MicroRNA in Solid Tumor and Hematological Diseases

MicroRNA in Solid Tumor and Hematological Diseases

Editor

Francesca Lovat

MDPI • Basel • Beijing • Wuhan • Barcelona • Belgrade • Manchester • Tokyo • Cluj • Tianjin

Editor
Francesca Lovat
Ohio State University
USA

Editorial Office
MDPI
St. Alban-Anlage 66
4052 Basel, Switzerland

This is a reprint of articles from the Special Issue published online in the open access journal *Biomedicines* (ISSN 2227-9059) (available at: https://www.mdpi.com/journal/biomedicines/special_issues/miRNA_tumor_hematol).

For citation purposes, cite each article independently as indicated on the article page online and as indicated below:

LastName, A.A.; LastName, B.B.; LastName, C.C. Article Title. *Journal Name* **Year**, *Volume Number*, Page Range.

ISBN 978-3-0365-3088-8 (Hbk)
ISBN 978-3-0365-3089-5 (PDF)

Contents

biomedicines

MDPI

Editorial

Editorial to the Special Issue "MicroRNA in Solid Tumor and Hematological Diseases"

Francesca Lovat

Department of Cancer Biology and Genetics and Comprehensive Cancer Center, The Ohio State University, Columbus, OH 43210, USA; francesca.lovat@osumc.edu

Citation: Lovat, F. Editorial to the Special Issue "MicroRNA in Solid Tumor and Hematological Diseases". *Biomedicines* **2021**, *9*, 1678. https://doi.org/10.3390/biomedicines9111678

Received: 5 November 2021
Accepted: 8 November 2021
Published: 12 November 2021

In the last two decades, the roles of microRNAs in the biology and progression of human cancer have been extensively studied; at present, these small non-coding RNAs are considered powerful gene regulators. microRNAs are involved in almost all biological and cellular processes, such as proliferation, differentiation and apoptosis [1]. Aberrant microRNA expression in cancer cells and in their microenvironment, and in body fluids such as serum or urine, has been linked to transformation, stemness, metastasis, resistance to chemotherapy and immune modulation in different tumor types, both solid tumor and hematological diseases.

This Special Issue includes eleven papers—five original manuscripts, six reviews and one systematic review—investigating the biological role of microRNAs in different tumor types and potential diagnostic and clinical approaches.

Lin and colleagues [2] presented a detailed overview of the role and regulatory mechanisms of microRNAs that control their dysregulated expression, focusing on solid tumors, including colorectal cancer, lung cancer, breast cancer, and liver cancer. Artemaki and co-workers [3] dissected the role of microRNAs in normal B-cell development and their deregulation in B-cell non-Hodgkin lymphomas (NHLs). Korac and colleagues [4] reported interesting details of miR-7's roles in cancer biology and development. Mainly described as a tumor-suppressor, miR-7 can act as an oncomiR, underlining how microRNA expression is tissue- and microenvironment-specific. Okada and colleagues [5] reported the tumor-suppressive role of the miR-139 duplex (miR-139-5p and miR-139-3p) in renal cell carcinoma (RCC). The miR-139 duplex can modulate and potentially silence different oncogenes involved in the pathogenesis of RCC. Lee and colleagues [6] described the role of microRNAs in the regulation of Hippo-YAP/TAZ signaling in liver cancer. MicroRNAs, functioning as oncogenes and/or tumor-suppressors, can directly or indirectly modulate the Hippo-YAP/TAZ signaling pathway, leading to the development and progression of hepatic cancer.

The discovery of the potential role of microRNAs as a biomarker in the detection of the early stage of the disease represents one of the most promising developments in microRNA research. It is well documented that microRNAs are released into body fluids, and their detection represents a powerful noninvasive and sensitive method for early diagnosis. Aita and colleagues [7] analyzed serum from patients with early pancreatic ductal adenocarcinoma (PDAC) to identify a microRNA signature, to ensure early diagnosis and potentially predict prognosis. Kovynev and colleagues [8] identified a set of microRNAs that were able to detect the minimal residual disease (MRD) of acute leukemia (AL), regardless of myeloid (AML) or lymphoblastic leukemia (ALL) origin, compared with the hematopoietic conditions induced by non-tumor pathologies (NTPs) used as controls. In addition to early disease detection, microRNAs serve as useful candidate biomarkers to stratify patients with primary resistance to a specific targeted therapy and those who have developed acquired resistance. Angerilli and colleagues [9] reported an accurate description of microRNAs as mediators of resistance to a specific targeted therapy used in the treatment of gastrointestinal tumors, focusing on anti-EGFR, anti-HER2 and anti-VEGF antibodies, small-molecule tyrosine kinase inhibitors and immune checkpoint inhibitors. Dias and co-workers [10] described a microRNA signature capable of modulating the L-type amino acid transporter 1 (LAT1) and alanine–serine–cysteine transporter 2 (ASCT2) expression in colorectal cancer (CRC). Cancer cells require an increased intake of amino acid to maintain their proliferation rate; therefore, dysregulation of these two amino acid transporters plays an important role in CRC development. This microRNA set could represent an inhibitory tool for a potential therapeutic approach. Ravegnini and colleagues [11] presented a systematic review of circulating microRNAs that were correlated with therapy in epithelial ovarian cancers (EOCs). The final analysis pinpointed the miR-200 family as the potential biomarker in EOC The miR-200 family has been described as being involved in the epithelial–mesenchymal transition (EMT) pathway, which promotes EOC progression and metastasis.

The results from studies of circulating microRNAs are often not reproducible due to a lack of accurate quantification of these molecules. The lack of standardization in the protocols has been identified as a possible cause of this issue, particularly the use of different strategies to normalize microRNA expression. Oto and colleagues [12] reported that miR-29c-3p represents the most stably expressed microRNA, and, therefore, the best normalizer in urine of bladder cancer (BC) patients. The discovery of this noninvasive stable reference will support future microRNA analysis among urine samples of BC patients.

All the research presented in this Special Issue provides an overview of the potential use of microRNAs to understand tumor biology and explain their roles as biomarkers to modulate the biological pathways that are critical for cancer development and progression, as well as providing useful insights to predict clinical outcomes or responses to therapy We are only at the beginning of the journey to understanding and clinically applying the knowledge of microRNAs. We hope that the studies presented in this Special Issue can spark additional interest and trigger new, insightful research in this field in rapid expansion.

Funding: This research did not receive external funding.

Institutional Review Board Statement: Not applicable.

Informed Consent Statement: Not applicable.

Data Availability Statement: Not applicable.

Acknowledgments: As Guest Editor, I would like to thank all the authors for their outstanding research contributions to this Special Issue, and the reviewers for their support in appraising the manuscripts. Thanks also to the *Biomedicines* editorial staff members, particularly Eastern Zhuang, for their assistance.

Conflicts of Interest: The authors declare no conflict of interest.

References

1. Lovat, F.; Valeri, N.; Croce, C.M. MicroRNAs in the pathogenesis of cancer. *Semin. Oncol.* **2011**, *38*, 724–733. [CrossRef] [PubMed]
2. Lin, Y.-C.; Chen, T.-H.; Huang, Y.-M.; Wei, P.-L.; Lin, J.-C. Involvement of microRNA in Solid Cancer: Role and Regulatory Mechanisms. *Biomedicines* **2021**, *9*, 343. [CrossRef] [PubMed]
3. Artemaki, P.I.; Letsos, P.A.; Zoupa, I.C.; Katsaraki, K.; Karousi, P.; Papageorgiou, S.G.; Pappa, V.; Scorilas, A.; Kontos, C.K. The Multifaceted Role and Utility of MicroRNAs in Indolent B-Cell Non-Hodgkin Lymphomas. *Biomedicines* **2021**, *9*, 333. [CrossRef]
4. Korać, P.; Antica, M.; Matulić, M. MiR-7 in Cancer Development. *Biomedicines* **2021**, *9*, 325. [CrossRef] [PubMed]
5. Okada, R.; Goto, Y.; Yamada, Y.; Kato, M.; Asai, S.; Moriya, S.; Ichikawa, T.; Seki, N. Regulation of Oncogenic Targets by the Tumor-Suppressive miR-139 Duplex (miR-139-5p and miR-139-3p) in Renal Cell Carcinoma. *Biomedicines* **2020**, *8*, 599. [CrossRef] [PubMed]
6. Lee, N.-H.; Kim, S.J.; Hyun, J. MicroRNAs Regulating Hippo-YAP Signaling in Liver Cancer. *Biomedicines* **2021**, *9*, 347. [CrossRef] [PubMed]
7. Aita, A.; Millino, C.; Sperti, C.; Pacchioni, B.; Plebani, M.; De Pittà, C.; Basso, D. Serum miRNA Profiling for Early PDAC Diagnosis and Prognosis: A Retrospective Study. *Biomedicines* **2021**, *9*, 845. [CrossRef] [PubMed]
8. Kovynev, I.B.; Titov, S.E.; Ruzankin, P.S.; Agakishiev, M.M.; Veryaskina, Y.A.; Nedel'ko, V.M.; Pospelova, T.I.; Zhimulev, I.F. Profiling 25 Bone Marrow microRNAs in Acute Leukemias and Secondary Nonleukemic Hematopoietic Conditions. *Biomedicines* **2020**, *8*, 607. [CrossRef] [PubMed]

9. Angerilli, V.; Galuppini, F.; Businello, G.; Dal Santo, L.; Savarino, E.; Realdon, S.; Guzzardo, V.; Nicolè, L.; Lazzarin, V.; Lonardi, S.; et al. MicroRNAs as Predictive Biomarkers of Resistance to Targeted Therapies in Gastrointestinal Tumors. *Biomedicines* **2021**, *9*, 318. [CrossRef] [PubMed]
10. Dias, F.; Almeida, C.; Teixeira, A.L.; Morais, M.; Medeiros, R. LAT1 and ASCT2 Related microRNAs as Potential New Therapeutic Agents against Colorectal Cancer Progression. *Biomedicines* **2021**, *9*, 195. [CrossRef] [PubMed]
11. Ravegnini, G.; De Iaco, P.; Gorini, F.; Dondi, G.; Klooster, I.; De Crescenzo, E.; Bovicelli, A.; Hrelia, P.; Perrone, A.M.; Angelini, S. Role of Circulating miRNAs in Therapeutic Response in Epithelial Ovarian Cancer: A Systematic Revision. *Biomedicines* **2021**, *9*, 1316. [CrossRef] [PubMed]
12. Oto, J.; Plana, E.; Fernández-Pardo, Á.; Cana, F.; Martínez-Sarmiento, M.; Vera-Donoso, C.D.; España, F.; Medina, P. Identification of miR-29c-3p as a Robust Normalizer for Urine microRNA Studies in Bladder Cancer. *Biomedicines* **2020**, *8*, 447. [CrossRef] [PubMed]

biomedicines

MDPI

Article

Serum miRNA Profiling for Early PDAC Diagnosis and Prognosis: A Retrospective Study

Ada Aita [1,†], Caterina Millino [2,†], Cosimo Sperti [3], Beniamina Pacchioni [2], Mario Plebani [1], Cristiano De Pittà [2,*] and Daniela Basso [1,*]

1 Department of Medicine-DIMED, University of Padova, 35128 Padova, Italy; ada.aita@unipd.it (A.A.); mario.plebani@unipd.it (M.P.)
2 Department of Biology, University of Padova, 35131 Padova, Italy; caterina.millino@unipd.it (C.M.); beniamina.pacchioni@unipd.it (B.P.)
3 Department of Surgical, Oncological and Gastroenterological Sciences-DiSCOG, University of Padova, 35128 Padova, Italy; cosimo.sperti@unipd.it
* Correspondence: cristiano.depitta@unipd.it (C.D.P.); daniela.basso@unipd.it (D.B.)
† These Authors contributed equally to this work.

Abstract: Background: Tumor stage predicts pancreatic cancer (PDAC) prognosis, but prolonged and short survivals have been described in patients with early-stage tumors. Circulating microRNA (miRNA) are an emerging class of suitable biomarkers for PDAC prognosis. Our aim was to identify whether serum miRNA signatures predict survival of early-stage PDAC. Methods: Serum RNA from archival 15 stage I-III PDAC patients and 4 controls was used for miRNAs expression profile (Agilent microarrays). PDAC patients with comparable age, gender, diabetes, jaundice and surgery were classified according to survival: less than 14 months (7/15 pts, group A) and more than 22 months (8/15 pts, group B). Bioinformatic data analysis was performed by two-class Significance Analysis of Microarray (SAM) algorithm. Binary logistic regression analyses considering PDAC diagnosis and outcome as dependent variables, and ROC analyses were also performed. Results: 2549 human miRNAs were screened out. At SAM, 76 differentially expressed miRNAs were found among controls and PDAC (FDR = 0.4%), the large majority (50/76, 66%) of them being downregulated in PDAC with respect to controls. Six miRNAs were independently correlated with early PDAC, and among these, *hsa-miR-6821-5p* was associated with the best ROC curve area in distinguishing controls from early PDAC. Among the 71 miRNAs differentially expressed between groups A and B, the most significant were *hsa-miR-3135b* expressed in group A only, *hsa-miR-6126* and *hsa-miR-486-5p* expressed in group B only. Eight miRNAs were correlated with the presence of lymph-node metastases; among these, *hsa-miR-4669* is of potential interest. *hsa-miR-4516*, increased in PDAC and found as an independent predictor of survival, has among its putative targets a series of gens involved in key pathways of cancer progression and dissemination, such as Wnt and p53 signalling pathways. Conclusions: A series of serum miRNAs was identified as potentially useful for the early diagnosis of PDAC, and for establishing a prognosis.

Keywords: PDAC; miRNAs; survival; expression profiling

Citation: Aita, A.; Millino, C.; Sperti, C.; Pacchioni, B.; Plebani, M.; De Pittà, C.; Basso, D. Serum miRNA Profiling for Early PDAC Diagnosis and Prognosis: A Retrospective Study. *Biomedicines* **2021**, *9*, 845. https://doi.org/10.3390/biomedicines9070845

Academic Editor: Francesca Lovat

Received: 26 April 2021
Accepted: 15 July 2021
Published: 20 July 2021

1. Introduction

The incidence of pancreatic ductal adenocarcinoma (PDAC) is increasing worldwide, this disease being among the first 15 leading forms of cancer in men and women, with its highest incidence rates in Europe, Northern America, and Australia/New Zealand [1,2].

Despite advances made in the last few decades in understanding PDAC biology and developing new targeted therapies [3–6], PDAC patients continue to have a dismal prognosis, with a 5-year survival rate of less than 20% [7,8]. Tumor stage is the main prognostic determinant, early-stage tumors being associated with a longer survival than that of locally advanced or metastatic tumors [9]. However, rapidly evolving tumors with metastatic spread are relatively frequent, even when PDAC is diagnosed at an early stage. Differences in tumor biology probably underlie the different clinical evolution of tumors that, at diagnosis, appear similar in their anatomic extension (i.e., the same TNM characteristics).

PDAC might harbor several mutations of the cell genome, and a vast heterogeneity among tumors has been described [10,11]. Despite this complexity, some clustered recurrent mutations have allowed the classification of PDAC into four main types associated with different prognoses [12], but also the classification of neoplastic cysts with implications for the clinical decision, thus allowing patients to be spared unnecessary surgery [13] On a ground of high genetic heterogeneity across tumors, mutations of some few genes are strongly correlated with PDAC, namely those of the *KRAS* oncogene and of *TP53*, *CDKN2A* and *SMAD4* tumor suppressor genes [14]. Among these genetic alterations, the most significantly associated with prognosis is *SMAD4* loss, found in about 60% of PDAC cases [15,16].

Although the above gene mutations alone or combined might determine the oncogenic and metastatic phenotype, molecules involved in regulating gene expression, namely microRNA (miRNAs), long non-coding RNA and circular RNA, have been the focus of attention in the last few decades due to their potential role in understanding tumor biology, and also in making the diagnosis, establishing the prognosis and providing tailored therapy [17–20]. miRNAs are small (20–25 nucleotide sequences) non-coding single-stranded RNAs that, by binding complementary mRNA sequences, might repress mRNA translation or enhance mRNA degradation, thus contributing to the regulation of a wide range of cellular activities. Any single mRNA molecule might be regulated by several miRNAs, and any miRNA might regulate several different mRNA molecules [21]. The expression of the key PDAC-associated oncogene *KRAS*, as well as that of genes encoding proteins entering in the core signaling pathways of PDAC, such as JAK/STAT, Wnt/β-catenin or TGF-β, evidences a number of regulating miRNAs with expression levels that are considered targetable by mimic or inhibitor compounds as new potential therapeutic strategies [14,18,22,23]. Moreover, miRNAs represent a great challenge for diagnosis, since they might be released by the tumor into biological fluids, such as plasma/serum or ascitic fluid [24,25], and because they are stable, thus limiting the impact of pre-analytical variability on the results. In the PDAC setting, several studies in the literature have evaluated different miRNAs panels as diagnostic and/or prognostic tools by analyzing tissue expression [26,27], serum levels [24], blood levels [28] and ascitic fluid levels [25]. New biomarkers of PDAC should aid early diagnosis and, as prognostic indices in early-stage cases, should be helpful in distinguishing between patients with long or short life expectancies, thus allowing a more tailored therapeutic approach [7,29].

The aim of the present retrospective study was, using microarray analysis of 2.549 miRNAs, to identify in sera those miRNAs able to diagnose early PDAC and to predict long or short survival after curative surgery.

2. Materials and Methods

2.1. Patients and Samples

The exploratory cohort comprised archival sera of 15 PDAC patients selected from a large retrospective series for microarray analysis. To ascertain the effects of survival on results, the initial selection criteria were: available data on survival allowing classification into two groups with short (<one year) and long (>two years) survival, tumor stage, tumor site, treatment, metabolic and biochemical characteristics being common to the two groups. On these bases the following criteria were established: (1) early-stage tumor; (2) tumor of the pancreas head; (3) R0 surgical treatment; (4) presence of diabetes mellitus or reduced glucose tolerance; (5) absence of jaundice, considering bilirubin as a potential interfering compound. Since only six cases met these criteria, we extended criteria to allow the potential presence of jaundice and tumor stage III besides stages I and II.

In addition, archival sera of four subjects with mild gastritis but without any present or previous evidence of cancer (2 males, 2 females, age range 62–74 years), were used as controls.

Another independent validation cohort of patients was selected from our archive. This included a reference group of 8 patients with mild gastritis without any present or previous evidence of cancer (7 males, 1 female, age range 42–79 years) and a series of 24 PDAC patients (6 males, 18 females, age range 48–83 years). PDAC stage was: IA = 4 patients; IIB = 11 patients; III = 8 patients; IV = 1 patient.

All sera were stored in aliquots at −80 °C and never subjected to thaw–freeze cycles before the analyses.

2.2. RNA Extraction

Total RNA, including miRNAs, was extracted from the archival sera (600 μL) using the miRNeasy Serum/Plasma Advanced Kit (Qiagen, Hilden, Germany) according to the manufacturer's instructions. miRNAs were quantified by Qubit microRNA assay kit (Thermo Fisher Scientific, Waltham, MA, USA). The RNA amount ranged from 3.5 to 18.9 ng/μL.

RNA integrity, and the content of miRNAs (%) in each sample were assessed by capillary electrophoresis with the Small RNA LabChip using the Agilent Bioanalyzer 2100 (Agilent Technologies, Santa Clara, CA, USA). Only samples with a quantity of ≥4000 pg were used for microarray analysis.

2.3. MicroRNA Expression Profiling

miRNAs expression profiles were obtained using "Agilent SurePrintG3 Human miRNA v.21 (8x60K)" microarray (Agilent Technologies, Santa Clara, CA, USA), which allows the detection of 2549 known human (miRBase Release 21.0) and 76 viral miRNAs (GEO Platform N. GPL24741).

Every slide contains eight individual microarrays, with 60,000 features each, including 2164 controls, used to estimate fluorescence background and background variance. Each miRNA was targeted by 16 to 20 array-probes of different sizes. Total RNA (4000 pg) was labelled with pCp Cy3, according to the Agilent's protocol, and unincorporated dyes were removed with MicroBioSpin6 columns (BioRad, Hercules, CA, USA) [30]. Probes were hybridized at 55 °C for 22 h using the Agilent's hybridization oven, which is suitable for bubble-mixing and microarray hybridization processes. Slides, washed by Agilent Gene expression wash buffers 1 and 2, were examined using an Agilent microarray scanner (model G2565CA) at 100% and 5% sensitivity settings. Agilent Feature Extraction software version 12.0.0.7 was used for image analysis of miRNA expression arrays. Raw miRNA data are available in the U.S. National Centre for Biotechnology Information Gene Expression Omnibus (GEO, http://www.ncbi.nlm.nih.gov/geo (accessed on 16 March 2021)) database with the Accession N. GSE168996.

2.4. Statistical and Functional Analysis of miRNA Expression Data

Inter-array normalization of miRNA expression levels was performed with cyclic Lowess for miRNA [31], the average of replicates being used. Feature Extraction software (Agilent Technologies, Santa Clara, CA, USA) was employed to obtain spot quality measures for evaluating the quality and the reliability of the hybridization. In particular, the flag "glsFound" (set to 1 if the spot had an intensity value significantly different from that of the local background, 0 otherwise) was used to filter out unreliable probes: a flag equal to 0 was noted as "not available" (NA). In order to make a robust and unbiased statistical analysis, probes with a high proportion of NA values were removed from the dataset. NA (47%) was used as a threshold in the filtering process, a total of 142 available human miRNAs being obtained. Cluster analysis with the average linkage method and Pearson correlation and profile similarity searches were performed with the Multi Experiment Viewer 4.9.1 (tMev) of the TM4 Microarray Software Suite [32]. All heat maps were obtained by morpheus software1 (https://software.broadinstitute.org/morpheus (accessed on 24 June 2021), Broad Institute, Cambridge, MA, USA).

Differentially expressed miRNAs were identified with two-class Significance Analysis of Microarray (SAM) algorithm [33] with default settings. SAM, which uses a permutation-based multiple testing algorithm, associates a variable false discovery rate (FDR) with the significant genes. FDR, which refers to the percentage of error that can occur in the identification of the statistically significant differentially expressed miRNAs in multiple comparisons, can be manually adjusted.

miRTarBase [34], a current and curated collection of miRNA-target interactions with experimental support, was used to predict target genes of differentially expressed miRNAs between PCDAC patients and controls. Biological pathway analysis of putative target genes was performed using DAVID v. 6.8 [35], which combines web tools such as gene functional classification, functional annotation chart or clustering and functional annotation table.

2.5. Reverse Transcription and Quantitative PCR (qRT-PCR) of miRNAs

Before starting the miRNA extraction procedure, 1 µL of Qiagen RNA Spike-In mix (UniSp2, UniSp4, and UniSp5) was added to 200 µL of serum (starting volume) to control the sample-to-sample variation in the RNA isolation procedure. Subsequently, total RNA, including miRNAs, was extracted using the miRNeasy Serum/Plasma Advanced Kit (Qiagen) according to the manufacturer's instructions. cDNA was synthesized using the miRCURY LNA RT Kit (Qiagen, Hilden, Germany) starting from 75 ng of total RNA in 10 µL with the addition of 1 µL of UniSp6 as exogenous miRNA spiked-in control. PCR was performed in a 10 µL volume containing 5 µL 2x miRCURY SYBR GREEN Master Mix (Qiagen, Hilden, Germany), 1 µL cDNA, and 1 µL of one of the following miRCURY LNA PCR primer sets (Qiagen): *hsa-miR-4516* (ID YP02112882), *hsa-miR-6089* (ID YP02109969), and *UniSp4* (ID YP00203953). The qPCR reactions were performed in an ABI7500 Real-Time PCR System (Applied Biosystems, Foster City, CA, USA). The qPCR cycling conditions were 95 °C for 2 min and 40 cycles (95 °C for 10 s and 56 °C for 1 min). Three replicates of each sample and control were amplified for each real-time PCR reaction. The relative expression levels between samples were calculated using the comparative delta Ct (threshold cycle number) method ($2^{-\Delta\Delta Ct}$), implemented in the 7500 Real Time PCR System software.

2.6. Statistical Analysis of Data

The non-parametric Kruskal–Wallis test, receiver operating characteristic curves (ROC) and binary logistic regression analyses were performed by Stata 13.1 (StataCorp, 4905 Lakeway Drive, TX, USA).

3. Results

3.1. Serum miRNA Expression Signatures for Early PDAC Diagnosis

Table 1 shows the clinical and biochemical characteristics of the 15 PDAC patients of the exploratory cohort, seven being short-term (<14 months), and eight long-term (>22 months) survivors. All 15 patients had a confirmed histological diagnosis of PDAC with grading, and all underwent radical surgery (R0) and had diabetes mellitus or reduced glucose tolerance diagnosed within 5 months (concurrent) or more than 5 months prior to PDAC diagnosis.

We analysed the expression of serum miRNAs in the 15 PDAC patients and in four controls using a 2549-miR microarrays platform. At SAM two-class analysis, 76 differentially expressed miRNAs were found among controls and PDAC, with an FDR of 0.4%. Interestingly, the large majority (50/76 samples, 66%) of differentially expressed miRNAs were downregulated in the sera of PDAC patients with respect to controls (Supplementary Table S1). A total of 7 (*hsa-miR-6126, hsa-miR-2392, hsa-miR-4327, hsa-miR-939-5p, hsa-miR-4655-3p, hsa-miR-371b-5p, hsa-miR-3135b*) out of the 76 miRNAs were undetectable in controls, but were expressed, albeit at low levels, only in PDAC patients' sera.

Table 1. Clinical and biochemical characteristics of PDAC patients included in the study.

Patient Nr.	Age	Gender	G	T	N	M	Stage	Survival (months)	Diabetes	CA 19-9kU/L	Total Bilirubin µmol/L	P-Glucose mmol/L	Survival
713	57	M	2	2	0	0	IB	113	DM (C)	131	10	5.0	L
708	70	M	1	3	0	0	IIA	86	DM (P)	1	8	10.4	L
386	59	M	2	3	1	0	IIB	75	RGT (C)	20	92	6.1	L
664	79	F	2	3	1	0	IIB	56	RGT (C)	na	39	4.7	L
642	79	M	2	3	0	0	IIA	45	RGT (C)	92	39	7.4	L
310	78	F	2	3	1	0	IIB	29	DM (P)	6969	135	8.8	L
703	72	F	2	3	1	0	IIB	25	DM (P)	5309	12	8.9	L
585	63	M	2	3	1	0	IIB	23	DM (P)	435	13	8.2	L
584	58	F	3	3	1	0	IIB	13	RGT (C)	106	81	5.2	S
354	63	M	2	4	1	0	III	11	DM (C)	519	94	5.9	S
301	65	M	3	3	1	0	IIB	11	RGT (C)	48	65	7.1	S
699	67	F	2	3	1	0	IIB	9	DM (C)	30	22	6.4	S
517	61	F	3	4	0	0	III	8	DM (C)	212	9	6.2	S
607	68	M	3	3	0	0	IIA	6	RGT (C)	29	7	5.3	S
550	84	F	2	3	0	0	IIA	2	DM (P)	105	39	8.1	S

Diabetes: DM = Diabetes Mellitus; RGT = Reduced Glucose Tolerance; C = concurrent diabetes diagnosis (≤5 months); P = previous diabetes diagnosis (>5 months); Survival: L = Long (>22 months); S = Short (<14 months).

Microarray data of any patient and control were then used for further analyses. We performed binary logistic regression analysis considering PDAC diagnosis as a dependent variable, and any of the 76 differentially expressed miRNAs between controls and PDAC (Supplementary Table S1), with age and gender, as predictor variables. Six out of the seventy-six miRNAs were significantly correlated with PDAC diagnosis, independently from age and gender: *hsa-miR-6089* ($\chi^2 = 8.58$, $p = 0.050$), *hsa-miR-4466* ($\chi^2 = 9.19$, $p = 0.047$), *hsa-miR-6821-5p* ($\chi^2 = 12.05$, $p = 0.048$), *hsa-miR-4669* ($\chi^2 = 7.62$, $p = 0.050$), *hsa-miR-1202* ($\chi^2 = 6.42$, $p = 0.048$), *hsa-miR-574-3p* ($\chi^2 = 9.03$, $p = 0.049$). The same binary logistic regression analysis performed with CA 19-9, age, and gender as predictors, was not significant ($\chi^2 = 5.81$, $p = 0.232$).

To further evaluate whether miRNAs outperform with respect to CA 19-9 in the early detection of PDAC, we compared receiver operating characteristic (ROC) curves of the 76 above reported miRNAs with the ROC curve of CA 19-9 considering the two groups of controls and PDAC. Those miRNAs showing an area under the ROC curve higher than that of CA 19-9 (0.8083 ± 0.0984, 95%CI: 0.61551–1.00000) are reported in Table 2.

Table 2. Serum miRNAs and early PDAC diagnosis. miRNAs showing an area under the ROC curve (AUC) higher than that of CA 19-9 in distinguishing controls from PDAC are reported. SE: standard error; 95% CI: 95% confidence interval; Increase/Decrease: miRNA levels increased or decreased in PDAC with respect to controls. Bold face: miRNAs significantly correlated with PDAC diagnosis at binary logistic regression analysis.

		ROC		Asymptotic Normal		Increase/ Decrease
	Obs	**AUC**	**SE**	**95% Conf. Interval**		
hsa-miR-7110-5p	19	0.8167	0.1360	0.55014	1.00000	Decrease
hsa-miR-3135-b	19	0.8333	0.0630	0.70987	0.95680	Increase
hsa-miR-4669	19	0.8333	0.1346	0.56959	1.00000	Decrease
hsa-miR-7107-5p	19	0.8333	0.1225	0.59329	1.00000	Decrease
hsa-miR-574-3p	19	0.8333	0.1225	0.59329	1.00000	Decrease
hsa-miR-1275	19	0.8333	0.1188	0.60049	1.00000	Decrease
hsa-miR-4466	19	0.8500	0.1351	0.58529	1.00000	Increase
hsa-miR-3679-5p	19	0.8500	0.1055	0.64316	1.00000	Decrease
hsa-miR-2392	19	0.8667	0.0591	0.75085	0.98249	Increase
hsa-miR-4655-3p	19	0.8667	0.0591	0.75085	0.98249	Increase
hsa-miR-6089	19	0.8667	0.1196	0.63230	1.00000	Increase
hsa-miR-5100	19	0.8667	0.0970	0.67646	1.00000	Increase
hsa-miR-6749-5p	19	0.8667	0.0858	0.69858	1.00000	Decrease
hsa-miR-4687-3p	19	0.8833	0.0922	0.70258	1.00000	Increase
hsa-miR-1915-3p	19	0.8833	0.0922	0.70258	1.00000	Increase
hsa-miR-6125	19	0.8833	0.1043	0.67896	1.00000	Increase
hsa-miR-1202	19	0.8833	0.0812	0.72426	1.00000	Decrease
hsa-miR-8485	19	0.8833	0.1043	0.67896	1.00000	Decrease
hsa-miR-6126	19	0.9000	0.0535	0.79524	1.00000	Increase
hsa-miR-939-5p	19	0.9000	0.0535	0.79524	1.00000	Increase
hsa-miR-6800-5p	19	0.9000	0.0796	0.74399	1.00000	Increase
hsa-miR-4516	19	0.9167	0.0684	0.78255	1.00000	Increase
hsa-miR-6869-5p	19	0.9167	0.0891	0.74206	1.00000	Increase
hsa-miR-6850-5p	19	0.9167	0.0746	0.77041	1.00000	Increase
hsa-miR-4327	19	0.9333	0.0454	0.84430	1.00000	Increase
hsa-miR-371b-5p	19	0.9333	0.0454	0.84430	1.00000	Increase
hsa-miR-6821-5p	19	0.9667	0.0403	0.88761	1.00000	Increase

3.2. *miRNA Expression Profiles and PDAC Survival Rates*

The circulating miRNAs expression profiles of controls were compared with those from patients with long (eight samples) and with short (seven samples) survival in order to identify miRNA expression signatures associated with different prognoses.

An unsupervised hierarchical clustering analysis, by using the list of differentially expressed genes between PDAC patients and controls, enabled the clear separation of controls and PDAC patients with long (Figure 1a) and short (Figure 1b) survival, respectively.

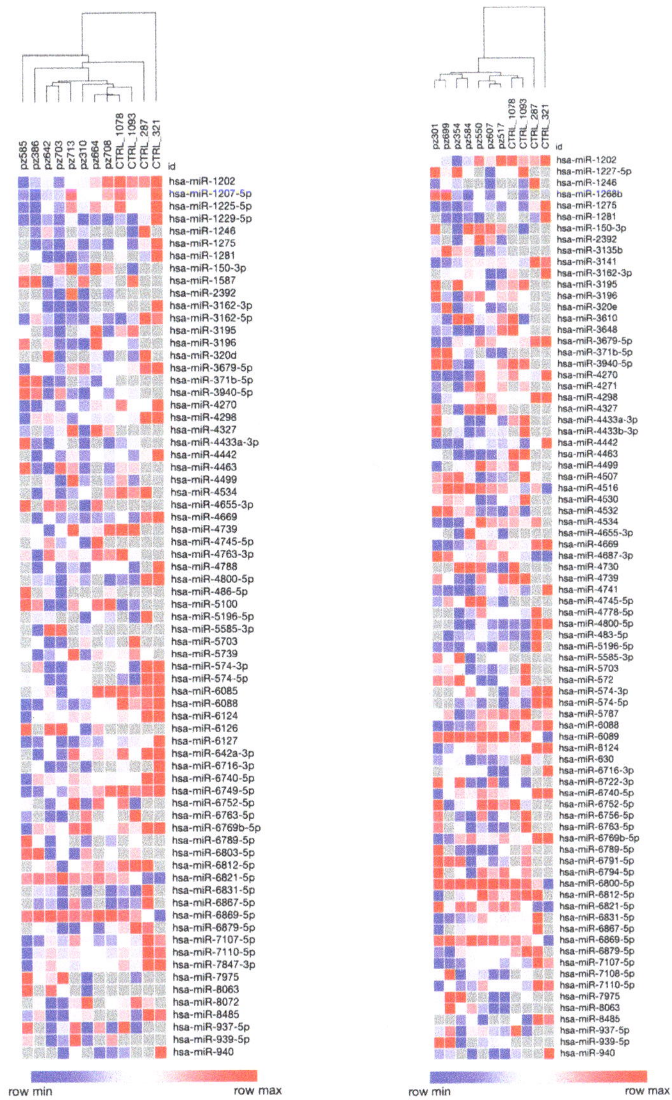

(a) (b)

Figure 1. Cluster analysis of differentially expressed miRNAs between PDAC patients and controls. Heat maps representing differentially expressed miRNAs of two different comparisons: long survival patients (**a**) and short survival (**b**) patients with respect to controls. Both dendrograms show the capability of circulating miRNA signatures to clearly separate PDAC patients with respect to controls. A color-coded scale for the normalized expression values is used: red and blue represent high and low expression levels. A complete list of differentially expressed genes identified by SAM two-class algorithm is provided in the Supplementary Materials (Tables S2 and S3).

A total of 71 and 80 differentially expressed miRNAs between long-term and short-term survivors with respect to the same controls were identified by SAM two-class analysis with an FDR < 3%. We observed an overall downregulation of miRNAs both in long-term (47/71, 66% differentially expressed miRNAs) and short-term (56/80, 70% differentially expressed miRNAs) survivors with respect to controls (Supplementary Tables S2 and S3). Among the above-mentioned miRNAs expressed in sera of PDAC patients, but not in those of controls, *hsa-miR-3135b* was detected only in patients with a short survival, whereas *hsa-miR-6126* and *hsa-miR-486-5p* were specifically observed in long-term survivors.

A SAM two-class analysis to compare short-term and long-term PDAC survivors to identify prognosis-specific signatures of circulating miRNAs, enabled the identification of 71 differentially expressed miRNAs with an extremely high FDR (62%), thus indicating that circulating miRNA expression profiles do not allow the correct stratification of PDAC patients according to their different prognoses. However, although almost all differentially expressed miRNAs (69/71, 97%) did not attain statistical significance, they were downregulated in PDAC patients with a poor prognosis with respect to those with a longer survival (Supplementary Table S4).

We performed binary logistic regression analysis considering survival of PDAC patients (long or short) as dependent variable, and miRNAs, age, and gender as predictor variables. In the analyses entered the 76 miRNAs differentially expressed between controls and PDAC (Supplementary Table S1), and the series of 71 miRNAs differentially expressed between PDAC patients with long and short survival (Supplementary Table S4). None of the evaluated miRNA reached the statistical significance.

To further ascertain whether serum miRNAs could predict a PDAC disease that is more likely with an adverse outcome, we evaluated the presence, if any, of an association between miRNAs and lymph node metastases, which was not found for CA 19-9 (χ^2 = 4.282, p = 0.1175). All 76 differentially expressed miRNAs between PDAC and controls were analyzed by Kruskal–Wallis rank test comparing controls with PDAC patients with or without lymph-node metastases. This resulted in 11 significant miRNAs, reported in Table 3.

Table 3. The Table reports results of Kruskal–Wallis rank test considering lymph node metastases as variable defining groups and any individual miRNA as predictor variable.

	χ^2	p
hsa-miR-1202	10.861	0.0044
hsa-miR-3679-5p	8.061	0.0178
hsa-miR-6088	11.782	0.0028
hsa-miR-6791-5p	6.027	0.0491
hsa-miR-1915-3p	10.861	0.0044
hsa-miR-371b-5p	11.982	0.0025
hsa-miR-4669	12.430	0.0020
hsa-miR-4499	9.682	0.0079
hsa-miR-4442	8.219	0.0164
hsa-miR-7107-5p	9.482	0.0087
hsa-miR-4800-5p	6.685	0.0353

Multiple comparison reached a significant difference for eight miRNAs, which expression levels are depicted in Figure 2.

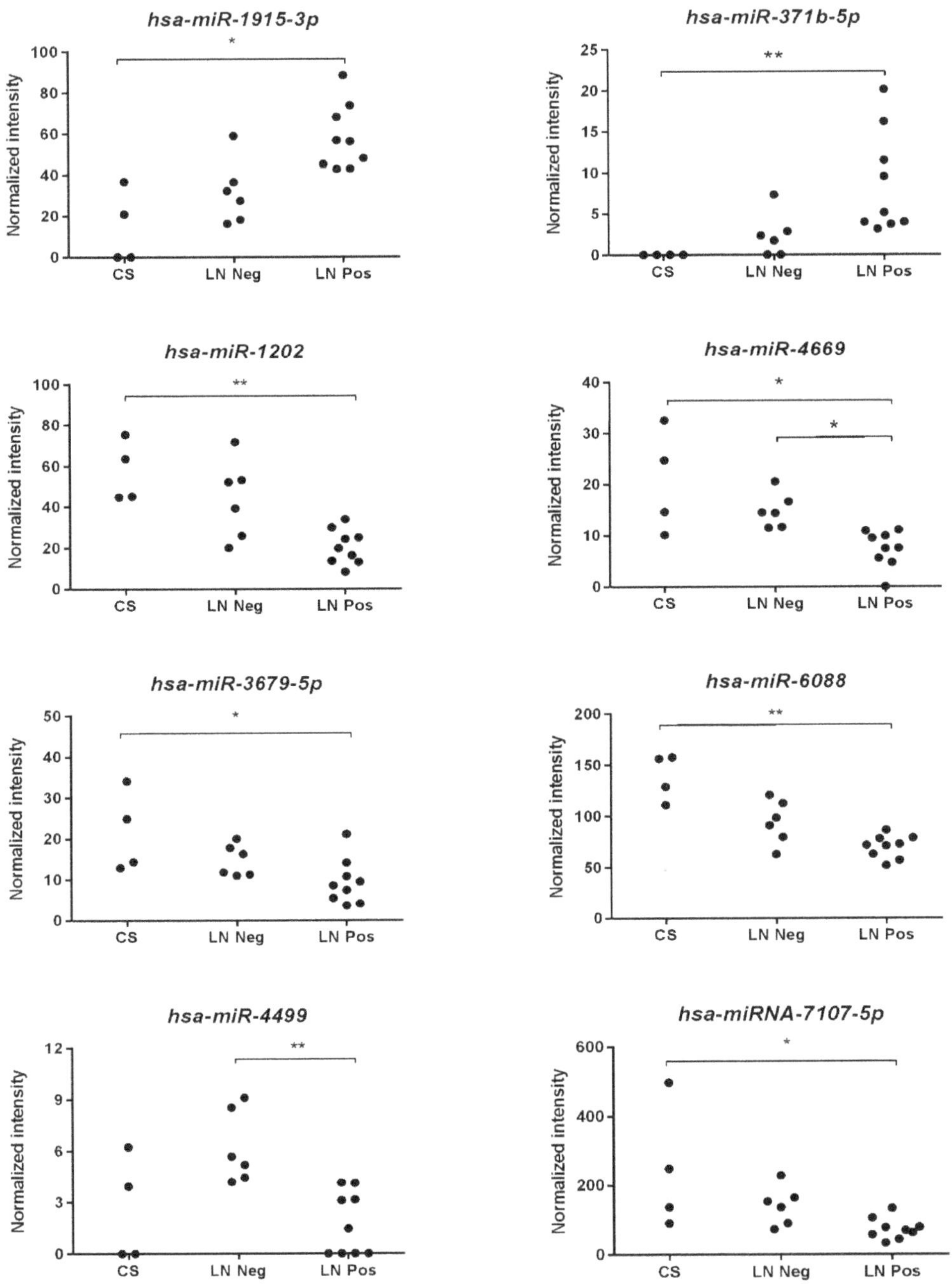

Figure 2. Expression levels of miRNAs significantly predicting adverse outcome. Normalized intensity values of miRNAs expression levels among controls (CS), PDAC patients with (LN pos) and without (LN neg) lymph node metastases have been compared. *: $p < 0.05$; **: $p < 0.005$.

3.3. Validation of Selected miRNAs in an Independent Cohort of PDAC Patients Using qRT-PCR

To validate the microarray expression data with an independent technique, we performed a qRT-PCR of two miRNAs (*hsa-miR-4516* and *hsa-miR-6089*) upregulated in PDAC with respect to controls, in a novel validation cohort of 24 PDAC patients and 8 controls Figure 3 shows the individual results of these two miRNAs, which was significantly different between PDAC and controls for *hsa-miR-4516* (Kruskal–Wallis rank test: $p = 0.0330$), not for *hsa-miR-6089* ($p = 0.4862$), although values tended to be higher in PDAC than in controls.

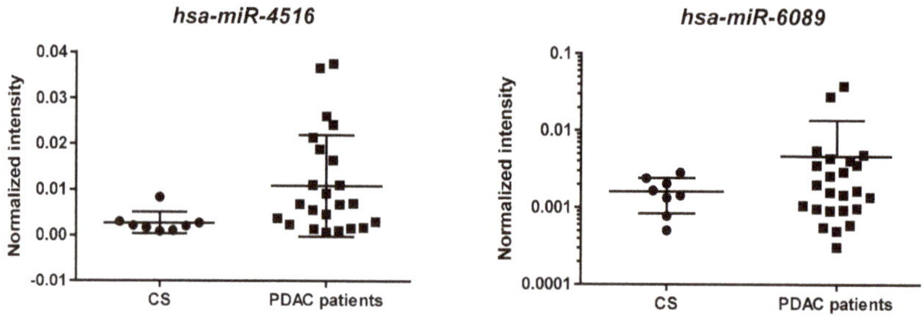

(**a**) (**b**)

Figure 3. Relative quantification of *hsa-miR-4516* (**a**) and *hsa-miR-6089* (**b**) serum levels in an independent cohort of PDAC patients. The relative serum expression levels of *hsa-miR-4516* and *hsa-miR-6089* measured in 8 controls (CS) and in 24 PDAC patients are shown.

We then ascertained whether these two miRNAs were correlated with survival, which ranged from 3 months to 10 years, with a median of 23 months. Cox regression analysis (Table 4) was performed, including the two evaluated miRNAs, age and gender, tumor stage, CA 19-9 and hemoglobin as predictors. Survival was significantly correlated with tumor stage, but also with *hsa-miR-4516* and hemoglobin, not with CA 19-9.

Table 4. Cox regression analysis in the validation cohort. Dependent variable: PDAC patients' survival.

Predictors	HR	SE	z	*p*	95% CI	
Sex	11.2116	15.25747	1.78	0.076	0.7785689	161.4499
Age	1.088652	0.0546704	1.69	0.091	0.9866048	1.201255
hsa-miR-4516	1.59×10^{-46}	7.36×10^{-45}	−2.28	0.023	6.58×10^{-86}	3.85×10^{-07}
hsa-miR-6089	2.91×10^{34}	1.24×10^{36}	1.86	0.063	0.0129917	6.53×10^{70}
CA 19-9	1.000101	0.0000555	1.81	0.070	0.9999919	1.000209
Hemoglobin	0.9515861	0.0202986	−2.33	0.020	0.9126218	0.9922141
Stage	20.64211	22.01492	2.84	0.005	2.552365	166.9418

3.4. Identification and Functional Analysis of Putative miRNAs Targets

miRTarBase (http://miRTarBase.mbc.nctu.edu.tw (accessed on 16 June 2021)) was used to predict putative targets of the 76 differentially expressed miRNAs between controls and PDAC patients (long and short survival). A list of 11.129 miRNA-target interactions (Table S5) were identified for a total of 5.183 putative target genes. In fact, a single miRNA has hundreds of putative targets because it has pleiotropic effects on different target genes based on the biological condition being explored. KEGG pathway analysis of putative targets was performed by using the DAVID web tool to identify the main biological pathways in which are involved the differentially expressed miRNAs among PDAC patients and controls. Among the 53 most enriched KEGG pathways (EASE < 0.05, Table S6), it is interesting to note that we found a large majority of putative targets involved in biological processes associated with cancer. The "Pancreatic cancer" pathway (Table 5) was one of the most enriched and this confirms the involvement of differentially expressed serum miRNAs in PDAC pathogenesis and/or progression. We also observed a statistically significant enrichment of target genes in biological processes that are considered highly important for cancer progression and dissemination, such as the "Wnt signalling pathway" (Table S7), the "p53 signalling pathway" (Table S8) and the "TGF-beta signalling pathway" (Table S9).

Table 5. List of miRNA-target interactions enriched in the "Pancreatic cancer" pathway. Bold face: the interactions between *miR-4516* (validated in qRT-PCR) and its target genes.

Entrez ID	Symbol	Gene Name	miRNA	Log2 (PDAC Patients/Control)	Experimental Evidence
208	*AKT2*	*AKT serine/threonine kinase 2*	*hsa-miR-2861*	0.60	Luciferase reporter assay; Western blot
9459	*ARHGEF6*	*Rac/Cdc42 guanine nucleotide exchange factor 6*	*hsa-miR-6127*	−1.90	PAR-CLIP
598	*BCL2L1*	*BCL2 like 1*	*hsa-miR-6127*	−1.90	PAR-CLIP
598	*BCL2L1*	*BCL2 like 1*	*hsa-miR-7110-5p*	−1.11	PAR-CLIP
598	*BCL2L1*	*BCL2 like 1*	*hsa-miR-4739*	−0.98	PAR-CLIP
598	*BCL2L1*	*BCL2 like 1*	*hsa-miR-5787*	−0.74	PAR-CLIP
598	*BCL2L1*	*BCL2 like 1*	*hsa-miR-6879-5p*	−0.66	PAR-CLIP
598	*BCL2L1*	*BCL2 like 1*	*hsa-miR-6756-5p*	−0.30	PAR-CLIP
598	*BCL2L1*	*BCL2 like 1*	*hsa-miR-6752-5p*	0.02	PAR-CLIP
598	*BCL2L1*	*BCL2 like 1*	*hsa-miR-6791-5p*	0.09	PAR-CLIP
598	*BCL2L1*	*BCL2 like 1*	*hsa-miR-371b-5p*	null	PAR-CLIP
598	*BCL2L2*	*BCL2 like 1*	*hsa-miR-630*	−1.34	Immunoblot;Luciferase reporter assay;qRT-PCR
595	*CCND1*	*cyclin D1*	*hsa-miR-574-5p*	−2.63	PAR-CLIP
595	*CCND1*	*cyclin D1*	*hsa-miR-3648*	−1.98	PAR-CLIP
595	*CCND1*	*cyclin D1*	*hsa-miR-5196-5p*	−1.69	PAR-CLIP
595	*CCND1*	*cyclin D1*	*hsa-miR-7107-5p*	−1.27	PAR-CLIP
595	*CCND1*	*cyclin D1*	*hsa-miR-6867-5p*	−0.84	PAR-CLIP
595	*CCND1*	*cyclin D1*	*hsa-miR-3940-5p*	0.01	qRT-PCR;Western blot
595	*CCND1*	*cyclin D1*	*hsa-miR-2861*	0.60	Luciferase reporter assay;Western blot
595	*CCND1*	*cyclin D1*	*hsa-miR-2392*	null	PAR-CLIP
1019	*CDK4*	*cyclin dependent kinase 4*	*hsa-miR-3135b*	null	HITS-CLIP
1021	*CDK6*	*cyclin dependent kinase 6*	*hsa-miR-6716-3p*	−2.51	PAR-CLIP
1021	*CDK6*	*cyclin dependent kinase 6*	*hsa-miR-8485*	−1.81	HITS-CLIP
1021	*CDK6*	*cyclin dependent kinase 6*	*hsa-miR-7847-3p*	−0.82	PAR-CLIP
1021	*CDK6*	*cyclin dependent kinase 6*	*hsa-miR-4534*	−0.48	PAR-CLIP
1021	*CDK6*	*cyclin dependent kinase 6*	*hsa-miR-5739*	−0.14	PAR-CLIP
1021	*CDK6*	*cyclin dependent kinase 6*	*hsa-miR-371b-5p*	null	PAR-CLIP
1869	*E2F1*	*E2F transcription factor 1*	*hsa-miR-940*	−3.46	PAR-CLIP
1869	*E2F1*	*E2F transcription factor 1*	*hsa-miR-6763-5p*	−1.09	PAR-CLIP
1869	*E2F1*	*E2F transcription factor 1*	*hsa-miR-939-5p*	null	PAR-CLIP
1870	*E2F2*	*E2F transcription factor 2*	*hsa-miR-4669*	−0.88	HITS-CLIP
1870	*E2F2*	*E2F transcription factor 2*	*hsa-miR-4433b-3p*	−0.40	PAR-CLIP
1871	*E2F3*	*E2F transcription factor 3*	*hsa-miR-6124*	−0.94	PAR-CLIP
1871	*E2F3*	*E2F transcription factor 3*	*hsa-miR-6867-5p*	−0.84	PAR-CLIP
1871	*E2F3*	*E2F transcription factor 3*	*hsa-miR-4778-5p*	−0.33	PAR-CLIP
1871	*E2F3*	*E2F transcription factor 3*	*hsa-miR-4271*	−0.19	PAR-CLIP
1950	*EGF*	*epidermal growth factor*	*hsa-miR-940*	−3.46	PAR-CLIP
1950	*EGF*	*epidermal growth factor*	*hsa-miR-4433a-3p*	−0.17	PAR-CLIP
1956	*EGFR*	*epidermal growth factor receptor*	*hsa-miR-574-3p*	−1.57	Luciferase reporter assay;Microarray;qRT-PCR;Western blot
1956	*EGFR*	*epidermal growth factor receptor*	*hsa-miR-2861*	0.60	Luciferase reporter assay;Western blot
2064	*ERBB2*	*erb-b2 receptor tyrosine kinase 2*	*hsa-miR-4270*	−0.97	HITS-CLIP

Table 5. *Cont.*

Entrez ID	Symbol	Gene Name	miRNA	Log2 (PDAC Patients/Control)	Experimental Evidence
8517	*IKBKG*	*inhibitor of kappa light polypeptide gene enhancer in B-cells, kinase gamma*	*hsa-miR-6127*	−1.90	PAR-CLIP
8517	*IKBKG*	*inhibitor of kappa light polypeptide gene enhancer in B-cells, kinase gamma*	*hsa-miR-8485*	−1.81	HITS-CLIP
8517	*IKBKG*	*inhibitor of kappa light polypeptide gene enhancer in B-cells, kinase gamma*	*hsa-miR-4739*	−0.98	HITS-CLIP
3845	*KRAS*	*KRAS proto-oncogene, GTPase*	*hsa-miR-671-5p*	−0.30	PAR-CLIP
5594	*MAPK1*	*mitogen-activated protein kinase 1*	*hsa-miR-6763-5p*	−1.09	PAR-CLIP
5594	*MAPK1*	*mitogen-activated protein kinase 1*	*hsa-miR-6831-5p*	−0.91	PAR-CLIP
5594	*MAPK1*	*mitogen-activated protein kinase 1*	*hsa-miR-6088*	−0.80	PAR-CLIP
5594	*MAPK1*	*mitogen-activated protein kinase 1*	*hsa-miR-4271*	−0.19	PAR-CLIP
5594	*MAPK1*	*mitogen-activated protein kinase 1*	*hsa-miR-4433a-3p*	−0.17	PAR-CLIP
5594	*MAPK1*	*mitogen-activated protein kinase 1*	*hsa-miR-6869-5p*	1.18	PAR-CLIP
5594	*MAPK1*	*mitogen-activated protein kinase 1*	*hsa-miR-3135b*	null	HITS-CLIP
5602	*MAPK10*	*mitogen-activated protein kinase 10*	*hsa-miR-574-5p*	−2.63	HITS-CLIP
5602	*MAPK10*	*mitogen-activated protein kinase 10*	*hsa-miR-6867-5p*	−0.84	HITS-CLIP
5599	*MAPK8*	*mitogen-activated protein kinase 8*	*hsa-miR-371b-5p*	null	PAR-CLIP
5293	*PIK3CD*	*phosphatidylinositol-4,5-bisphosphate 3-kinase catalytic subunit delta*	*hsa-miR-4433a-3p*	−0.17	PAR-CLIP
5294	*PIK3CG*	*phosphatidylinositol-4,5-bisphosphate 3-kinase catalytic subunit gamma*	*hsa-miR-8485*	−1.81	PAR-CLIP
5295	*PIK3R1*	*phosphoinositide-3-kinase regulatory subunit 1*	*hsa-miR-8485*	−1.81	PAR-CLIP
5295	*PIK3R1*	*phosphoinositide-3-kinase regulatory subunit 1*	*hsa-miR-1202*	−0.94	PAR-CLIP
5296	*PIK3R2*	*phosphoinositide-3-kinase regulatory subunit 2*	*hsa-miR-3135b*	null	PAR-CLIP
5879	*RAC1*	*ras-related C3 botulinum toxin substrate 1*	*hsa-miR-574-3p*	−1.57	Luciferase reporter assay;Microarray;qRT-PCR;Western blot
5879	*RAC1*	*ras-related C3 botulinum toxin substrate 1*	*hsa-miR-6763-5p*	−1.09	PAR-CLIP
5879	*RAC1*	*ras-related C3 botulinum toxin substrate 1*	*hsa-miR-6124*	−0.94	PAR-CLIP
5879	*RAC1*	*ras-related C3 botulinum toxin substrate 1*	*hsa-miR-939-5p*	null	PAR-CLIP
5881	*RAC3*	*ras-related C3 botulinum toxin substrate 3*	*hsa-miR-5703*	−1.70	PAR-CLIP
5881	***RAC3***	*ras-related C3 botulinum toxin substrate 3*	***hsa-miR-4516***	1.51	PAR-CLIP
5888	*RAD51*	*RAD51 recombinase*	*hsa-miR-940*	−3.46	PAR-CLIP
5888	*RAD51*	*RAD51 recombinase*	*hsa-miR-7847-3p*	−0.82	PAR-CLIP

Table 5. *Cont.*

Entrez ID	Symbol	Gene Name	miRNA	Log2 (PDAC Patients/Control)	Experimental Evidence
5888	*RAD51*	*RAD51 recombinase*	*hsa-miR-1915-3p*	0.72	HITS-CLIP
5888	*RAD51*	*RAD51 recombinase*	*hsa-miR-7975*	1.09	HITS-CLIP
5888	*RAD51*	*RAD51 recombinase*	*hsa-miR-371b-5p*	null	HITS CLIP
5894	*RAF1*	*Raf-1 proto-oncogene, serine/threonine kinase*	*hsa-miR-4534*	−0.48	PAR-CLIP
5894	*RAF1*	*Raf-1 proto-oncogene, serine/threonine kinase*	*hsa-miR-6789-5p*	−0.27	PAR-CLIP
5970	*RELA*	*RELA proto-oncogene, NF-kB subunit*	*hsa-miR-3162-3p*	−2.72	PAR-CLIP
5970	*RELA*	*RELA proto-oncogene, NF-kB subunit*	*hsa-miR-4534*	−0.48	PAR-CLIP
4087	*SMAD2*	*SMAD family member 2*	*hsa-miR-8485*	−1.81	HITS-CLIP
4087	*SMAD2*	*SMAD family member 2*	*hsa-miR-937-5p*	−0.38	HITS-CLIP
4089	*SMAD4*	*SMAD family member 4*	*hsa-miR-574-5p*	−2.63	HITS-CLIP
4089	*SMAD4*	*SMAD family member 4*	*hsa-miR-574-3p*	−1.57	Luciferase reporter assay;qRT-PCR;Western blot
4089	*SMAD4*	*SMAD family member 4*	*hsa-miR-6867-5p*	−0.84	HITS-CLIP
4089	*SMAD4*	*SMAD family member 4*	*hsa-miR-371b-5p*	null	PAR-CLIP
6774	*STAT3*	*signal transducer and activator of transcription 3*	*hsa-miR-4270*	−0.97	PAR-CLIP
6774	***STAT3***	*signal transducer and activator of transcription 3*	***hsa-miR-4516***	1.51	Luciferase reporter assay;Microarray;qRT-PCR;Western blot
7040	*TGFB1*	*transforming growth factor beta 1*	*hsa-miR-574-3p*	−1.57	Luciferase reporter assay;qRT-PCR;Western blot
7046	*TGFBR1*	*transforming growth factor beta receptor 1*	*hsa-miR-6831-5p*	−0.91	HITS-CLIP
7048	*TGFBR2*	*transforming growth factor beta receptor 2*	*hsa-miR-940*	−3.46	PAR-CLIP
7048	*TGFBR2*	*transforming growth factor beta receptor 2*	*hsa-miR-574-5p*	−2.63	HITS-CLIP
7048	*TGFBR2*	*transforming growth factor beta receptor 2*	*hsa-miR-630*	−1.34	Microarray
7048	*TGFBR2*	*transforming growth factor beta receptor 2*	*hsa-miR-6867-5p*	−0.84	HITS-CLIP
7157	*TP53*	*tumor protein p53*	*hsa-miR-6127*	−1.90	PAR-CLIP
7157	*TP53*	*tumor protein p53*	*hsa-miR-5703*	−1.70	PAR-CLIP
7157	*TP53*	*tumor protein p53*	*hsa-miR-7110-5p*	−1.11	PAR-CLIP
7157	*TP53*	*tumor protein p53*	*hsa-miR-937-5p*	−0.38	PAR-CLIP
7157	*TP53*	*tumor protein p53*	*hsa-miR-6756-5p*	−0.30	PAR-CLIP
7157	*TP53*	*tumor protein p53*	*hsa-miR-4271*	−0.19	PAR-CLIP
7157	*TP53*	*tumor protein p53*	*hsa-miR-6752-5p*	0.02	PAR-CLIP
7157	***TP53***	*tumor protein p53*	***hsa-miR-4516***	1.51	PAR-CLIP
7422	*VEGFA*	*vascular endothelial growth factor A*	*hsa-miR-8485*	−1.81	PAR-CLIP;HITS-CLIP
7422	*VEGFA*	*vascular endothelial growth factor A*	*hsa-miR-6769b-5p*	−0.70	PAR-CLIP
7422	*VEGFA*	*vascular endothelial growth factor A*	*hsa-miR-6756-5p*	−0.30	PAR-CLIP

4. Discussion

PDAC is the only cancer type with a survival rate that has not improved in the last 40 years [8]. Since most patients are diagnosed when they have metastases, they can only be considered candidates for palliative treatment. Radical surgery for patients with locally confined tumors might not provide long-term survival, the median survival interval following diagnosis ranging from eight to ten months and early tumor relapse occurring in most patients [29]. Several factors affect the survival rates, such as cancer type, tumor stage at diagnosis, treatment modality, age, sex, overall health, lifestyle and differences in healthcare systems [36,37].

Although survival duration is mainly related to tumor stage, both long-term and short-term survivals have been described in patients diagnosed with early-stage tumors [29,36,37] Recent studies have reported the prognostic utility of circulating miRNAs profiling in several malignancies due to their altered expression during tumorigenesis and their stability in body fluids [17,18]. In sera, miRNAs could be either entrapped in exosomes or they could circulate bound to proteins, such as Ago2 [38]. Although circulating exosomes are important vehicles of tumor-derived miRNAs and a potential source for biomarker identification, the overall RNA yield that could be obtained after exosomes isolation is generally lower than that from pooled sera, thus compromising test sensitivity [39]. For this reason, we choose to analyse whole sera.

In the present retrospective study, we performed whole miRNAs microarray analysis in the sera of cancer-free controls and of selected early-stage PDAC patients with different survival rates (<14 months or >22 months after curative surgery) in order to establish whether miRNAs measurement can enable early disease diagnosis and predict survival in a minimally invasive way. Unsupervised cluster analysis allowed a distinction between PDAC patients and controls. A number of 76 miRNAs were identified as differently expressed between PDAC patients (short and long-term survivors) with respect to controls, and the large majority of them were downregulated in PDAC patients with respect to controls, in line with findings reported in the current literature. In fact, the overall downregulation of miRNAs, emerging as a common hallmark of cancer, contributes to the malignant phenotype and a poor prognosis [40]. Our results not only corroborate this assumption, but also indicate that this is an early and detectable finding, since our patients' series comprised only early-stage PDAC. Furthermore, seven circulating miRNAs (*hsa-miR-6126*, *hsa-miR-2392*, *hsa-miR-4327*, *hsa-miR-939-5p*, *hsa-miR-4655-3p*, *hsa-miR-371b-5p*, *hsa-miR-3135b*), expressed in low levels, were found only in the sera of PADC patients and not in controls, thus suggesting their potential role as an emerging class of suitable biomarkers for this cancer type.

Further studies are required to investigate the effect of these selected miRNAs, which were until now poorly evaluated in the PDAC setting, although some of them have been studied in other cancer types. Our findings are in agreement with those of Mazza et al., who demonstrated that *hsa-miR-6126* was one of the most significantly increased miRNAs in the plasma of PDAC patients [41], and the same *hsa-miR-6126* has been found in the tumor tissues of five patients with stage II colon cancer but not in cancer-free tissues [42]. *hsa-miR-939-5p*, of potential interest for its role in invasive cancer, is reportedly upregulated in lung adenocarcinomas [43], hepatocellular carcinomas [44,45], ovarian cancer tissue [46] and in the promotion of blood vessel invasion in breast cancer [47]. The recent study by Shen et al. [48] shed some light on the role of this *hsa-miR-939-5p* in PDAC biology and prognosis. By studying human pancreatic cancer tissues and cell lines, the authors demonstrated that high expression levels of this miRNA are associated with a poor prognosis in patients, while promoting in vitro tumor cell migration and invasion by targeting the Rho GTPase, activating protein 4.

Another series of six miRNAs was found to be significantly and independently correlated with early PDAC, namely *hsa-miR-6089*, *hsa-miR-4466*, *hsa-miR-6821-5p*, *hsa-miR-4669*, *hsa-miR-1202* and *hsa-miR-574-3p*. Among these, *hsa-miR-6821-5p* was associated with the best ROC curve area in distinguishing controls from early PDAC, supporting its potential role in early diagnosis. This result fits well with findings of Keller et al. [49], who demonstrated that *hsa-miR-6821-5p* is one of the most deregulated miRNAs in serum for decades prior lung, breast, or colon cancer diagnosis. The same study emphasized *hsa-miR-4687-3p* and *hsa-miR-574-3p* as potential pre-diagnostic markers and, in agreement, we identified these miRNAs among those most significant in distinguishing early PDAC from controls. All the above-described miRNAs outperformed with respect to the established PDAC biomarker CA 19-9, which levels did not significantly differ between controls and patients. This finding is not surprising, considering its low sensitivity for early tumor diagnosis [50].

When miRNA expression data of PDAC patients were analysed in order to identify a miRNA transcriptional signature characterizing early-stage PDAC patients with different prognoses, unsupervised cluster analysis did not enable a distinction between short-term and long-term survivors, but SAM two-class analysis (high FDR: 62.5%) allowed the identification of 71 differentially expressed miRNAs. The majority of survival-related differentially expressed miRNAs were downregulated in short-term survival with respect to long-term survival patients, thus confirming the association between cancer progression and miRNA downregulation, as reported above [40].

We first focused our attention on the three miRNAs that were expressed exclusively in the sera of PDAC patients with long-term (*hsa-miR-486-5p* and *hsa-miR-6126*) or short-term (*hsa-miR-3135b*) survival, but not in controls.

The finding of *hsa-miR-486-5p* expression exclusively in long-term survival patients supports the hypothesis that tumor progression determines a progressive decline for this miRNA expression, as occurs in colorectal cancer [51]. While no data in the literature report an association between *hsa-miR-3135b* and cancer progression and/or survival, *hsa-miR-6126* has been recently identified by Mazza et al. in PDAC plasma with levels increasing with metastases, but not correlated with survival [41]. The differences between our results and the results of Mazza et al. [41] might depend on the different characteristics of the patients studied: our series was mainly represented by stage I-II, while the series of Mazza et al. by stage III-IV PDAC cases.

To obtain further insights into the association between serum miRNAs and outcome, we verified whether miRNAs were correlated with the presence or absence of lymph-node metastases, known to be associated with worse outcome [52]. Eight miRNAs were identified among the most significant (Figure 2). Few data are present in the literature on these miRNAs, but *hsa-miR-1202*, which decreased in the presence of lymph-node metastases, was described to similarly decline with the progression of cervical cancer in tissue samples [53] and it was shown to exert an anti-tumor effect in vitro on HCC cells lines [54]. The *hsa-miRNA-4669*, which also declines in the presence of lymph-node metastases, was also described to diminish in sera of colorectal cancer patients [55].

One upregulated miRNA, namely hsa-miR-4516, was confirmed by qRT-PCR in an independent PDAC patients' validation cohort to have increased expression levels in PDAC with respect to controls. This miRNA was also an independent predictor of survival, which was correlated as expected with tumor stage, but also with hemoglobin levels, not with CA 19-9. Furthermore, the putative targets of this miRNA were statistically significantly enriched in the target's prediction analysis performed with miRTarBase, and they are involved in key pathways of cancer progression and dissemination such as the "Wnt signalling pathway" (*CCND2*, *RAC3*, *TP53*, *WNT8B*, *FBXW11*, *SOX17*, *TBL1XR1*) and the "p53 signalling pathway" (*CDKN1A*, *SFN*, *TP53*, *RPRM*).

Overall, our results support the importance of exploring the use of liquid biopsies in PDAC patients, managing the clinical utility of circulating miRNAs as novel diagnostic and prognostic biomarkers for this disease.

The main limitation of our study is related to the few numbers of the studied patients in both the exploratory and validation cohorts. This limitation derives from the choice to study early tumors with short or particularly long survival that imposed strict selection criteria (cases of early-stage cancer, without jaundice, with diabetes or reduced glucose tolerance, and with survival data including long-term survival). Considering that at diagnosis, PDAC patients with stage I–II are less than 20% [56], this left few cases within a large retrospective cohort meeting the requirements. On the other hand, our findings were not biased by tumor stage and allowed to consider the identified circulating miRNAs as early biomarkers.

5. Conclusions

In conclusion, the findings made in the present study suggest that new circulating miRNAs are a potentially useful biomarker for making an early diagnosis of PDAC and for establishing the prognosis. These miRNAs include *hsa-miR-3135b, hsa-miR-6126, hsa-miR-486-5p, hsa-miR-6821-5p, hsa-miR-4669* and *hsa-miR-4516* that, after future and independent validation, could be measured in the sera of PDAC patients, thus enabling improved management and prolonging survival.

Supplementary Materials: The following are available online at https://www.mdpi.com/article/10.3390/biomedicines9070845/s1. Table S1: Differentially expressed miRNAs between PDAC patients and controls. Table S2: Differentially expressed miRNAs between long survival PDAC patients and controls. Table S3: Differentially expressed miRNAs between short survival PDAC patients and controls. Table S4: Differentially expressed miRNAs between short and long survival PDAC patients. Table S5: miRNA's putative targets prediction by miRTarBase. Table S6: KEGG pathways. Table S7: Wnt signalling pathway. Table S8: p53 signalling pathway. Table S9. TGF-beta signalling pathway.

Author Contributions: Conceptualization, A.A., D.B. and C.D.P.; methodology, A.A., C.M. and B.P.; software, C.D.P.; validation, D.B. and C.D.P.; formal analysis, C.D.P.; investigation, A.A., C.M., C.S. and B.P.; resources, D.B.; data curation, A.A., D.B. and C.D.P.; writing—original draft preparation, A.A. and D.B.; writing—review and editing, C.M. and C.D.P.; visualization, C.S.; supervision, M.P.; project administration, D.B. and C.D.P.; funding acquisition, M.P. All authors have read and agreed to the published version of the manuscript.

Funding: This work was supported by the European Commission under the 7th Framework Programme [grant agreement no. 306031]. Under the auspices of the WIRSUNG ONLUS Association, Padova, Italy.

Institutional Review Board Statement: The study was conducted according to the guidelines of the Declaration of Helsinki and approved by the Ethics Committee of the University-Hospital of Padova (protocol code 2872P, date of approval: 20 February 2013).

Informed Consent Statement: Informed consent was obtained from all subjects involved in the study.

Data Availability Statement: Raw miRNA data are available in the U.S. National Centre for Biotechnology Information Gene Expression Omnibus (GEO, http://www.ncbi.nlm.nih.gov/geo (accessed on 16 March 2021)) database with the Accession N. GSE168996.

Acknowledgments: We would like to thank Paolo Martini (Dept. Molecular and Translational Medicine, University of Brescia, 25123 Brescia, Italy) and Gabriele Sales (Dept. Biology, University of Padova, 35131 Padova, Italy) for help regarding bioinformatics programming, statistical analysis, and miRNA target predictions.

Conflicts of Interest: The authors declare no conflict of interest. The funders had no role in the design of the study; in the collection, analyses, or interpretation of data; in the writing of the manuscript, or in the decision to publish the results.

References

1. Sung, H.; Ferlay, J.; Siegel, R.L.; Laversanne, M.; Soerjomataram, I.; Jemal, A.; Bray, F. Global cancer statistics 2020: GLOBOCAN estimates of incidence and mortality worldwide for 36 cancers in 185 countries. *CA Cancer J. Clin.* **2021**, *71*, 209–249. [CrossRef] [PubMed]
2. Siegel, R.L.; Miller, K.D.; Fuchs, H.E.; Jemal, A. Cancer Statistics, 2021. *CA Cancer J. Clin.* **2021**, *71*, 7–33. [CrossRef] [PubMed]
3. Bardeesy, N.; Depinho, R.A. Pancreatic cancer biology and genetics. *Nat. Rev. Cancer* **2002**, *2*, 897–909. [CrossRef] [PubMed]
4. Ryan, D.P.; Hong, T.S.; Bardeesy, N. Pancreatic adenocarcinoma. *N. Engl. J. Med.* **2014**, *371*, 1039–1049. [CrossRef]
5. Hezel, A.F.; Kimmelman, A.C.; Stanger, B.Z.; Bardeesy, N.; Depinho, R.A. Genetics and biology of pancreatic ductal adenocarcinoma. *Genes Dev.* **2006**, *20*, 1218–1249. [CrossRef] [PubMed]
6. Al-Share, B.; Hammad, N.; Diab, M. Pancreatic adenocarcinoma: Molecular drivers and the role of targeted therapy. *Cancer Metastasis Rev.* **2021**, *40*, 355–371. [CrossRef] [PubMed]
7. Gheorghe, G.; Bungau, S.; Ilie, M.; Behl, T.; Vesa, C.M.; Brisc, C.; Bacalbasa, N.; Turi, V.; Costache, R.S.; Diaconu, C.C. Early diagnosis of pancreatic cancer: The key for survival. *Diagnostics* **2020**, *10*, 869. [CrossRef] [PubMed]
8. DeSantis, C.E.; Lin, C.C.; Mariotto, A.B.; Siegel, R.L.; Stein, K.D.; Kramer, J.L.; Alteri, R.; Robbins, A.S.; Jemal, A. Cancer treatment and survivorship statistics, 2014. *CA Cancer J. Clin.* **2014**, *64*, 252–271. [CrossRef] [PubMed]

9. Cong, L.; Liu, Q.; Zhang, R.; Cui, M.; Zhang, X.; Gao, X.; Guo, J.; Dai, M.; Zhang, T.; Liao, Q.; et al. Tumor size classification of the 8th edition of TNM staging system is superior to that of the 7th edition in predicting the survival outcome of pancreatic cancer patients after radical resection and adjuvant chemotherapy. *Sci. Rep.* **2018**, *8*, 10383–10389. [CrossRef] [PubMed]
10. Waddell, N.; Pajic, M.; Patch, A.-M.; Chang, D.K.; Kassahn, K.S.; Bailey, P.; Johns, A.L.; Miller, D.; Nones, K.; Quek, K.; et al. Whole genomes redefine the mutational landscape of pancreatic cancer. *Nature* **2015**, *518*, 495–501. [CrossRef] [PubMed]
11. Witkiewicz, A.K.; McMillan, E.A.; Balaji, U.; Baek, G.; Lin, W.-C.; Mansour, J.; Mollaee, M.; Wagner, K.-U.; Koduru, P.; Yopp, A.; et al. Whole-exome sequencing of pancreatic cancer defines genetic diversity and therapeutic targets. *Nat. Commun.* **2015**, *6*, 6744. [CrossRef]
12. Bailey, P.; Chang, D.K.; Nones, K.; Johns, A.L.; Patch, A.-M.; Gingras, M.-C.; Miller, D.K.; Christ, A.N.; Bruxner, T.J.C.; Quinn, M.C.; et al. Genomic analyses identify molecular subtypes of pancreatic cancer. *Nature* **2016**, *531*, 47–52. [CrossRef] [PubMed]
13. Springer, S.; Wang, Y.; Molin, M.D.; Masica, D.L.; Jiao, Y.; Kinde, I.; Blackford, A.; Raman, S.P.; Wolfgang, C.L.; Tomita, T.; et al. A combination of molecular markers and clinical features improve the classification of pancreatic cysts. *Gastroenterology* **2015**, *149*, 1501–1510. [CrossRef]
14. Knudsen, E.S.; O'Reilly, E.M.; Brody, J.R.; Witkiewicz, A.K. Genetic diversity of pancreatic ductal adenocarcinoma and opportunities for precision medicine. *Gastroenterology* **2016**, *150*, 48–63. [CrossRef] [PubMed]
15. Oshima, M.; Okano, K.; Muraki, S.; Haba, R.; Maeba, T.; Suzuki, Y.; Yachida, S. Immunohistochemically detected expression of 3 major genes (CDKN2A/p16, TP53, and SMAD4/DPC4) strongly predicts survival in patients with resectable pancreatic cancer. *Ann. Surg.* **2013**, *258*, 336–346. [CrossRef] [PubMed]
16. Iacobuzio-Donahue, C.A.; Fu, B.; Yachida, S.; Luo, M.; Abe, H.; Henderson, C.M.; Vilardell, F.; Wang, Z.; Keller, J.W.; Banerjee, P.; et al. DPC4 gene status of the primary carcinoma correlates with patterns of failure in patients with pancreatic cancer. *J. Clin. Oncol.* **2009**, *27*, 1806–1813. [CrossRef]
17. Berindan-Neagoe, I.; Monroig, P.D.C.; Pasculli, B.; Calin, G.A. MicroRNAome genome: A treasure for cancer diagnosis and therapy. *CA Cancer J. Clin.* **2014**, *64*, 311–336. [CrossRef]
18. Sarwar, A.; Wang, B.; Su, Q.; Zhang, Y. MiRNAs directly targeting the key intermediates of biological pathways in pancreatic cancer. *Biochem. Pharmacol.* **2020**, *70*, 114357.
19. Rong, Z.; Xu, J.; Shi, S.; Tan, Z.; Meng, Q.; Hua, J.; Liu, J.; Zhang, B.; Wang, W.; Yu, X.; et al. Circular RNA in pancreatic cancer: A novel avenue for the roles of diagnosis and treatment. *Theranostics* **2021**, *11*, 2755–2769. [CrossRef]
20. Pandya, G.; Kirtonia, A.; Sethi, G.; Pandey, A.K.; Garg, M. The implication of long non-coding RNAs in the diagnosis, pathogenesis and drug resistance of pancreatic ductal adenocarcinoma and their possible therapeutic potential. *Biochim. Biophys. Acta Rev. Cancer* **2020**, *1874*, 188423. [CrossRef]
21. Terrinoni, A.; Calabrese, C.; Basso, D.; Aita, A.; Caporali, S.; Plebani, M.; Bernardini, S. The circulating miRNAs as diagnostic and prognostic markers. *Clin. Chem. Lab. Med.* **2019**, *57*, 932–953. [CrossRef]
22. Hao, J.; Zhang, S.; Zhou, Y.; Hu, X.; Shao, C. MicroRNA 483-3p suppresses the expression of DPC4/Smad4 in pancreatic cancer. *FEBS Lett.* **2011**, *585*, 207–213. [CrossRef]
23. Blahna, M.T.; Hata, A. Regulation of miRNA biogenesis as an integrated component of growth factor signaling. *Curr. Opin. Cell Biol.* **2013**, *25*, 233–240. [CrossRef]
24. Yang, W.; Ju, H.-Y.; Tian, X.-F. Hsa-miR-4730 as a new and potential diagnostic and prognostic indicators for pancreatic cancer. *Eur. Rev. Med. Pharmacol. Sci.* **2020**, *24*, 8801–8811.
25. Hata, T.; Mizuma, M.; Masuda, K.; Chiba, K.; Ishida, M.; Ohtsuka, H.; Nakagawa, K.; Morikawa, T.; Kamei, T.; Unno, M. MicroRNA-593-3p expression in peritoneal lavage fluid as a prognostic marker for pancreatic cancer patients undergoing staging laparoscopy. *Ann. Surg. Oncol.* **2021**, *70*, 7–11.
26. Wolfe, A.R.; Wald, P.; Webb, A.; Sebastian, N.; Walston, S.; Robb, R.; Chen, W.; Vedaie, M.; Dillhoff, M.; Frankel, W.L.; et al. A microRNA-based signature predicts local-regional failure and overall survival after pancreatic cancer resection. *Oncotarget* **2020**, *11*, 913–923. [CrossRef] [PubMed]
27. Wu, X.; Huang, J.; Yang, Z.; Zhu, Y.; Zhang, Y.; Wang, J.; Yao, W. MicroRNA-221-3p is related to survival and promotes tumour progression in pancreatic cancer: A comprehensive study on functions and clinicopathological value. *Cancer Cell Int.* **2020**, *20*, 443. [CrossRef] [PubMed]
28. Schultz, N.A.; Dehlendorff, C.; Jensen, B.V.; Bjerregaard, J.K.; Nielsen, K.R.; Bojesen, S.E.; Calatayud, D.; Nielsen, S.E.; Yilmaz, M.; Holländer, N.H.; et al. MicroRNA biomarkers in whole blood for detection of pancreatic cancer. *JAMA* **2014**, *311*, 392–404. [CrossRef] [PubMed]
29. Ma, J.; Jemal, A. The rise and fall of cancer mortality in the USA: Why does pancreatic cancer not follow the trend? *Future Oncol.* **2013**, *9*, 917–919. [CrossRef] [PubMed]
30. Wang, H.; Ach, R.A.; Curry, B. Direct and sensitive miRNA profiling from low-input total RNA. *RNA* **2007**, *13*, 151–159. [CrossRef]
31. Risso, D.; Massa, M.S.; Chiogna, M.; Romualdi, C. A modified LOESS normalization applied to microRNA arrays: A comparative evaluation. *Bioinformatics* **2009**, *25*, 2685–2691. [CrossRef]
32. Saeed, A.I.; Bhagabati, N.K.; Braisted, J.C.; Liang, W.; Sharov, V.; Howe, E.A.; Li, J.; Thiagarajan, M.; White, J.A.; Quackenbush, J. TM4 microarray software suite. *Methods Enzymol.* **2006**, *411*, 134–193.
33. Tibshirani, R.; Hastie, T.; Narasimhan, B.; Chu, G. Diagnosis of multiple cancer types by shrunken centroids of gene expression. *Proc. Natl. Acad. Sci. USA* **2002**, *99*, 6567–6572. [CrossRef]

34. Hsu, S.D.; Lin, F.M.; Wu, W.Y.; Liang, C.; Huang, W.C.; Chan, W.L.; Tsai, W.T.; Chen, G.Z.; Lee, C.J.; Chiu, C.M.; et al. miRTarBase A database curates experimentally validated microRNA-target interactions. *Nucleic Acids Res.* **2011**, *39*, D163–D169. [CrossRef]
35. Sherman, B.T.; Lempicki, R.A. Systematic and integrative analysis of large gene lists using DAVID bioinformatics resources. *Nat Protoc.* **2009**, *4*, 44–57.
36. Vega, E.A.; Kutlu, O.C.; Salehi, O.; James, D.; Alarcon, S.V.; Herrick, B.; Krishnan, S.; Kozyreva, O.; Conrad, C. Preoperative chemotherapy for pancreatic cancer improves survival and R0 rate even in early stage I. *J. Gastrointest. Surg.* **2020**, *24*, 2409–2415 [CrossRef]
37. Shapiro, M.; Chen, Q.; Huang, Q.; Boosalis, V.A.; Yoon, C.H.; Saund, M.S.; Whang, E.E.; Gold, J.S. Associations of socioeconomic variables with resection, stage, and survival in patients with early-stage pancreatic cancer. *JAMA Surg.* **2016**, *151*, 338–345 [CrossRef] [PubMed]
38. Turchinovich, A.; Weiz, L.; Langheinz, A.; Burwinkel, B. Characterization of extracellular circulating microRNA. *Nucleic Acids Res.* **2011**, *39*, 7223–7233. [CrossRef]
39. Endzeliņš, E.; Berger, A.; Melne, V.; Bajo-Santos, C.; Soboļevska, K.; Ābols, A.; Rodriguez, M.; Šantare, D.; Rudņickiha, A.; Lietuvietis, V.; et al. Detection of circulating miRNAs: Comparative analysis of extracellular vesicle-incorporated miRNAs and cell-free miRNAs in whole plasma of prostate cancer patients. *BMC Cancer* **2017**, *17*, 730. [CrossRef] [PubMed]
40. Williams, M.; Cheng, Y.Y.; Blenkiron, C.; Reid, G. Exploring mechanisms of microRNA downregulation in cancer. *Microrna* **2017**, *6*, 2–16. [CrossRef] [PubMed]
41. Mazza, T.; Gioffreda, D.; Fontana, A.; Biagini, T.; Carella, M.; Palumbo, O.; Maiello, E.; Bazzocchi, F.; Andriulli, A.; Tavano, F. Clinical significance of circulating miR-1273g-3p and miR-122-5p in pancreatic cancer. *Front. Oncol.* **2020**, *10*, 44. [CrossRef] [PubMed]
42. Gungormez, C.; Gumushan Aktas, H.; Dilsiz, N.; Borazan, E. Novel miRNAs as potential biomarkers in stage II colon cancer: Microarray analysis. *Mol. Biol. Rep.* **2019**, *46*, 4175–4183. [CrossRef] [PubMed]
43. Rani, S.; Gately, K.; Crown, J.; O'Byrne, K.; O'Driscoll, L. Global analysis of serum microRNAs as potential biomarkers for lung adenocarcinoma. *Cancer Biol. Ther.* **2013**, *14*, 1104–1112. [CrossRef] [PubMed]
44. Fornari, F.; Ferracin, M.; Trerè, D.; Milazzo, M.; Marinelli, S.; Galassi, M.; Venerandi, L.; Pollutri, D.; Patrizi, C.; Borghi, A.; et al. Circulating microRNAs, miR-939, miR-595, miR-519d and miR-494, identify cirrhotic patients with HCC. *PLoS ONE* **2015**, *10*, e0141448. [CrossRef]
45. Chen, F.; Ni, X.; Chen, L.; Wang, X.; Xu, J. miR-939-3p promotes epithelial-mesenchymal transition and may be used as a prognostic marker in hepatocellular carcinoma. *Oncol. Lett.* **2020**, *19*, 2727–2732. [CrossRef] [PubMed]
46. Ying, X.; Li-ya, Q.; Feng, Z.; Yin, W.; Ji-hong, L. MiR-939 promotes the proliferation of human ovarian cancer cells by repressing APC2 expression. *Biomed Pharm.* **2015**, *71*, 64–69. [CrossRef]
47. Di Modica, M.; Regondi, V.; Sandri, M.; Iorio, M.V.; Zanetti, A.; Tagliabue, E.; Casalini, P.; Triulzi, T. Breast cancer-secreted miR-939 downregulates VE-cadherin and destroys the barrier function of endothelial monolayers. *Cancer Lett.* **2017**, *384*, 94–100. [CrossRef]
48. Shen, Y.; Chen, G.; Gao, H.; Li, Y.; Zhuang, L.; Meng, Z.; Liu, L. miR-939-5p contributes to the migration and invasion of pancreatic cancer by targeting ARHGAP4. *OncoTargets Ther.* **2020**, *13*, 389–399. [CrossRef]
49. Keller, A.; Fehlmann, T.; Backes, C.; Kern, F.; Gislefoss, R.; Langseth, H.; Rounge, T.B.; Ludwig, N.; Meese, E. Competitive learning suggests circulating miRNA profiles for cancers decades prior to diagnosis. *RNA Biol.* **2020**, *17*, 1416–1426. [CrossRef]
50. Luo, G.; Jin, K.; Deng, S.; Cheng, H.; Fan, Z.; Gong, Y.; Qian, Y.; Huang, Q.; Ni, Q.; Liu, C.; et al. Roles of CA19-9 in pancreatic cancer: Biomarker, predictor and promoter. *Biochim. Biophys. Acta Rev. Cancer* **2020**, *1875*, 188409. [CrossRef]
51. Pisano, A.; Griñan-Lison, C.; Farace, C.; Fiorito, G.; Fenu, G.; Jiménez, G.; Scognamillo, F.; Peña-Martin, J.; Naccarati, A.; Pröll, J.; et al. The inhibitory role of miR-486-5p on CSC phenotype has diagnostic and prognostic potential in colorectal cancer. *Cancers* **2020**, *12*, 3432. [CrossRef] [PubMed]
52. Pederzoli, P.; Bassi, C.; Falconi, M.; Pedrazzoli, S. Does the extent of lymphatic resection affect the outcome in pancreatic cancer? *Digestion* **1997**, *58*, 536–541. [CrossRef] [PubMed]
53. Yang, X.; Yan, Z.; Yang, H.; Ni, H.; Zhang, L.; Wang, Y. Clinical value of combined detection of miR-1202 and miR-195 in early diagnosis of cervical cancer. *Oncol. Lett.* **2019**, *17*, 3387–3391. [CrossRef] [PubMed]
54. Du, B.; Zhang, P.; Tan, Z.; Xu, J. MiR-1202 suppresses hepatocellular carcinoma cells migration and invasion by targeting cyclin dependent kinase 14. *Biomed Pharm.* **2017**, *96*, 1246–1252. [CrossRef]
55. Wang, Y.N.; Chen, Z.H.; Chen, W.C. Novel circulating microRNAs expression profile in colon cancer: A pilot study. *Eur. J. Med. Res.* **2017**, *22*, 51. [CrossRef] [PubMed]
56. Hidalgo, M.; Cascinu, S.; Kleeff, J.; Labianca, R.; Löhr, J.M.; Neoptolemos, J.; Real, F.X.; Van Laethem, J.L.; Heinemann, V. Addressing the challenges of pancreatic cancer: Future directions for improving outcomes. *Pancreatology* **2015**, *15*, 8–18. [CrossRef] [PubMed]

biomedicines

MDPI

Article

Profiling 25 Bone Marrow microRNAs in Acute Leukemias and Secondary Nonleukemic Hematopoietic Conditions

Igor B. Kovynev [1], Sergei E. Titov [2,3], Pavel S. Ruzankin [4,5], Mechti M. Agakishiev [1,6], Yuliya A. Veryaskina [2,7,*], Viktor M. Nedel'ko [4,5], Tatiana I. Pospelova [1] and Igor F. Zhimulev [2]

1 Department of Therapy, Hematology and Transfusiology, Novosibirsk State Medical University, 630091 Novosibirsk, Russia; kovin_gem@mail.ru (I.B.K.); m_agakishiev@mail.ru (M.M.A.); depart04@mail.ru (T.I.P.)
2 Laboratory of Molecular Genetics, Department of the Structure and Function of Chromosomes, Institute of Molecular and Cellular Biology, Siberian Branch of the Russian Academy of Sciences, 630090 Novosibirsk, Russia; titovse78@gmail.com (S.E.T.); zhimulev@mcb.nsc.ru (I.F.Z.)
3 AO Vector-Best, 630117 Novosibirsk, Russia
4 Sobolev Institute of Mathematics, Siberian Branch of Russian Academy of Sciences, 630090 Novosibirsk, Russia; ruzankin@math.nsc.ru (P.S.R.); nedelko@math.nsc.ru (V.M.N.)
5 Department of Mathematics and Mechanics, Novosibirsk State University, 630090 Novosibirsk, Russia
6 Department of Hematology, City Clinical Hospital #2, 630051 Novosibirsk, Russia
7 Laboratory of Gene Engineering, Institute of Cytology and Genetics, Siberian Branch of the Russian Academy of Sciences, 630090 Novosibirsk, Russia
* Correspondence: microrna@inbox.ru

Received: 5 November 2020; Accepted: 12 December 2020; Published: 14 December 2020

Abstract: Introduction: The standard treatment of acute leukemias (AL) is becoming more efficacious and more selective toward the mechanisms via which to suppress hematologic cancers. This tendency in hematology imposes additional requirements on the identification of molecular-genetic features of tumor clones. MicroRNA (miRNA, miR) expression levels correlate with cytogenetic and molecular subtypes of acute leukemias recognized by classification systems. The aim of this work is analyzing the miRNA expression profiles in acute myeloblastic leukemia (AML) and acute lymphoblastic leukemia (ALL) and hematopoietic conditions induced by non-tumor pathologies (NTP). Methods: A total of 114 cytological samples obtained by sternal puncture and aspiration biopsy of bone marrow (22 ALLs, 44 AMLs, and 48 NTPs) were analyzed by real-time PCR regarding preselected 25 miRNAs. For the classification of the samples, logistic regression was used with balancing of comparison group weights. Results: Our results indicated potential feasibility of (i) differentiating ALL+AML from a nontumor hematopoietic pathology with 93% sensitivity and 92% specificity using miR-150:miR-21, miR-20a:miR-221, and miR-24:nf3 (where nf3 is a normalization factor calculated from threshold cycle values of miR-103a, miR-191, and miR-378); (ii) diagnosing ALL with 81% sensitivity and 81% specificity using miR-181b:miR-100, miR-223:miR-124, and miR-24:nf3; and (iii) diagnosing AML with 81% sensitivity and 84% specificity using miR-150:miR-221, miR-100:miR-24, and miR-181a:miR-191. Conclusion: The results presented herein allow the miRNA expression profile to de used for differentiation between AL and NTP, no matter what AL subtype.

Keywords: acute myeloblastic leukemia; acute lymphoblastic leukemia; microRNA

1. Introduction

Acute leukemia (AL) is a hematologic cancer arising from early hematopoietic precursors that have undergone malignant transformation and for that reason lost their ability to differentiate into mature blood cells. From a clinical perspective, the progression of this type of hematologic cancer may be evident on a hemogram as pancytopenia due to the withdrawal of normal lineages from bone marrow by tumor blasts, as pronounced tumor toxicity, as secondary lesions to internal organs due to blastic infiltration, and as hematopoietic deficiencies (anemia, hemorrhagic syndrome, and infection-based agranulocytosis). Special epidemiological features of ALs make these pathologies some of the most important types of hematologic cancers that are determinants of treatment efficacy of hematologic cancers [1].

The standard treatment of ALs is becoming more efficacious and more selective toward the mechanisms via which to suppress hematologic cancers. Cells that remain in the BM after chemotherapy are believed to be responsible for relapses. This is why protocols for an early and accurate detection of the residual leukemic cells (minimal residual disease (MRD)) are important. MRD monitoring may have importance for clinical decision making, as it allows survival rates and relapse risk to be accurately assessed [2].

Classical cell-based methods used to ascertain relapsed AL, including flow immunocytofluorometry, are well proven and reliable. However, these methods are no aids to monitor early molecular-genetic events in the genomes of the pre-cancerous precursors that have implications for AL progression. This is why the only privilege that a physician armed with standard laboratory methods for detecting a relapse has at the moment is to observe what clonal selection has done rather than to have full control of the process.

Residual leukemic cells are not seldom present in very low numbers, and so their detection requires more sensitive methods [3]. Real-time PCR and droplet digital PCR are some of them. However, MRD detection can be complicated by a phenomenon known as 'clonal evolution'. Thus, analysis of all molecular markers at diagnosis and first relapse may show that the predominant clone is not always of one and the same origin [4]. There is little doubt that the use of any MRD detection method requires a biomarker or a combination of biomarkers that can clearly differentiate between normal and cancer cells. This is especially difficult in AL due to a high genetic instability of tumor cells' genome and, as a consequence, a substantial clonal heterogeneity of this disease.

On the other hand, patients who are neither genetically nor epigenetically predisposed to tumor progression/relapse have to rely on standard chemotherapeutic relapse prevention protocols and are exposed to undue risk of side effects and even death.

Analysis of molecular-genetic factors in the pathogenesis of leukemias has allowed to discover a new regulatory mechanism underlying an abnormal function of the key genes responsible for differentiation into myeloid and lymphoid cell types. This mechanism involves microRNAs (miRNAs, miRs), i.e., short noncoding RNAs exerting regulatory action on the expression of target genes, both transcriptionally and translationally [5].

It has been demonstrated that miRNA expression levels correlate with cytogenetic and molecular AL subtypes recognized by classification systems and determine many properties of tumor blasts [5,6]. Aberrations in miRNA expression profiles in AL have been demonstrated [7–9]. Today, there is little doubt that miRNAs correlate strongly with the efficacy of standard chemotherapy for AL and clinical outcomes of hematologic cancers [10].

The aim of this work is analyzing the miRNA expression profiles in acute myeloblastic leukemia (AML), acute lymphoblastic leukemia (ALL), and hematopoietic conditions induced by non-tumor pathologies (NTP). Data obtained will allow us to identify the miRNAs that can be used as high-sensitivity biomarkers to detect MRD with.

2. Experimental Section

Clinical material. A total of 114 cytological samples were obtained by sternal puncture and aspiration biopsy of bone marrow on the posterior iliac spine. All the cases were AL patients at the Novosibirsk Municipal Hematological Center before treatment initiation. Cytological material was obtained in compliance with Russian laws and regulations, written informed consent was obtained from each patient, and all the data were depersonalized. The study protocol No.15 of May 25, 2020 was approved by the Ethics Committee of Novosibirsk State Medical University.

The types of hematologic cancers included in the study population were ALL (22 specimens) and AML (44 specimens). The characteristics of the groups are shown in Supplementary Table S1. Work with healthy bone marrow donors for allogeneic transplantation is beyond the competence of our clinic. Taken together, we decided to go with a control group composed of people who had no hematologic cancer, but had indications for bone marrow examination to exclude one. They were people with secondary anemic and cytopenic conditions, in whom leukemias were not confirmed by myelography (NTP (48 specimens)). The characteristics of the NTP group are shown in Table 1.

Table 1. Clinical data of non-cancerous blood diseases (NTP) (n = 48).

Characteristic	n (%)
Gender	
Male	23 (48)
Female	25 (52)
Age	
>60 years	11 (23)
<60 years	37 (77)
Median hemoglobin, g/L	90
Median WBC count, ×109/L	6.7
Median ANC, /dL	5
Median platelet count, ×109/L	200.5
Sybtype	
Iron-deficiency anemia	28 (58)
hemolytic anemia	3 (6)
B12 deficiency anemia	5 (10)
chronic disease anemia	6 (13)
immune thrombocytopenia	5 (10)
aplastic anemia	1 (2)

Abbreviations: WBC = white blood cell; ANC = absolute neutrophil count.

Selecting miRNAs. MiRNAs were chosen based on literature data. The experimental analysis involved 25 miRNAs: miR-100-5p, -124-3p, -126-3p, -128-3p, -146a-5p, -150-5p, -155-5p, -18a-5p, -181a-5p, -181b-5p, -196b-5p, -20a-5p, -21-5p, -210-3p, -221-3p, -223-3p, -24-3p, -26a-5p, -29b-3p, -451a, -9-5p, -92a-3p, -96-5p, -99a-5p, and let-7a [10–15]. Reference miRNAs were miR-378-3p, -191-5p, and -103a-3p, which were selected by means of our original data and literature data [16]. In some classification variables, the geometric mean of threshold cycle (Ct) values of the three reference miRNAs was employed for normalization as proposed by Vandesompele [17].

Total nucleic acid isolation. Nucleic acid was isolated and, as described by Titov et al., a dried cytological smear was washed into a microcentrifuge tube with three 200 μL portions of guanidine lysis buffer [18]. The sample was vigorously mixed and incubated in a thermal shaker for 15 min at 65 °C. Next, an equal volume of isopropanol was added. The reaction solution was thoroughly mixed and kept at room temperature for 5 min. After centrifugation for 10 min at 14000 g, the supernatant was discarded, and the pellet was washed with 500 μL of 70% ethanol and 300 μL of acetone. Finally, the RNA was dissolved in 200 μL of deionized water. If not analyzed immediately, RNA samples were stored at 20 °C.

Oligonucleotide primers and probes. All the oligonucleotides, including fluorescently labeled ones, were synthesized by AO Vector-Best (Novosibirsk, Russia). The oligonucleotides were chosen using an online tool, PrimerQuest (https://eu.idtdna.com/). For each miRNA, several sets of oligonucleotides were chosen, from which those with the highest real-time PCR efficiency were selected. PCR efficiency was assessed by constructing a standard curve for serial dilutions of synthetic miRNA analogs

(OOO Biosan, Novosibirsk, Russia) of known concentration. Depending on the system, the E value varied from 91.5% to 99.8%. The sequences of the oligonucleotides are given in Supplementary Table S2.

Detection of miRNA by real-time PCR. Mature miRNAs were detected via the method proposed by Chen et al. [19]. For each miRNA, reverse transcription was carried out, followed by real-time PCR as described by Titov et al. [18]. Reverse transcription and PCR for each sample involved one replicate each. Concentrations of some miRNAs were normalized to "nf3," which is the geometric mean of Ct values of the three reference miRNAs (miR-103a-3p, miR-191-5p, and miR-378-3p), by the $2^{-\Delta Ct}$ method [20]. Concentrations of some other miRNAs were normalized to another miRNA (instead of nf3), as described below. In other words, nf3 served as the normalization factor for some classification variables.

Classifying samples. To investigate the association between miRNA concentrations and outcomes in NTP, ALL, and AML, classification variables were created that represent binary logarithms of pairwise ratios of miRNA concentrations; for example, the classification variable miR-150:miR-378 denotes Ct(miR-150) minus Ct(miR-378). The paper by Ivanov et al. explains why it is worthwhile to use ratios corresponding to pairs of markers (where an oncogenic or tumor suppressor miRNA is normalized to another marker miRNA) rather than stand-alone markers (individual miRNAs) normalized to housekeeping genes [21]. Above-mentioned normalization factor nf3 was employed to create some classification variables (i.e., to calculate some ratios), for example, miR-150:nf3 was equal to Ct(miR-150) − Ct(nf3). The reference miRNAs miR-103a-3p, miR-191-5p, and miR-378-3p were included in our analysis as nonreference miRNAs too, even though they are components of nf3. For instance, variables miR-150:miR-378 and miR-191:nf3 were utilized in the classification analysis. A total of 406 of such classification variables based on miRNA Ct values were tested in this analysis.

Primary analysis. For each classification variable, the following comparisons were made: NTP vs. others, ALL vs. others, AML vs. others, and ALL vs. AML. The comparisons were carried out by the exact Mann–Whitney test. In accordance with the Bonferroni approach to multiple comparisons, differences with *p*-values less than 0.05/(4 × 406) were considered statistically significant. In addition to the *p*-values, we calculated the following prediction accuracy measures for leave-one-out cross-validation: accuracy, sensitivity, specificity, and receiver-operating characteristic (ROC) area under the curve (AUC) with DeLong's confidence interval. Predicted values were computed via logistic regression with balancing of the comparison groups' weights. The accuracy, sensitivity, specificity, and ROC AUC values in the leave-one-out cross-validation were calculated for the threshold value of 0.5.

Secondary analysis. To assess the possibility of improving prediction accuracy, we examined all logistic regression models based on two or three of the created variables. In total, for each of the four comparisons, (406 × 405/2) + (406 × 405 × 404/6) = 11,153,835 models with two or three regressors were tested. Additionally, we tried the following machine learning methods: support vector machine, linear discriminant analysis, and boosting, but their performance was not better than that of logistic regression, and therefore their results are omitted here.

The computations were made in the R software, v.3.6.3 (R Core Team).

3. Results

3.1. *Comparing miRNA Concentrations among ALL, AML, and NTP Samples*

Relative concentrations of miRNAs in different sample types were determined by RT-PCR (Figure 1). As an example, the scatterplot of miR-150:nf3 and miR-221:nf3 for individual patients is displayed in Figure 2.

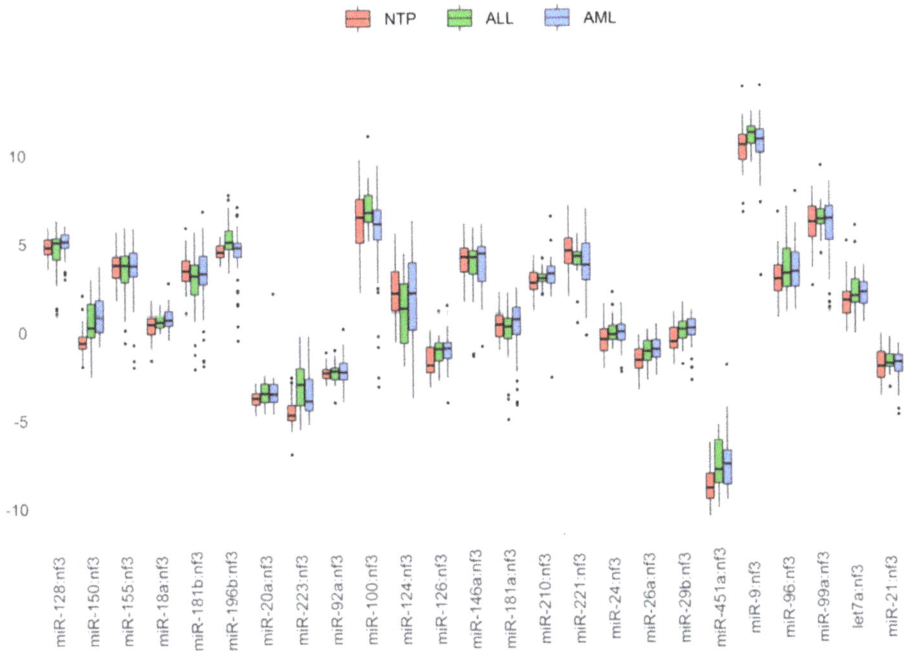

Figure 1. Boxplot for binary logarithms of some miRNA concentrations, those that were normalized to nf3. The boxes depict medians with the 1st and 3rd quartiles.

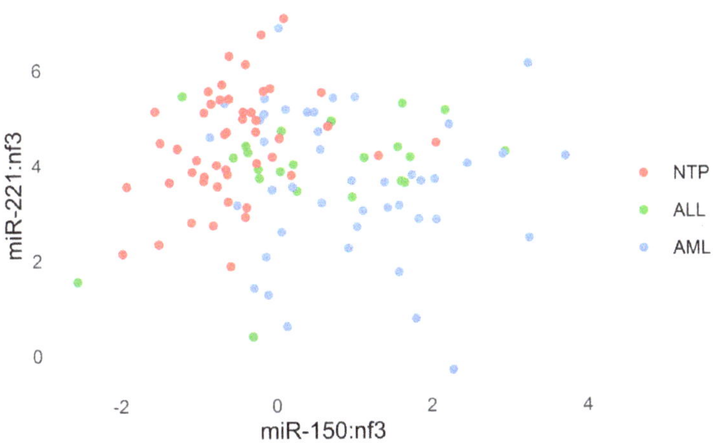

Figure 2. The scatterplot of miR-150:nf3 and miR-221:nf3 for individual patients.

3.2. Sample Classification

Although significant differences in the concentrations of seven miRNAs were found between either ALL or AML and NTP samples, none of the miRNAs could serve as a single marker (Figure 1). For this reason, when classifying samples, we used ratios of concentrations for pairs of miRNAs (some of these ratios involved reference miRNAs and/or nf3) in univariate classification models or combinations of such ratios in multivariable classification models.

The results of the comparison between NTP and the others are given in Table 2; those for ALL vs. others, in Table 3; the results on AML vs. others in Table 4, and the comparison between ALL and AML is presented in Table 5. The tables list univariate models for the variables for which $p \times 4 \times 406 < 0.05$ and those classification variables that are based on normalization to nf3. For the sake of convenience, in the tables, the p-values are multiplied by the number of comparisons 4×406 to allow the reader to compare the "normalized" p-values with the common threshold of 0.05, which is equivalent to comparing the "raw" p-values with $0.05/(4 \times 406)$. Additionally, some logistic regression models based on two or three variables with the best cross-validation accuracy are presented in the tables.

According to Table 2, which shows the results of the comparison between NTP and the others, the best prediction accuracy (93.8% sensitivity with 92.4% specificity) was manifested by the classification model based on covariates miR-150:miR-21, miR-20a:miR-221, and miR-24:nf3. Among univariate models, miR-150:miR-378 had the best prediction accuracy (87.5% sensitivity with 72.7% specificity).

Table 2. Comparison of NTP vs. others.

	p-Value × 4 × 406	CV Accuracy	CV Sensitivity	CV Specificity	CV AUC
miR-150:miR-21 + miR-20a:miR-221 + miR-24:nf3		0.930	0.938	0.924	0.949 (0.910, 0.989)
miR-150:miR-223 + miR-150:miR-221 + miR-126:miR-191		0.921	0.917	0.924	0.950 (0.910, 0.991)
miR-150:miR-223 + miR-150:nf3 + miR-126:miR-221		0.921	0.917	0.924	0.959 (0.926, 0.993)
miR-150:miR-223 + miR-223:miR-221 + miR-126:miR-191		0.921	0.917	0.924	0.950 (0.910, 0.991)
miR-150:miR-223 + miR-223:nf3 + miR-126:miR-221		0.921	0.917	0.924	0.959 (0.926, 0.993)
miR-150:miR-221 + miR-223:miR-221 + miR-126:miR-191		0.921	0.917	0.924	0.950 (0.910, 0.991)
miR-150:nf3 + miR-20a:miR-221 + miR-24:miR-103a		0.921	0.917	0.924	0.946 (0.900, 0.991)
miR-150:nf3 + miR-223:nf3 + miR-126:miR-221		0.921	0.917	0.924	0.959 (0.926, 0.993)
miR-150:miR-146a + miR-155:miR-221 + miR-24:miR-378		0.921	0.938	0.909	0.951 (0.910, 0.993)
miR-150:miR-221 + miR-196b:miR-99a + miR-24:nf3		0.921	0.958	0.894	0.950 (0.909, 0.992)
miR-223:miR-378 + miR-221:miR-24 + miR-29b:nf3		0.921	0.958	0.894	0.943 (0.895, 0.990)
miR-223:nf3 + miR-221:miR-24		0.886	0.917	0.864	0.919 (0.865, 0.972)

Table 2. *Cont.*

	p-Value × 4 × 406	CV Accuracy	CV Sensitivity	CV Specificity	CV AUC
miR-150:miR-221 + miR-24:miR-378		0.877	0.917	0.848	0.931 (0.886, 0.976)
miR-223:miR-221 + miR-126:miR-103a		0.877	0.875	0.879	0.929 (0.881, 0.977)
miR-223:miR-221 + miR-126:miR-191		0.877	0.896	0.864	0.932 (0.888, 0.975)
miR-223:miR-221 + miR 29b:nf3		0.877	0.875	0.879	0.930 (0.880, 0.980)
miR-150:miR-378	0.00000000071	0.789	0.875	0.727	0.863 (0.794, 0.932)
miR-150:nf3	0.0000000024	0.772	0.854	0.712	0.857 (0.785, 0.929)
miR-221:miR-24	0.000000044	0.746	0.792	0.712	0.836 (0.764, 0.908)
miR-223:miR-221	0.00000056	0.719	0.792	0.667	0.820 (0.744, 0.895)
miR-150:miR-221	0.00000099	0.728	0.792	0.682	0.817 (0.741, 0.893)
miR-150:miR-191	0.0000021	0.746	0.812	0.697	0.811 (0.730, 0.892)
miR-223:miR-378	0.0000045	0.693	0.792	0.621	0.806 (0.726, 0.886)
miR-150:miR-92a	0.000043	0.746	0.792	0.712	0.788 (0.703, 0.873)
miR-150:miR-103a	0.000045	0.719	0.771	0.682	0.785 (0.699, 0.871)
miR-128:miR-150	0.000068	0.693	0.729	0.667	0.782 (0.698, 0.867)
miR-150:miR-146a	0.00011	0.719	0.792	0.667	0.779 (0.695, 0.863)
miR-150:miR-181a	0.00011	0.711	0.792	0.652	0.778 (0.695, 0.861)
miR-451a:miR-103a	0.00015	0.684	0.667	0.697	0.775 (0.690, 0.860)
miR-150:miR-181b	0.00017	0.711	0.812	0.636	0.773 (0.687, 0.860)
miR-92a:miR-451a	0.00032	0.667	0.688	0.652	0.769 (0.682, 0.855)
miR-221:miR-26a	0.00045	0.728	0.771	0.697	0.764 (0.675, 0.853)
miR-126:miR-221	0.00059	0.728	0.750	0.712	0.764 (0.676, 0.852)
miR-150:miR-21	0.00099	0.719	0.729	0.712	0.759 (0.668, 0.849)
miR-451a:nf3	0.0015	0.649	0.667	0.636	0.753 (0.665, 0.842)
miR-451a:miR-21	0.0016	0.667	0.792	0.576	0.753 (0.662, 0.844)
miR-181b:miR-223	0.0019	0.640	0.833	0.500	0.751 (0.662, 0.841)
miR-221:miR-451a	0.0020	0.693	0.750	0.652	0.751 (0.663, 0.840)
miR-451a:miR-378	0.0032	0.667	0.750	0.606	0.745 (0.656, 0.834)
miR-221:miR-9	0.0034	0.702	0.771	0.652	0.747 (0.656, 0.837)
miR-221:miR-29b	0.0042	0.711	0.729	0.697	0.747 (0.656, 0.838)
miR-26a:miR-378	0.0050	0.675	0.646	0.697	0.741 (0.649, 0.833)
miR-150:miR-20a	0.0054	0.693	0.688	0.697	0.744 (0.653, 0.835)
miR-29b:miR-378	0.0069	0.711	0.708	0.712	0.740 (0.647, 0.832)
miR-150:miR-18a	0.0074	0.693	0.729	0.667	0.739 (0.647, 0.831)
miR-24:miR-378	0.0091	0.667	0.625	0.697	0.735 (0.638, 0.833)
miR-150:miR-99a	0.011	0.702	0.750	0.667	0.732 (0.639, 0.825)
miR-451a:let7a	0.016	0.684	0.771	0.621	0.729 (0.637, 0.822)
miR-223:nf3	0.017	0.675	0.792	0.591	0.725 (0.633, 0.818)

Table 2. *Cont.*

	p-Value × 4 × 406	CV Accuracy	CV Sensitivity	CV Specificity	CV AUC
miR-150:miR-196b	0.017	0.667	0.792	0.576	0.732 (0.638, 0.827)
miR-126:miR-378	0.019	0.675	0.667	0.682	0.729 (0.633, 0.825)
miR-150:miR-155	0.026	0.684	0.792	0.606	0.727 (0.633, 0.821)
miR-20a:miR-451a	0.026	0.632	0.646	0.621	0.724 (0.629, 0.820)
miR-451a:miR-191	0.032	0.649	0.667	0.636	0.722 (0.628, 0.815)
miR-29b:nf3	0.18	0.684	0.646	0.712	0.698 (0.597, 0.799)
miR-126:nf3	0.41	0.649	0.646	0.652	0.686 (0.581, 0.791)
miR-26a:nf3	0.70	0.623	0.604	0.636	0.681 (0.581, 0.781)
miR-20a:nf3	>1	0.588	0.604	0.576	0.636 (0.535, 0.737)
miR-210:nf3	>1	0.632	0.583	0.667	0.628 (0.521, 0.735)
let7a:nf3	>1	0.570	0.625	0.530	0.627 (0.519, 0.735)
miR-221:nf3	>1	0.570	0.604	0.545	0.624 (0.520, 0.728)
miR-24:nf3	>1	0.579	0.542	0.606	0.628 (0.520, 0.735)
miR-196b:nf3	>1	0.614	0.562	0.652	0.621 (0.517, 0.724)
miR-18a:nf3	>1	0.570	0.542	0.591	0.614 (0.508, 0.720)
miR-96:nf3	>1	0.561	0.646	0.500	0.608 (0.504, 0.713)
miR-9:nf3	>1	0.614	0.500	0.697	0.597 (0.489, 0.705)
miR-128:nf3	>1	0.596	0.521	0.652	0.543 (0.433, 0.652)
miR-124:nf3	>1	0.553	0.583	0.530	0.553 (0.448, 0.659)
miR-21:nf3	>1	0.553	0.542	0.561	0.534 (0.420, 0.648)
miR-181b:nf3	>1	0.579	0.667	0.515	0.553 (0.447, 0.659)
miR-92a:nf3	>1	0.544	0.562	0.530	0.490 (0.383, 0.597)
miR-155:nf3	>1	0.491	0.583	0.424	0.462 (0.356, 0.569)
miR-100:nf3	>1	0.456	0.521	0.409	0.500 (0.389, 0.611)
miR-181a:nf3	>1	0.465	0.562	0.394	0.523 (0.416, 0.630)
miR-99a:nf3	>1	0.439	0.521	0.379	0.591 (0.486, 0.696)
miR-146a:nf3	>1	0.465	0.604	0.364	0.511 (0.404, 0.618)

In this table, sensitivity is the proportion of NTP patients that are correctly identified as such, and specificity is the proportion of ALL+AML patients correctly identified as such.

According to Table 3, which presents the results of the comparison between ALL and the others, the model based on covariates miR-181b:miR-100, miR-223:miR-124, and miR-24:nf3 had the best prediction accuracy (81.8% sensitivity with 81.5% specificity).

Table 3. Comparison of ALL vs. others.

	p-Value × 4 × 406	CV Accuracy	CV Sensitivity	CV Specificity	CV AUC
miR-181b:miR-100 + miR-223:miR-124 + miR-24:nf3		0.816	0.818	0.815	0.796 (0.679, 0.914)
miR-155:miR-124 + miR-181b:miR-100 + miR-223:miR-103a		0.807	0.818	0.804	0.839 (0.736, 0.941)

Table 3. *Cont.*

	p-Value × 4 × 406	CV Accuracy	CV Sensitivity	CV Specificity	CV AUC
miR-155:miR-378 + miR-181b:miR-223 + miR-100:miR-210		0.807	0.818	0.804	0.829 (0.732, 0.926)
miR-196b:miR-124 + miR-92a:miR-24 + miR-100:miR-181a		0.807	0.818	0.804	0.850 (0.763, 0.936)
miR-155:miR-378 + miR-181b:miR-196b + miR-223:miR-146a		0.807	0.773	0.815	0.768 (0.646, 0.890)
miR-155:miR-378 + miR-181b:miR-223 + miR-196b:miR-24		0.807	0.773	0.815	0.772 (0.663, 0.882)
miR-155:miR-100 + miR-196b:miR-124 + miR-223:miR-92a		0.807	0.727	0.826	0.801 (0.689, 0.914)
miR-181b:miR-223 + miR-196b:miR-103a + miR-100:miR-124		0.807	0.727	0.826	0.796 (0.674, 0.918)
miR-196b:miR-124 + miR-223:miR-26a + miR-99a:miR-378		0.807	0.727	0.826	0.792 (0.673, 0.911)
miR-196b:nf3 + miR-223:miR-181a + miR-126:miR-210		0.807	0.727	0.826	0.726 (0.590, 0.862)
miR-223:miR-26a + miR-100:miR-181a + miR-451a:miR-103a		0.807	0.727	0.826	0.777 (0.658, 0.896)
miR-196b:miR-181a + miR-92a:miR-24 + miR-221:miR-21		0.807	0.682	0.837	0.719 (0.584, 0.855)
miR-155:miR-92a + miR-181b:miR-196b + miR-221:miR-451a		0.807	0.545	0.870	0.698 (0.567, 0.829)
miR-196b:miR-181a + miR-20a:miR-451a + let7a:miR-21		0.807	0.545	0.870	0.643 (0.488, 0.798)
miR-155:miR-92a + miR-181b:miR-126 + miR-196b:miR-181a		0.807	0.500	0.880	0.650 (0.496, 0.803)
miR-181b:miR-223 + miR-196b:miR-103a		0.789	0.682	0.815	0.727 (0.596, 0.858)
miR-18a:miR-451a + miR-181a:miR-24		0.789	0.545	0.848	0.647 (0.502, 0.793)
miR-181b:miR-223 + miR-221:miR-9		0.789	0.545	0.848	0.727 (0.616, 0.838)
miR-128:miR-21 + miR-223:miR-181a		0.789	0.500	0.859	0.617 (0.452, 0.781)
miR-196b:nf3	>1	0.649	0.591	0.663	0.703 (0.570, 0.836)
miR-223:nf3	>1	0.684	0.591	0.707	0.686 (0.548, 0.824)
miR-100:nf3	>1	0.596	0.591	0.598	0.643 (0.527, 0.759)

Table 3. *Cont.*

	p-Value × 4 × 406	CV Accuracy	CV Sensitivity	CV Specificity	CV AUC
miR-9:nf3	>1	0.570	0.682	0.543	0.625 (0.502, 0.748)
miR-451a:nf3	>1	0.596	0.500	0.620	0.595 (0.453, 0.737)
miR-124:nf3	>1	0.588	0.545	0.598	0.595 (0.459, 0.732)
miR-150:nf3	>1	0.614	0.455	0.652	0.570 (0.439, 0.700)
miR-29b:nf3	>1	0.526	0.591	0.511	0.583 (0.450, 0.715)
miR-126:nf3	>1	0.570	0.636	0.554	0.564 (0.436, 0.691)
miR-24:nf3	>1	0.535	0.455	0.554	0.574 (0.449, 0.699)
miR-21:nf3	>1	0.518	0.500	0.522	0.564 (0.439, 0.689)
miR-181a:nf3	>1	0.588	0.364	0.641	0.558 (0.428, 0.688)
miR-181b:nf3	>1	0.623	0.409	0.674	0.558 (0.409, 0.707)
miR-20a:nf3	>1	0.596	0.545	0.609	0.509 (0.361, 0.658)
let7a:nf3	>1	0.570	0.455	0.598	0.549 (0.409, 0.689)
miR-96:nf3	>1	0.579	0.455	0.609	0.535 (0.384, 0.686)
miR-26a:nf3	>1	0.518	0.545	0.511	0.506 (0.372, 0.641)
miR-146a:nf3	>1	0.596	0.364	0.652	0.502 (0.366, 0.637)
miR-99a:nf3	>1	0.500	0.636	0.467	0.523 (0.408, 0.637)
miR-18a:nf3	>1	0.491	0.455	0.500	0.502 (0.379, 0.625)
miR-155:nf3	>1	0.561	0.364	0.609	0.499 (0.348, 0.650)
miR-128:nf3	>1	0.588	0.364	0.641	0.489 (0.334, 0.644)
miR-221:nf3	>1	0.500	0.318	0.543	0.602 (0.482, 0.722)
miR-210:nf3	>1	0.482	0.500	0.478	0.650 (0.536, 0.763)
miR-92a:nf3	>1	0.447	0.364	0.467	0.629 (0.507, 0.752)

In this table, sensitivity is the proportion of ALL patients that are correctly identified as such, and specificity is the proportion of NTP+AML patients correctly identified as such.

According to Table 4, which lists findings of the comparison between AML and the others, the best prediction accuracy (81.8% sensitivity with 84.3% specificity) was shown the model based on the covariates miR-150:miR-221, miR-100:miR-24, and miR-181a:miR-191. Among univariate models, miR-150:miR-191 had the best prediction accuracy (68.2% sensitivity with 78.6% specificity).

Table 4. Comparison of AML vs. others.

	p-Value × 4 × 406	CV Accuracy	CV Sensitivity	CV Specificity	CV AUC
miR-150:miR-221 + miR-100:miR-24 + miR-181a:miR-191		0.833	0.818	0.843	0.868 (0.803, 0.934)
miR-150:miR-221 + miR-100:miR-124 + miR-26a:nf3		0.825	0.841	0.814	0.882 (0.821, 0.944)
miR-150:miR-100 + miR-181a:miR-221 + miR-24:nf3		0.825	0.818	0.829	0.872 (0.807, 0.937)
miR-150:miR-100 + miR-181a:nf3 + miR-221:miR-24		0.825	0.818	0.829	0.882 (0.819, 0.944)

Table 4. *Cont.*

	p-Value × 4 × 406	CV Accuracy	CV Sensitivity	CV Specificity	CV AUC
miR-150:miR-21 + miR-18a:miR-92a + miR-26a:miR-191		0.825	0.795	0.843	0.831 (0.754, 0.908)
miR-150:nf3 + miR-20a:miR-92a + miR-100:miR-124		0.825	0.795	0.843	0.843 (0.765, 0.921)
miR-223:miR-100 + miR-146a:miR-103a + miR-221:miR-451a		0.825	0.795	0.843	0.836 (0.755, 0.917)
miR-223:miR-103a + miR-100:miR-451a + miR-146a:miR-221		0.825	0.795	0.843	0.823 (0.741, 0.905)
miR-100:miR-126 + miR-146a:miR-221 + miR-26a:miR-21		0.825	0.773	0.857	0.822 (0.736, 0.908)
miR-128:miR-221 + miR-20a:miR-100 + miR-24:nf3		0.825	0.773	0.857	0.812 (0.726, 0.898)
miR-150:miR-221 + miR-196b:miR-24 + miR-100:miR-99a		0.825	0.773	0.857	0.819 (0.730, 0.907)
miR-181b:miR-100 + miR-146a:miR-103a + miR-221:miR-451a		0.825	0.773	0.857	0.834 (0.752, 0.915)
miR-150:miR-221 + miR-26a:miR-103a		0.798	0.795	0.800	0.837 (0.763, 0.911)
miR-150:miR-191 + miR-124:miR-221		0.781	0.773	0.786	0.789 (0.700, 0.878)
miR-150:miR-21 + miR-26a:miR-191		0.781	0.750	0.800	0.823 (0.745, 0.901)
miR-150:miR-191 + miR-210:miR-21		0.781	0.727	0.814	0.803 (0.719, 0.887)
miR-150:miR-191 + miR-181b:miR-124		0.781	0.705	0.829	0.780 (0.689, 0.872)
miR-150:miR-100 + miR-26a:miR-378		0.781	0.682	0.843	0.808 (0.728, 0.889)
miR-150:nf3	0.000022	0.737	0.636	0.800	0.794 (0.712, 0.876)
miR-150:miR-191	0.000057	0.746	0.682	0.786	0.786 (0.699, 0.874)
miR-150:miR-378	0.00016	0.711	0.614	0.771	0.775 (0.690, 0.860)
miR-150:miR-100	0.0011	0.702	0.659	0.729	0.757 (0.665, 0.849)
miR-150:miR-221	0.0015	0.719	0.659	0.757	0.756 (0.659, 0.854)
miR-150:miR-103a	0.0072	0.719	0.682	0.743	0.741 (0.646, 0.837)
miR-150:miR-196b	0.0072	0.684	0.568	0.757	0.741 (0.645, 0.838)
miR-150:miR-21	0.0076	0.684	0.682	0.686	0.739 (0.645, 0.833)
miR-150:miR-92a	0.020	0.711	0.636	0.757	0.729 (0.631, 0.828)
miR-221:miR-24	0.022	0.711	0.636	0.757	0.728 (0.630, 0.825)
miR-26a:miR-21	0.028	0.711	0.636	0.757	0.726 (0.624, 0.829)
miR-221:miR-26a	0.042	0.693	0.636	0.729	0.718 (0.613, 0.822)

Table 4. *Cont.*

	p-Value × 4 × 406	CV Accuracy	CV Sensitivity	CV Specificity	CV AUC
miR-100:miR-26a	0.043	0.667	0.659	0.671	0.721 (0.624, 0.819)
miR-451a:nf3	> 1	0.649	0.614	0.671	0.675 (0.575, 0.775)
miR-26a:nf3	> 1	0.596	0.614	0.586	0.649 (0.547, 0.751)
miR-29b:nf3	> 1	0.623	0.705	0.571	0.634 (0.527, 0.740)
miR-126:nf3	> 1	0.570	0.568	0.571	0.630 (0.526, 0.734)
miR-210:nf3	> 1	0.623	0.636	0.614	0.629 (0.520, 0.737)
miR-221:nf3	> 1	0.605	0.568	0.629	0.613 (0.500, 0.725)
miR-128:nf3	> 1	0.544	0.636	0.486	0.604 (0.497, 0.712)
miR-100:nf3	> 1	0.588	0.477	0.657	0.608 (0.501, 0.715)
miR-223:nf3	> 1	0.596	0.477	0.671	0.591 (0.483, 0.699)
miR-18a:nf3	> 1	0.570	0.591	0.557	0.592 (0.483, 0.700)
miR-20a:nf3	> 1	0.570	0.500	0.614	0.584 (0.474, 0.694)
let7a:nf3	> 1	0.570	0.523	0.600	0.573 (0.467, 0.679)
miR-24:nf3	> 1	0.579	0.591	0.571	0.564 (0.456, 0.673)
miR-96:nf3	> 1	0.561	0.477	0.614	0.558 (0.449, 0.666)
miR-181a:nf3	> 1	0.421	0.182	0.571	0.777 (0.685, 0.869)
miR-155:nf3	> 1	0.342	0.227	0.414	0.867 (0.797, 0.936)
miR-9:nf3	> 1	0.500	0.341	0.600	0.572 (0.463, 0.682)
miR-92a:nf3	> 1	0.509	0.477	0.529	0.498 (0.380, 0.616)
miR-146a:nf3	> 1	0.491	0.318	0.600	0.629 (0.518, 0.740)
miR-99a:nf3	> 1	0.544	0.432	0.614	0.505 (0.388, 0.622)
miR-181b:nf3	> 1	0.570	0.455	0.643	0.486 (0.371, 0.602)
miR-21:nf3	> 1	0.465	0.432	0.486	0.586 (0.480, 0.693)
miR-196b:nf3	> 1	0.491	0.386	0.557	0.507 (0.393, 0.621)
miR-124:nf3	> 1	0.377	0.227	0.471	0.827 (0.745, 0.910)

In this table, sensitivity is the proportion of AML patients that are correctly identified as such, and specificity is the proportion of NTP+ALL patients correctly identified as such.

According to Table 5, which lists the results for the comparison between ALL and AML, the best prediction accuracy (86.4% sensitivity with 84.1% specificity) belonged to the classification model based on covariates miR-100:miR-124, miR-24:miR-26a, and miR-24:miR-9, the model based on covariates miR-100:miR-124, miR-24:miR-26a, and miR-26a:miR-9, and the model based on mirR-100:miR-124, miR-24:miR-9, and miR-26a:miR-9.

Table 5. Comparison of ALL vs. AML.

	p-Value × 4 × 406	CV Accuracy	CV Sensitivity	CV Specificity	CV AUC
miR-100:miR-124 + miR-24:miR-26a + miR-24:miR-9		0.848	0.864	0.841	0.893 (0.809, 0.976)
miR-100:miR-124 + miR-24:miR-26a + miR-26a:miR-9		0.848	0.864	0.841	0.893 (0.809, 0.976)

Table 5. *Cont.*

	p-Value × 4 × 406	CV Accuracy	CV Sensitivity	CV Specificity	CV AUC
miR-100:miR-124 + miR-24:miR-9 + miR-26a:miR-9		0.848	0.864	0.841	0.893 (0.809, 0.976)
miR-155:miR-181b + miR-100:miR-124 + miR-24:miR-26a		0.848	0.818	0.864	0.897 (0.810, 0.984)
miR-155:miR-124 + miR-181b:miR-100 + miR-24:miR-26a		0.848	0.818	0.864	0.893 (0.802, 0.983)
miR-20a:miR-9 + miR-100:miR-124 + miR-24:miR-26a		0.848	0.818	0.864	0.871 (0.781, 0.961)
miR-223:miR-124 + miR-92a:miR-100		0.773	0.773	0.773	0.794 (0.682, 0.907)
miR-100:miR-124 + miR-24:miR-26a		0.773	0.727	0.795	0.851 (0.756, 0.946)
miR-223:miR-124 + miR-100:miR-26a		0.773	0.682	0.818	0.818 (0.715, 0.921)
miR-100:nf3	>1	0.652	0.636	0.659	0.692 (0.561, 0.824)
miR-196b:nf3	>1	0.591	0.591	0.591	0.657 (0.514, 0.800)
miR-181a:nf3	>1	0.576	0.318	0.705	0.560 (0.416, 0.703)
miR-210:nf3	>1	0.621	0.727	0.568	0.572 (0.428, 0.716)
miR-150:nf3	>1	0.530	0.591	0.500	0.568 (0.416, 0.721)
miR-223:nf3	>1	0.652	0.591	0.682	0.565 (0.408, 0.722)
miR-124:nf3	>1	0.530	0.545	0.523	0.544 (0.399, 0.690)
miR-9:nf3	>1	0.530	0.682	0.455	0.569 (0.426, 0.712)
miR-128:nf3	>1	0.606	0.409	0.705	0.568 (0.409, 0.727)
miR-221:nf3	>1	0.606	0.682	0.568	0.544 (0.401, 0.688)
miR-26a:nf3	>1	0.561	0.545	0.568	0.507 (0.351, 0.664)
miR-21:nf3	>1	0.485	0.500	0.477	0.472 (0.325, 0.619)
miR-146a:nf3	>1	0.515	0.227	0.659	0.438 (0.292, 0.584)
miR-181b:nf3	>1	0.606	0.409	0.705	0.496 (0.341, 0.650)
miR-155:nf3	>1	0.545	0.364	0.636	0.505 (0.349, 0.661)
miR-18a:nf3	>1	0.530	0.636	0.477	0.485 (0.333, 0.636)
miR-99a:nf3	>1	0.500	0.636	0.432	0.485 (0.345, 0.624)
miR-92a:nf3	>1	0.470	0.455	0.477	0.548 (0.404, 0.691)
miR-24:nf3	>1	0.455	0.364	0.500	0.518 (0.367, 0.668)
let7a:nf3	>1	0.485	0.409	0.523	0.586 (0.434, 0.738)
miR-29b:nf3	>1	0.424	0.455	0.409	0.689 (0.555, 0.823)
miR-126:nf3	>1	0.455	0.409	0.477	0.753 (0.625, 0.881)
miR-96:nf3	>1	0.515	0.364	0.591	0.644 (0.500, 0.787)
miR-451a:nf3	>1	0.470	0.500	0.455	0.679 (0.540, 0.818)
miR-20a:nf3	>1	0.485	0.591	0.432	0.595 (0.448, 0.742)

In this table, sensitivity is the proportion of ALL patients that are correctly identified as such, and specificity is the proportion of AML patients correctly identified as such.

4. Discussion

The expression of miRNA is always different between tumors and healthy tissues or a secondary nontumor pathology. Additionally, differences in the miRNA profiles among different tumor types and among different stages of the same malignant tumor are not uncommon [22,23]. Because miRNAs are highly stable in tissues and body fluids, they appear to be promising diagnostic markers.

In this work, we compared the expression profiles of 25 miRNA in bone marrow samples from new cases of AML, ALL, and NTPs.

Our study indicates that each of the aforementioned hematopoietic bone marrow disorders may be identified through profiling of miRNAs. The results of our analysis show that some of our classification variables are statistically significantly associated with the comparison groups.

The results from our multivariable models, even though they may be subject to overfitting to some extent, cannot be explained by overfitting alone. For example, in the comparison of NTP with ALL+AML, the best accuracy (93.8% sensitivity with 92.4% specificity) was shown by the classification model based on covariates miR-150:miR-21, miR-20a:miR-221, and miR-24:nf3. In the exact Mann–Whitney test comparing the values predicted by logistic regression during the leave-one-out cross-validation between the comparison groups, we obtained a p-value $< 2.2 \times 10^{-16}$. On the other hand, dividing 0.05 by the total number of comparisons yields $0.05/(4 \times (11{,}153{,}835 + 406)) = 1.1 \times 10^{-9}$, which is at least 5×10^{6} times greater than the above p-value.

As for individual miRNAs, NTP was best discriminated from ALL+AML by miR-150 (85.4% sensitivity with 71.2% specificity for the corresponding variable, miR-150:nf3), ALL was best discriminated from AML+NTP by miRNA-223 (59.1% sensitivity with 70.7% sensitivity for the corresponding variable, miR-223:nf3), AML was best differentiated from ALL+NTP by miRNA-150 (63.6% sensitivity with 80.0% specificity for miR-150:nf3), and, finally, ALL was best differentiated from AML by miRNA-100 (63.6% sensitivity with 65.9% specificity for miR-100:nf3), although expression deregulation of these miRNAs has been described [14]. AML is a special case because changes in the expression of miRNA-126, -29b, and -26a are reported more often [24–27] in comparison with miRNA-150. In ALL, changes in the expression of miRNA-150 and -155 are detected more often [28–32] as compared to miRNA-223.

Inconsistency between miRNA profiling results of different studies is a common problem. This could be because different authors use dissimilar control groups; this is especially true for the research on hematologic cancers. Another source of inconsistency is a relatively large error in measurements of miRNA concentrations whether by RT-PCR or via microarray technologies. Because of the measurement errors, potentially significant under-three-fold changes in miRNA concentrations may not be taken into account accurately. Thus, in this study, the most reliable results, which at the same time were most consistent between different studies, were obtained only for miRNAs whose cancer-specific differences in concentration were the greatest.

Therefore, our results point to potential feasibility of (i) discriminating ALL+AML from NTP with 93% sensitivity of and 92% specificity using miR-150:miR-21, miR-20a:miR-221, and miR-24:nf3; (ii) diagnosing ALL with 80% sensitivity and 81% specificity by means of miR-196b:miR-221, miR-223:miR-378, and miR-100:miR-29b; and (iii) diagnosing AML with 80% sensitivity and 83% specificity with the help of miR-150:miR-100, miR-181a:miR-191, and miR-221:miR-26a.

The results presented herein allow the miRNA expression profile to be used for differentiation between AL and NTP, no matter what AL subtype.

Effective treatment for acute leukemias strongly depends on our understanding of the basics of their genesis. One of the fundamental mechanisms underlying leukemias is the phenomenon known as 'clonal selection'. This pathogenetic mechanism is in fact responsible for all the clinical features of tumor relapse to come. One of the problems where the data obtained can be helpful is the detection of MRD. There is little doubt that the use of any MRD detection method requires a biomarker or a combination of biomarkers that can clearly differentiate between normal and cancer cells. Data obtained allow miRNAs to be seen as promising biomarkers with sufficient sensitivity and specificity to detect MRD, no matter what clonal nature of the disease.

The aim of this publication is to present the results of the first, pilot stage of the research project. With reliance on a wide spectrum of miRNAs, we have confirmed the feasibility of using analyses of miRNA patterns for solving differential problems when analyzing bone marrow samples.

To this end, it is necessary to do testing with larger sample sizes. The obvious extension of this work will be to analyze miRNA expression profiles after consolidation chemotherapy and to see whether they correlate with the patient's clinical data.

Supplementary Materials: The following are available online at http://www.mdpi.com/2227-9059/8/12/607/s1, Table S1: Patient characteristics at the time of diagnosis, Table S2: The oligonucleotide sequences used in the study.

Author Contributions: I.B.K. contributed to the conception and design of the work, data collection, and writing of the manuscript draft. S.E.T. contributed to the conception and design of the work, participated in the experiments and data analysis, and wrote the manuscript draft. M.M.A. collected clinical samples. Y.A.V. conducted the experiments. P.S.R. and V.M.N. carried out statistical analysis and drafted the figure. T.I.P. and I.F.Z. contributed to the conception and design of the work, supervised this study, and reviewed and edited the manuscript. All authors have read and agreed to the published version of the manuscript.

Funding: The work of S.E.T. was financially supported by the Russian Science Foundation (project No. 20-14-00074). The work of Y.A.V. was supported by the Russian Foundation for Basic Research (project No. 19-34-60024). The work of P.S.R. was supported by the Mathematical Center in Akademgorodok under agreement No. 075-15-2019-1675 with the Ministry of Science and Higher Education of the Russian Federation. The work of V.M.N. was supported by a state contract of Sobolev Institute of Mathematics (Project No. 0314-2019-0015).

Acknowledgments: The authors thank the Siberian Supercomputer Center of the Siberian Branch of the Russian Academy of Sciences for providing supercomputer facilities.

Conflicts of Interest: The authors have no competing interests.

References

1. Kovynev, I.B.; Agakishiev, M.M.; Kolesnikova, M.A.; Zalavina, S.V.; Tarnovsky, R.V.; Soldatova, G.S.; Tairova, S.A.; Nechunaeva, I.N.; Pospelova, T.I. Epidemiology and Molecular Genetics of Acute Leukemia in a Major Industrial Center of Western Siberia. *Int. J. Adv. Biotechnol. Res.* **2019**, *2*, 260–266.
2. Cruz, N.M.; Mencia-Trinchant, N.; Hassane, D.C.; Guzman, M.L. Minimal residual disease in acute myelogenous leukemia. *Int. J. Lab. Hematol.* **2017**, *39* (Suppl. 1), 53–60. [CrossRef] [PubMed]
3. Brooimans, R.A.; van der Velden, V.H.J.; Boeckx, N.; Slomp, J.; Preijers, F.; Te Marvelde, J.G.; Van, N.M.; Heijs, A.; Huys, E.; van der Holt, B.; et al. Immunophenotypic measurable residual disease (MRD) in acute myeloid leukemia: Is multicentric MRD assessment feasible? *Leuk. Res.* **2019**, *76*, 39–47. [CrossRef] [PubMed]
4. Eckert, C.; Flohr, T.; Koehler, R.; Hagedorn, N.; Moericke, A.; Stanulla, M.; Kirschner-Schwabe, R.; Cario, G.; Stackelberg, A.; Bartram, C.R.; et al. Very early/early relapses of acute lymphoblastic leukemia show unexpected changes of clonal markers and high heterogeneity in response to initial and relapse treatment. *Leukemia* **2011**, *25*, 1305–1313. [CrossRef]
5. Cammarata, G.; Augugliaro, L.; Salemi, D.; La Rosa, C.A.M.; Dagnino, L.; Civiletto, G.; Messana, F.; Marfia, A.; Bica, M.G.; Cascio, L.; et al. Differential expression of specific microRNA and their targets in acute myeloid leukemia. *Am. J. Hematol.* **2010**, *85*, 331–339. [CrossRef]
6. Li, Z.; Lu, J.; Sun, M.; Mi, S.; Zhang, H.; Luo, R.T.; Chen, P.; Wang, Y.; Yan, M.; Qian, Z.; et al. Distinct microRNA expression profiles in acute myeloid leukemia with common translocations. *Proc. Natl. Acad. Sci. USA* **2008**, *105*, 15535–15540. [CrossRef]
7. Kluiver, J.; Kroesen, B.J.; Poppema, S.; van den Berg, A. The role of microRNAs in normal hematopoiesis and hematopoietic malignancies. *Leukemia* **2006**, *20*, 1931–1936. [CrossRef]
8. Fabbri, M.; Garzon, R.; Andreeff, M.; Kantarjian, H.M.; Garcia-Manero, G.; Calin, G.A. MicroRNAs and noncoding RNAs in hematological malignancies: Molecular, clinical and therapeutic implications. *Leukemia* **2008**, *22*, 1095–1105. [CrossRef]
9. Garzon, R.; Croce, C.M. MicroRNAs in normal and malignant hematopoiesis. *Curr. Opin. Hematol.* **2008**, *15*, 352–358. [CrossRef]

10. Trino, S.; Lamorte, D.; Caivano, A.; Laurenzana, I.; Tagliaferri, D.; Falco, G.; Del Vecchio, L.; Musto, P.; De Luca, L. MicroRNAs as New Biomarkers for Diagnosis and Prognosis, and as Potential Therapeutic Targets in Acute Myeloid Leukemia. *Int. J. Mol. Sci.* **2018**, *19*, 460. [CrossRef]
11. Wang, Y.; Li, Z.; He, C.; Wang, D.; Yuan, X.; Chen, J.; Jin, J. MicroRNAs expression signatures are associated with lineage and survival in acute leukemias. *Blood Cells Mol. Dis.* **2010**, *44*, 191–197. [CrossRef]
12. Luan, C.; Yang, Z.; Chen, B. The functional role of microRNA in acute lymphoblastic leukemia: Relevance for diagnosis, differential diagnosis, prognosis, and therapy. *OncoTargets Ther.* **2015**, *8*, 2903–2914.
13. Liao, Q.; Wang, B.; Li, X.; Jiang, G. miRNAs in acute myeloid leukemia. *Oncotarget* **2016**, *8*, 3666–3682. [CrossRef] [PubMed]
14. Yeh, C.; Moles, R.; Nicot, C. Clinical significance of microRNAs in chronic and acute human leukemia. *Mol. Cancer* **2016**, *15*, 37. [CrossRef]
15. Ultimo, S.; Martelli, A.M.; Zauli, G.; Vitale, M.; Calin, G.A.; Neri, L.M. Roles and clinical implications of microRNAs in acute lymphoblastic leukemia. *J. Cell. Physiol.* **2018**, *233*, 5642–5654. [CrossRef] [PubMed]
16. de Leeuw, D.C.; van den Ancker, W.; Denkers, F.; de Menezes, R.X.; Westers, T.M.; Ossenkoppele, G.J.; van de Loosdrecht, A.A.; Smit, L. MicroRNA profiling can classify acute leukemias of ambiguous lineage as either acute myeloid leukemia or acute lymphoid leukemia. *Clin. Cancer Res.* **2013**, *19*, 2187–2196. [CrossRef] [PubMed]
17. Vandesompele, J.; De Preter, K.; Pattyn, F.; Poppe, B.; Van Roy, N.; De Paepe, A.; Speleman, F. Accurate normalization of real-time quantitative RT-PCR data by geometric averaging of multiple internal control genes. *Genome Biol.* **2002**, *3*, 34. [CrossRef]
18. Titov, S.E.; Ivanov, M.K.; Karpinskaya, E.V.; Tsivlikova, E.V.; Shevchenko, S.P.; Veryaskina, Y.A.; Akhmerova, L.G.; Poloz, T.L.; Klimova, O.A.; Gulyaeva, L.F.; et al. miRNA profiling, detection of BRAF V600E mutation and RET-PTC1 translocation in patients from Novosibirsk oblast (Russia) with different types of thyroid tumors. *BMC Cancer* **2016**, *16*, 201. [CrossRef]
19. Chen, C.; Ridzon, D.A.; Broomer, A.J.; Zhou, Z.; Lee, D.H.; Nguyen, J.T.; Barbisin, M.; Xu, N.L.; Mahuvakar, V.R.; Andersen, M.R.; et al. Real-time quantification of microRNAs by stem–loop RT–PCR. *Nucleic Acids Res.* **2005**, *33*, 1–9. [CrossRef]
20. Livak, K.J.; Schmittgen, T.D. Analysis of relative gene expression data using real-time quantitative PCR and the 2-ΔΔCt method. *Methods* **2001**, *25*, 402–408. [CrossRef]
21. Ivanov, M.K.; Titov, S.E.; Glushkov, S.A.; Dzyubenko, V.V.; Malek, A.V.; Arkhangelskaya, P.A.; Samsonov, R.B.; Mikhetko, A.A.; Bakhidze, E.V.; Berlev, I.V.; et al. Detection of high-grade neoplasia in air-dried cervical PAP smears by a microRNA-based classifier. *Oncol. Rep.* **2018**, *39*, 1099–1111. [CrossRef] [PubMed]
22. Veryaskina, Y.A.; Titov, S.E.; Kometova, V.V.; Rodionov, V.V.; Zhimulev, I.F. Intratumoral Heterogeneity of Expression of 16 miRNA in Luminal Cancer of the Mammary Gland. *Non-Coding RNA* **2020**, *6*, 16. [CrossRef] [PubMed]
23. Veryaskina, Y.A.; Titov, S.E.; Kovynev, I.B.; Pospelova, T.I.; Zhimulev, I.F. The miRNA Profile in Non-Hodgkin's Lymphoma Patients with Secondary Myelodysplasia. *Cells* **2020**, *9*, 2318. [CrossRef] [PubMed]
24. Gong, J.N.; Yu, J.; Lin, H.S.; Zhang, X.H.; Yin, X.L.; Xiao, Z.; Wang, F.; Wang, X.S.; Su, R.; Shen, C.; et al. The role, mechanism and potentially therapeutic application of microRNA-29 family in acute myeloid leukemia. *Cell Death Differ.* **2014**, *21*, 100–112. [CrossRef]
25. Jongen-Lavrencic, M.; Sun, S.M.; Dijkstra, M.K.; Valk, P.J.; Lowenberg, B. MicroRNA expression profiling in relation to the genetic heterogeneity of acute myeloid leukemia. *Blood* **2008**, *111*, 5078–5085. [CrossRef]
26. Wang, S.; Aurora, A.B.; Johnson, B.A.; Qi, X.; McAnally, J.; Hill, J.A.; Richardson, J.A.; Bassel-Duby, R.; Olson, E.N. The endothelial-specific microRNA miR-126 governs vascular integrity and angiogenesis. *Dev. Cell* **2008**, *15*, 261–271. [CrossRef]
27. Cattaneo, M.; Pelosi, E.; Castelli, G.; Cerio, A.M.; D'angiò, A.; Porretti, L.; Rebulla, P.; Pavesi, L.; Russo, G.; Giordano, A.; et al. A miRNA Signature in Human Cord Blood Stem and Progenitor Cells as Potential Biomarker of Specific Acute Myeloid Leukemia Subtypes. *J. Cell. Physiol.* **2015**, *230*, 1770–1780. [CrossRef]
28. Havelange, V.; Garzon, R. MicroRNAs: Emerging Key Regulators of Hematopoiesis. *Am. J. Hematol.* **2010**, *85*, 935–942. [CrossRef]
29. Slavov, S.N.; Teixeira, H.L.G.; Rego, E.M. The role of micro-ribonucleic acids in normal hematopoiesis and leukemic T-lymphogenesis. *Braz. J. Med. Biol. Res.* **2010**, *43*, 619–626. [CrossRef]

30. Ghisi, M.; Corradin, A.; Basso, K.; Frasson, C.; Serafin, V.; Mukherjee, S.; Mussolin, L.; Ruggero, K.; Bonanno, L.; Guffanti, A.; et al. Margherita Ghisi, Alberto Corradin, Katia Basso, Chiara Frasson, Valentina Serafin, Subhamoy Mukherjee, Lara Mussolin, Katia Ruggero, Laura Bonanno, Alessandro Guffanti et al. Modulation of microRNA expression in human T-cell development: Targeting of NOTCH3 by miR-150. *Blood* **2011**, *117*, 7053–7062.
31. Seddiki, N.; Brezar, V.; Ruffin, N.; Levy, Y.; Swaminathan, S. Role of miR-155 in the regulation of lymphocyte immune function and disease. *Immunology.* **2014**, *142*, 32–38. [CrossRef] [PubMed]
32. He, Y.; Chevillet, J.R.; Liu, G.; Kim, T.K.; Wang, K. The effects of microRNA on the absorption, distribution, metabolism and excretion of drugs. *Br. J. Pharmacol.* **2015**, *172*, 2733–2747. [CrossRef] [PubMed]

Publisher's Note: MDPI stays neutral with regard to jurisdictional claims in published maps and institutional affiliations.

biomedicines

MDPI

Article

Regulation of Oncogenic Targets by the Tumor-Suppressive *miR-139* Duplex (*miR-139-5p* and *miR-139-3p*) in Renal Cell Carcinoma

Reona Okada [1], Yusuke Goto [1,2], Yasutaka Yamada [1,2], Mayuko Kato [1,2], Shunichi Asai [1], Shogo Moriya [3], Tomohiko Ichikawa [2] and Naohiko Seki [1,*]

1 Department of Functional Genomics, Chiba University Graduate School of Medicine, Chiba 260-8670, Japan; reonaokada@chiba-u.jp (R.O.); caxa1117@chiba-u.jp (Y.G.), yasutaka_yamada@dfci.harvard.edu (Y.Y.), mayukokato@chiba-u.jp (M.K.); cada5015@chiba-u.jp (S.A.)

2 Department of Urology, Chiba University Graduate School of Medicine, Chiba 260-8670, Japan; tomohiko_ichikawa@faculty.chiba-u.jp

3 Department of Biochemistry and Genetics, Chiba University Graduate School of Medicine, Chiba 260-8670, Japan; moriya.shogo@chiba-u.jp

* Correspondence: naoseki@faculty.chiba-u.jp; Tel.: +81-43-226-2971

Received: 20 November 2020; Accepted: 10 December 2020; Published: 12 December 2020

Abstract: We previously found that both the guide and passenger strands of the *miR-139* duplex (*miR-139-5p* and *miR-139-3p*, respectively) were downregulated in cancer tissues. Analysis of TCGA datasets revealed that low expression of *miR-139-5p* ($p < 0.0001$) and *miR-139-3p* ($p < 0.0001$) was closely associated with 5-year survival rates of patients with renal cell carcinoma (RCC). Ectopic expression assays showed that *miR-139-5p* and *miR-139-3p* acted as tumor-suppressive miRNAs in RCC cells. Here, 19 and 22 genes were identified as putative targets of *miR-139-5p* and *miR-139-3p* in RCC cells, respectively. Among these genes, high expression of *PLXDC1*, *TET3*, *PXN*, *ARHGEF19*, *ELK1*, *DCBLD1*, *IKBKB*, and *CSF1* significantly predicted shorter survival in RCC patients according to TCGA analyses ($p < 0.05$). Importantly, the expression levels of four of these genes, *PXN*, *ARHGEF19*, *ELK1*, and *IKBKB*, were independent prognostic factors for patient survival ($p < 0.05$). We focused on *PXN* (paxillin) and investigated its potential oncogenic role in RCC cells. *PXN* knockdown significantly inhibited cancer cell migration and invasion, possibly by regulating epithelial–mesenchymal transition. Involvement of the *miR-139-3p* passenger strand in RCC molecular pathogenesis is a new concept. Analyses of tumor-suppressive-miRNA-mediated molecular networks provide important insights into the molecular pathogenesis of RCC.

Keywords: *miR-139-5p*; *miR-139-3p*; renal cell carcinoma (RCC); microRNA; tumor suppressor; paxillin (*PXN*)

1. Introduction

Renal cell carcinoma (RCC) accounts for approximately 3% of adult malignancies, with more than 330,000 cases newly diagnosed in 2018 and more than 100,000 deaths annually [1]. Clear cell RCC is the most common histological subtype of RCC, accounting for approximately 75% of all cases [2]. Approximately 20–30% of RCC patients have metastatic lesions at diagnosis, and thus the 5-year survival rate of these patients is less than 20% [2,3]. In addition, more than 20% of patients develop metastases during the postoperative follow-up period [4]. Elucidation of the molecular mechanisms involved in distant metastasis of RCC will contribute to the development of new diagnostic and therapeutic strategies.

MicroRNAs (miRNAs) are classified as noncoding RNAs approximately 18–25 bases long that are found in a wide range of organisms, from plants to humans [5]. miRNAs bind to the 3′-UTR of their target genes (protein coding and noncoding) in a sequence-dependent manner to regulate their expression. miRNAs are involved in various intracellular processes [6,7]. A unique property of these molecules is that a single miRNA can control a vast number of genes in normal and diseased cells [6,7]. Therefore, aberrant expression of miRNAs disrupts intracellular transcriptional networks, which in turn causes human diseases. In cancer cells, aberrant expression of miRNAs is closely associated with cancer cell progression, metastasis, and drug resistance [8,9].

According to the original theory regarding miRNA biogenesis, the passenger strand of the miRNA duplex is degraded and therefore considered to have no function [6,7]. Our RNA-sequencing-based signatures refute this concept, and our recent studies showed downregulation of both the guide and passenger strands of miRNA duplexes (e.g., *miR-30c*, *miR-99a*, *miR-101*, *miR-143*, *miR-145*, and *miR-150*) in cancer tissues [10–12]. Furthermore, ectopic expression assays revealed that the passenger strand of some miRNAs, similar to the guide strand, functions as tumor-suppressive miRNAs, regulating many oncogenes intracellularly [13–16]. Involvement of miRNA passenger strands in the molecular pathogenesis of human cancers is a new concept, and additional research on passenger strands is needed. A recent study confirmed that both strands of miRNAs are functional and, despite having different seed sequences, cooperate to control molecular pathways across cancer types [17].

In this study, we focused on *miR-139-5p* (the guide strand) and *miR-139-3p* (the passenger strand) and investigated their functional significance. We searched for oncogenes that are controlled by *miR-139-3p* in RCC cells in public databases and identified 22 genes as putative targets of *miR-139-3p* regulation. Among these genes, high expression of *PXN*, *ELK1*, *ARHGEF19*, *DCBLD1*, *IKBKB*, and *CSF1* significantly predicted shorter survival in RCC patients according to The Cancer Genome Atlas (TCGA) analyses ($p < 0.05$). We discuss the oncogenic role of PXN in RCC cells.

2. Materials and Methods

2.1. Human RCC Cell Lines

The A498 and 786-O human RCC cell lines were used in this study and were obtained from the Japanese Collection of Research Bioresources Cell Bank. Cell maintenance was performed as we described previously [15].

2.2. RNA Extraction and Quantitative Reverse Transcription-Polymerase Chain Reaction

RNA was extracted from cell lines using TRIzol reagent (Invitrogen, Carlsbad, CA, USA) according to the manufacturer's protocol. For gene expression assays, reverse transcription of the RNA was performed using the High-Capacity cDNA Reverse Transcription Kit (Applied Biosystems, Waltham, MA, USA) and TaqMan Gene Expression Assays (Applied Biosystems), according to our previous studies [13–16]. The expression of *GAPDH* was evaluated as the internal control. We used the CFX Connect Real-Time PCR Detection System (Bio-Rad, Hercules, CA, USA). The TaqMan primers and probes used in this study are listed in Table S1.

2.3. Transfection of miRNAs, siRNAs, and Plasmid Vectors into RCC cells

Transfection of miRNAs and siRNAs into RCC cells was performed using Lipofectamine RNAiMAX reagent (Invitrogen) according to our previous studies [13–16]. miRNAs or siRNAs were added to a final concentration of 10 nM. Plasmid vectors were transfected into the cells using Lipofectamine 2000 (Invitrogen). The final concentration was 50 ng/well.

2.4. Functional Assays (Cell Proliferation, Migration, and Invasion Assays) in RCC cells

XTT assay for cell proliferation, wound healing assay for migration, and Matrigel chamber assay for invasion were performed using RCC cells as described previously [13–16]. The reagents used are listed in Table S1.

2.5. Identification of miR-139-5p and miR-139-3p Gene Targets in RCC Cells

The search strategy used to identify miRNA targets is summarized in Figure S1. The expression profiles of *miR-139-5p/miR-139-3p*-transfected A498 and 786-O cells from the Gene Expression Omnibus (GEO) GSE129043 dataset were used. GSE36895 datasets were also utilized (details below). The TargetScanHuman database (http://www.targetscan.org/vert_72/) was used to predict miRNA-binding sites.

2.6. In silico Analysis of RCC Public Databases

For comparison of gene expression levels between normal and cancer tissues, we utilized the GSE36895 datasets obtained from GEO. GSE36895 contains mRNA microarray data from clear cell RCC tumors, normal kidney tissues, and mouse tumor graft samples obtained using Affymetrix U133 Plus 2.0 whole-genome chips (Affymetrix, Santa Clara, CA, USA). Expression is shown as signal intensities, and for each gene with multiple probes, the mean intensity value was used. In addition, RNA sequencing data from TCGA—Kidney renal clear cell carcinoma (TCGA-KIRC) RNA sequencing datasets were utilized to re-evaluate gene expression levels in normal versus tumor samples.

2.7. Clinicopathological Analysis of RCC

For Kaplan–Meier analyses of overall survival, we downloaded TCGA clinical data (Firehose Legacy) from cBioportal (https://www.cbioportal.org). The patients were divided into two groups by median expression for each gene, according to the data collected from OncoLnc (http://www.oncolnc.org). R version 4.0.2 (R Foundation for Statistical Computing, Vienna, Austria) was used for the analyses.

Multivariate Cox regression analyses were also performed using TCGA-KIRC clinical data and survival data according to the expression level of each gene from OncoLnc to identify factors associated with RCC patient survival. In addition to gene expression, the tumor stage, pathological grade, and age group were evaluated as potential independent prognostic factors. The multivariate analyses were performed using JMP Pro 15.0.0 (SAS Institute Inc., Cary, NC, USA).

We performed gene set enrichment analysis (http://software.broadinstitute.org/gsea/index.jsp) to obtain lists of differentially expressed genes between high and low *PXN* expression groups in the TCGA-KIRC cohort.

2.8. Plasmid Construction and Dual-Luciferase Reporter Assays

psiCHECK-2 plasmid vectors (Promega, Madison, WI, USA) harboring the wild-type or a deletion sequence of the *miR-139-3p*-binding site within the 3′-UTR of *PXN* were prepared. The predicted binding site sequence was obtained from the TargetScanHuman database (release 7.2). Cells were co-transfected with *miR-139-3p* and the plasmid vectors for 36 h, after which firefly and *Renilla* luciferase activities in cell lysates were measured consecutively using the Dual-Luciferase Reporter Assay System (Promega). *Renilla* luciferase activities are expressed as normalized values to firefly luciferase activities. The dual-luciferase reporter assay procedure was described in our previous studies [13–16].

2.9. Western Blotting

Cell lysates were prepared 48 h after transfection with RIPA buffer (Nacalai Tesque, Chukyo-ku, Kyoto, Japan). Then, 20 µg of protein lysates were separated on 4–15% Mini-PROTEAN TGX Precast Gels (Bio-Rad), and transferred to Immun-Blot PVDF membranes (Bio-Rad). Blocking was performed with Blocking One (Nacalai Tesque) for 30 min. The antibodies used in this study are shown in Table S1.

2.10. Statistical Analyses

For comparisons among multiple groups, Dunnett's test was applied. The statistical analyses were performed using JMP Pro 15. Significance levels were set to $p < 0.05$ if not otherwise mentioned.

3. Results

3.1. Analysis of miR-139-5p and miR-139-3p Expression Levels in Clinical RCC Tissues and Their Clinical Significance

The expression levels of *miR-139-5p* and *miR-139-3p* were evaluated in RNA sequencing data from RCC tissue samples obtained from TCGA. *miR-139-5p* and *miR-139-3p* were significantly downregulated in RCC tissues compared with normal tissues ($p < 0.0001$ and $p < 0.0001$, respectively; Figure 1A). Cohort analyses using the TCGA-KIRC datasets revealed that low expression of *miR-139-5p* and *miR-139-3p* was associated with poorer survival in patients with RCC ($p < 0.0001$ and $p < 0.0001$, respectively; Figure 1B).

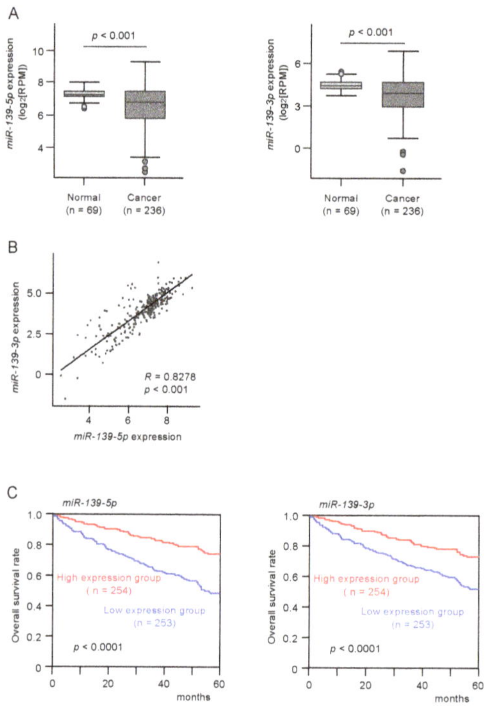

Figure 1. The expression and clinical significance of *miR-139-5p* and *miR-139-3p* in renal cell carcinoma (RCC) clinical specimens. (**A**) Expression of *miR-139-5p* and *miR-139-3p* were significantly reduced in The Cancer Genome Atlas—Kidney Renal Clear Cell Carcinoma (TCGA-KIRC) cancer specimens compared with adjacent normal specimens ($p < 0.001$). (**B**) Spearman's rank test showed positive correlations between the expression levels of *miR-139-5p* and *miR-139-3p* in TCGA-KIRC clinical specimens (R = 0.8278, $p < 0.001$). (**C**) Kaplan–Meier survival curves of patients from TCGA-KIRC cohort. Patients were divided into two groups according to the median expression levels of miR-139-5p or miR-139-3p: high- and low-expression groups. Both *miR-139-5p* and *miR-139-3p* expression levels were significantly associated with the 5-year survival rate of RCC patients ($p < 0.0001$).

3.2. Tumor-Suppressive Functions of miR-139-5p and miR-139-3p in RCC Cells

To investigate the tumor-suppressive functions of *miR-139-5p* and *miR-139-3p* in RCC cells, we assessed cell proliferation, migration, and invasion after ectopic transfection of *miR-139-5p* and *miR-139-3p* into A498 and 786-O cells. Ectopic expression of the two miRNAs did not significantly affect the proliferation of RCC cells (Figure 2A). In contrast, the expression of these miRNAs significantly inhibited the migration and invasive abilities of RCC cells (Figure 2B,C).

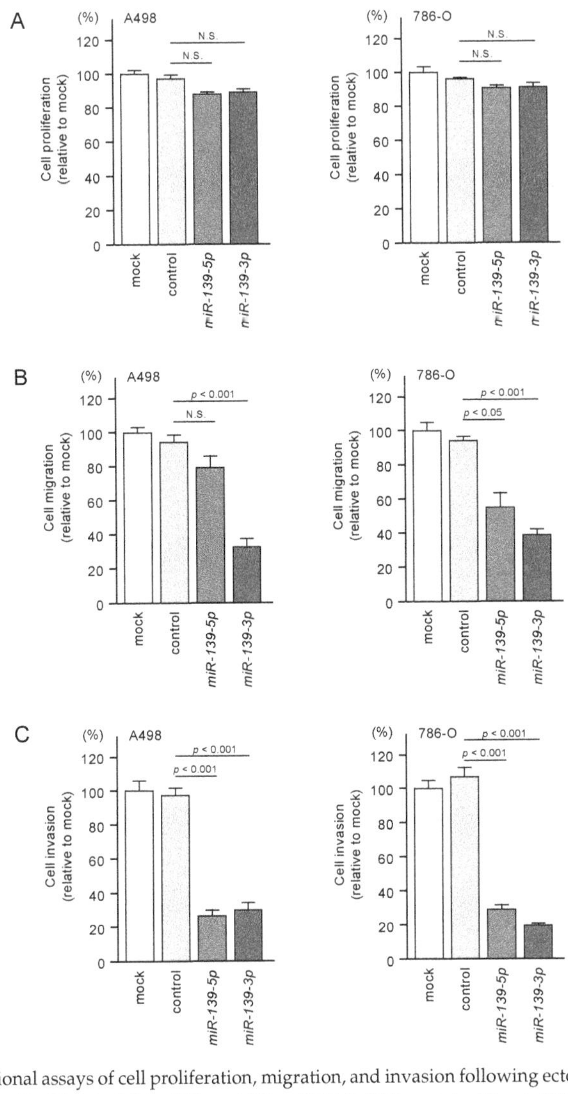

Figure 2. Functional assays of cell proliferation, migration, and invasion following ectopic expression of *miR-139-5p* and *miR-139-3p* in RCC cell lines A498 and 786-O. N.S.: not significant. (**A**) Cell proliferation was assessed using XTT assays. Data were collected 72 h after miRNA transfection. (**B**) Cell migration was assessed using wound healing assays. (**C**) Cell invasions were determined 48 h after seeding miRNA-transfected cells into chambers using Matrigel invasion assays.

3.3. Identification of Putative Oncogenic Targets Regulated by miR-139-3p and miR-139-3p in RCC Cells

To identify the genes regulated by *miR-139-5p* and *miR-139-3p* in RCC cells, we integrated three datasets. First, we obtained RNA microarray data from *miR-139-5p*- or *miR-139-3p*-transfected A498 or 786-O cells (GSE129043). Second, we used data from the TargetScanHuman database (release 7.2) to obtain annotated putative targets regulated by each miRNA strand. Third, we extracted genes highly expressed in clinical specimens of RCC from GSE36895.

A schematic of the strategy used to narrow down the gene list is shown in Figure S1. A total of 19 genes regulated by *miR-139-5p* and 22 by *miR-139-3p* were finally selected (Table 1A,B).

Table 1. A. Candidate target genes of *miR-139-5p*. **B.** Candidate target genes of *miR-139-3p*.

A. Candidate target genes of *miR-139-5p*

Gene Symbol	Gene Name	Entrez Gene ID	*miR-139-5p*-Transfected A498 Cells (log2 FC)	*miR-139-5p*-Transfected 786-O Cells (log2 FC)	GSE36895 (log2 FC)	Total Binding Sites	TCGA 5y OS *p*-Value
PLXDC1	plexin domain containing 1	57125	−0.615791	−1.455703	1.6528159	1	0.0098
TET3	tet methylcytosine dioxygenase 3	200424	−1.074818	−1.152835	0.7350367	1	0.0261
IRF4	interferon regulatory factor 4	3662	−2.491284	−1.563216	0.72956836	1	0.0575
RAB27B	RAB27B, member RAS oncogene family	5874	−0.620716	−0.970745	0.7823266	1	0.1244
FCHSD2	FCH and double SH3 domains 2	9873	−1.834753	−1.364281	0.5673638	1	0.1565
DMD	dystrophin	1756	−0.98734	−1.12498	0.46094477	1	0.3891
APOL6	apolipoprotein L, 6	80830	−1.363298	−0.574163	0.43652818	2	0.4733
AP1S2	adaptor-related protein complex 1, sigma 2 subunit	8905	−0.571341	−0.63469	0.57979524	1	0.5072
PTPRU	protein tyrosine phosphatase, receptor type, U	10076	−1.430874	−0.883479	0.8481535	1	0.6152
TRAT1	T cell receptor associated transmembrane adaptor 1	50852	−1.192173	−1.781637	1.9733018	2	0.7395
SLC39A14	solute carrier family 39 (zinc transporter), member 14	23516	−0.536152	−0.705082	1.0408258	1	0.8125
OTUD4	OTU deubiquitinase 4	54726	−1.720465	−1.59549	0.21931966	1	0.9338
CDCA7L	cell division cycle associated 7-like	55536	−1.478013	−0.564377	2.1733913	1	0.969
ZNF678	zinc finger protein 678	339500	−1.585917	−0.621832	0.25128952	2	0.0086 *
FGFBP2	fibroblast growth factor binding protein 2	83888	−0.830757	−1.317887	1.5742466	1	0.0058 *
ATP2B2	ATPase, Ca++ transporting, plasma membrane 2	491	−1.109889	−2.668194	1.3068246	3	0.0050 *
EML1	echinoderm microtubule associated protein like 1	2009	−0.56874	−0.541978	0.38891175	2	0.0014 *
PCSK5	proprotein convertase subtilisin/kexin type 5	5125	−1.751102	−0.577052	0.57546955	1	0.0006 *
FAM168A	family with sequence similarity 168, member A	23201	−1.390649	−0.959992	0.25754136	1	0.0002 *

B. Candidate target genes of *miR-139-3p*

Gene Symbol	Gene Name	Entrez Gene ID	*miR-139-3p*-Transfected A498 Cells (log2 FC)	*miR-139-3p*-Transfected 786-O Cells (log2 FC)	GSE36895 (log2 FC)	Total Binding Sites	TCGA 5y OS *p*-Value
PXN	paxillin	5829	−1.164181	−0.707167	0.481819	2	<0.0001
ARHGEF19	Rho guanine nucleotide exchange factor (GEF) 19	128272	−0.607603	−1.826996	1.1049173	1	<0.0001
ELK1	ELK1, member of ETS oncogene family	2002	−1.518329	−0.839984	0.5987212	2	0.0001
CSF1	colony stimulating factor 1 (macrophage)	1435	−0.952795	−0.537074	1.0153022	1	0.0124
IKBKB	inhibitor of kappa light polypeptide gene enhancer in B-cells, kinase beta	3551	−0.511766	−1.882033	0.23441868	1	0.0251
DCBLD1	discoidin, CUB and LCCL domain containing 1	285761	−1.398862	−0.668245	0.28628728	1	0.0285
SYT11	synaptotagmin XI	23208	−1.034462	−0.578617	0.4411527	1	0.0556
SERPINE1	serpin peptidase inhibitor, clade E (nexin, plasminogen activator inhibitor type 1), member 1	5054	−2.516404	−0.663638	1.8049024	3	0.0731
KDM6B	lysine (K)-specific demethylase 6B	23135	−0.62567	−0.733171	0.42303625	1	0.1019
RASSF5	Ras association (RalGDS/AF-6) domain family member 5	83593	−1.180982	−4.605049	1.5059676	1	0.3643
ACBD3	acyl-CoA binding domain containing 3	64746	−0.688812	−0.542166	0.65537864	1	0.3841
APOL6	apolipoprotein L, 6	80830	−1.350307	−0.858142	0.43652818	1	0.4733
EPN2	epsin 2	22905	−0.619177	−0.520903	0.29098234	1	0.581
GIT2	G protein-coupled receptor kinase interacting ArfGAP 2	9815	−0.64332	−0.544951	1.4113243	1	0.7179
KIF3C	kinesin family member 3C	3797	−0.694929	−0.956568	0.29198763	1	0.9148
ARAP2	ArfGAP with RhoGAP domain, ankyrin repeat and PH domain 2	116984	−1.161352	−1.258001	0.3943753	1	0.0940 *
RFX2	regulatory factor X, 2 (influences HLA class II expression)	5990	−0.937881	−0.580092	1.4766915	1	0.0671 *
RNF125	ring finger protein 125, E3 ubiquitin protein ligase	54941	−0.539067	−1.811576	0.6103558	1	0.0394 *
ARSK	arylsulfatase family, member K	153642	−1.505581	−0.887476	0.4327301	1	0.0219 *
STAG2	stromal antigen 2	10735	−0.527565	−0.563985	0.38703138	1	<0.0001 *
TNS1	tensin 1	7145	−0.764289	−0.609893	0.28155625	1	<0.0001 *

* Better prognosis in high-expression group.

3.4. Clinical Significance of miR-139 Target Genes in RCC Pathogenesis

Kaplan–Meier analyses of 5-year overall survival were performed according to high versus low expression of *miR-139* target genes. The high expression of two *miR-139-5p* (*PLXDC1* and *TET3*) and six *miR-139-3p* target genes (*PXN, ARHGEF19, ELK1, CSF1, IKBKB*, and *DCBLD1*) was found to be related to better 5-year overall survival rates of the patients ($p < 0.05$, Table 1A,B, and Figure 3).

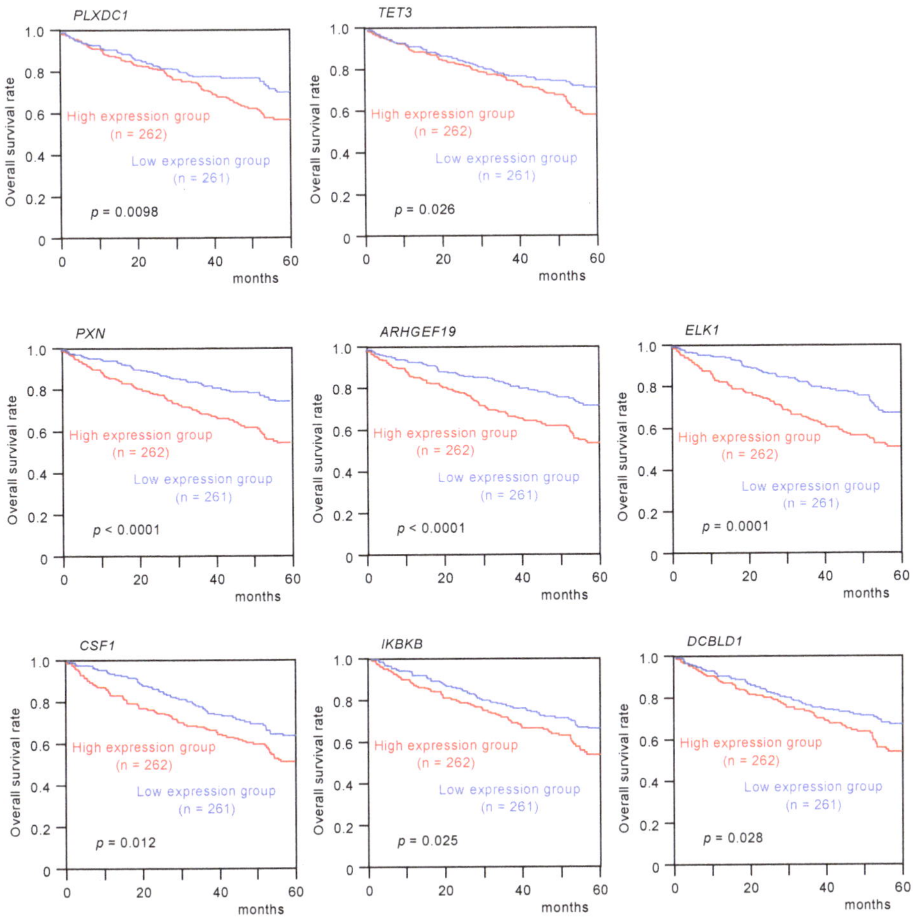

Figure 3. Clinical significance of *miR-139-5p* or *miR-139-3p* target genes in TCGA-KIRC database. High expression of two *miR-139-5p* target genes (*PLXDC1* and *TET3*) and six *miR-139-3p* target genes (*PXN*, *ARHGEF19*, *ELK1*, *CSF1*, *IKBKB*, and *DCBLD1*) were significantly associated with poor prognosis in patients with RCC. Kaplan–Meier curves for 5-year overall survival according to the expression of each *miR-139-3p* target gene are shown.

Furthermore, multivariate analyses identified four genes (*PXN*, *ARHGEF19*, *ELK1*, and *IKBKB*) as independent prognostic factors for patient survival ($p < 0.05$; Figure 4). We validated the expression levels of these target genes using TCGA-KIRC RNA sequencing dataset. All target genes except *ARHGEF19* were upregulated in cancer tissues in this cohort (Figure 5).

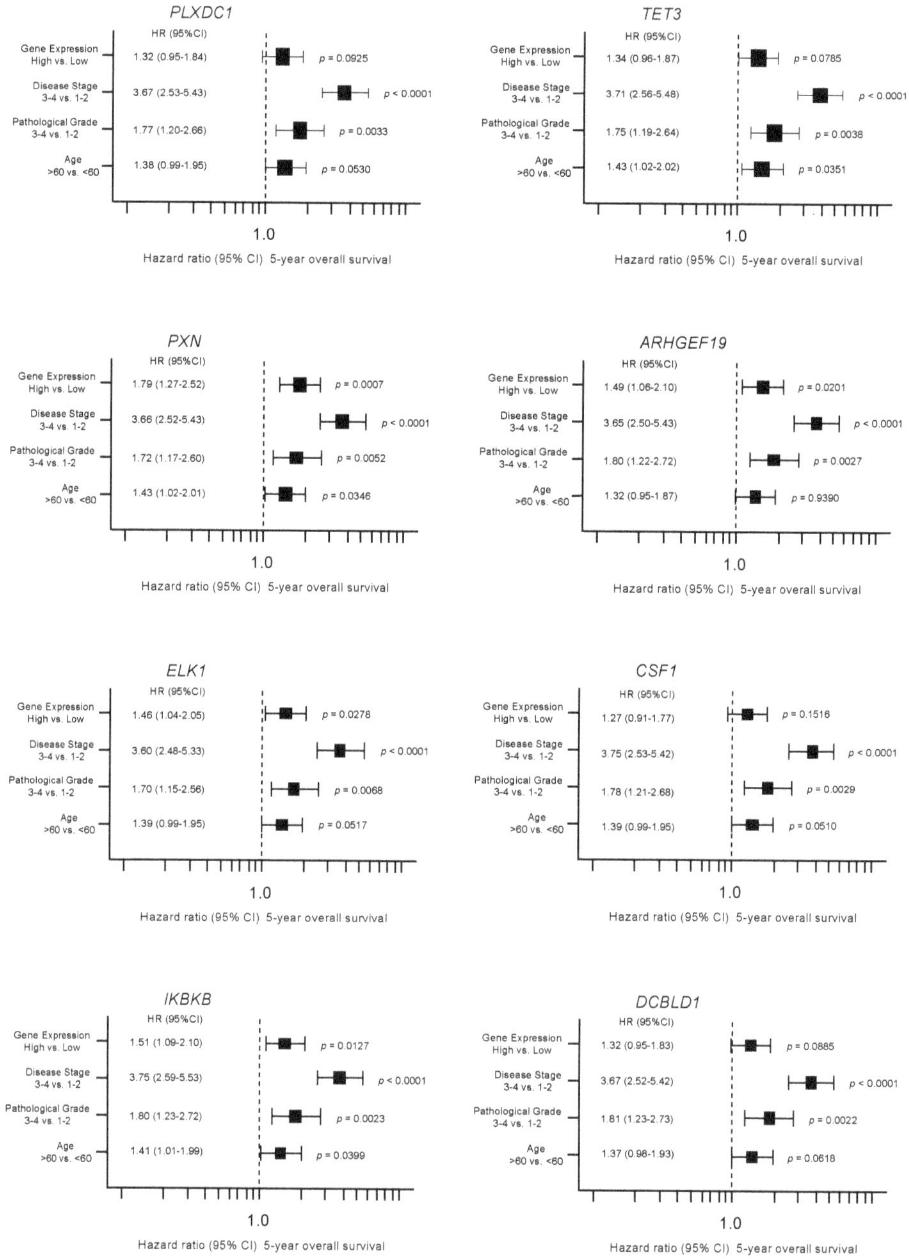

Figure 4. Forest plot showing the results of multivariate analyses of eight genes (*PLXDC1*, *TET3*, *PXN*, *ARHGEF19*, *ELK1*, *CSF1*, *IKBKB*, and *DCBLD1*). In addition to the gene expression level, the tumor stage, pathological grade, and age group were evaluated as potential independent factors associated with survival. Numbers of cases per each group are shown in Table S2.

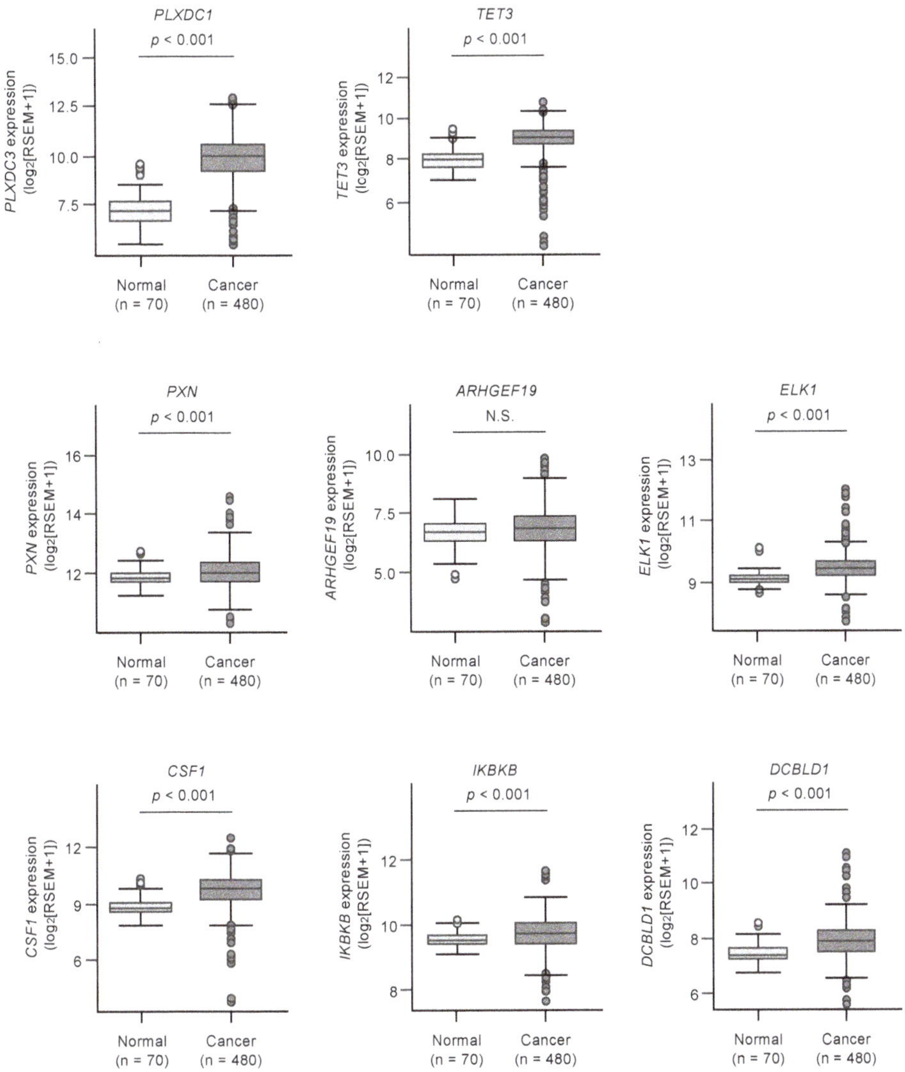

Figure 5. Expression levels of *miR-139-5p* or *miR-139-3p* target genes (*PLXDC1* and *TET3* for *miR-139-5p*, *PXN*, *ARHGEF19*, *ELK1*, *CSF1*, *IKBKB*, and *DCBLD1* for *miR-139-3p*) in TCGA-KIRC cohort. Seven genes (*PLXDC1*, *TET3*, *PXN*, *ELK1*, *CSF1*, *IKBKB*, and *DCBLD1*) were confirmed to be significantly upregulated in clinical specimens. N.S.: not significant.

For subsequent analyses, we focused on miR-139-3p (the passenger strand) target genes, according to our previous studies emphasizing the novel roles of miRNA passenger strands. Among these target genes, we focused on *PXN*, which showed the most significant association with RCC patient survival in the multivariate analysis and has been reported previously as an oncogene in other types of cancers [18,19].

3.5. Direct Regulation of PXN by miR-139-3p in RCC Cells

In RCC cells transfected with *miR-139-3p*, both PXN mRNA and protein levels were significantly downregulated (Figure 6A,B). Western blot images are shown in Figure S2. To validate that *miR-139-3p* binds directly to *PXN* to downregulate its expression, we performed dual-luciferase reporter assays. A498 and 786-O cells were co-transfected with plasmid vectors and *miR-139-3p*. We used two different plasmid vectors: one containing the partial wild-type sequence of the *miR-139-3p*-binding site predicted by TargetScanHuman database ("wild-type sequence" in Figure 6C) and the other containing this sequence lacking the binding site ("deletion-type sequence" in Figure 6C). Luciferase activity was significantly reduced in cells transfected with the wild-type sequence but not in cells transfected with the deletion sequence (Figure 6D). These results suggest that *miR-139-3p* directly binds to the 3′-UTR of *PXN*.

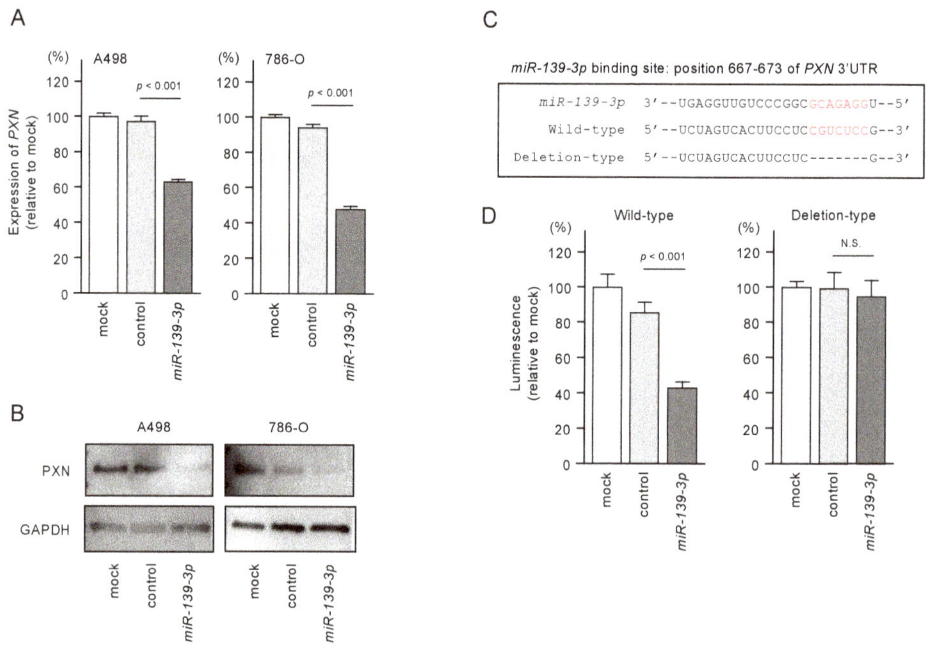

Figure 6. Expression of paxillin (PXN) was regulated directly by *miR-139-3p* in RCC cells. (**A**) Expression of *PXN* mRNA was significantly suppressed in *miR-139-3p*-transfected A498 and 786-O cells (48 h after transfection). Expression of *GAPDH* was used as an internal control. (**B**) Expression of PXN protein was reduced in *miR-139-3p*-transfected RCC cells (48 h after transfection). GAPDH was used as a loading control. (**C**) The TargetScanHuman database predicted one putative *miR-139-3p*-binding site in the 3′-UTR of *PXN*. (**D**) Dual-luciferase reporter assays showed decreased luminescence activity in RCC cells co-transfected with *miR-139-3p* together with a vector harboring the "wild-type sequence". Normalized data were calculated as *Renilla*/firefly luciferase activity ratios. N.S.: not significant.

3.6. PXN Knockdown Assays in RCC Cells

We assessed the oncogenic functions of PXN in RCC cells by performing knockdown using two small interfering RNAs (siRNAs) targeting *PXN*. The mRNA and protein levels of PXN were successfully downregulated by either siRNAs in A498 and 786-O cells (Figure 7A,B). Western blot images are shown in Figure S2.

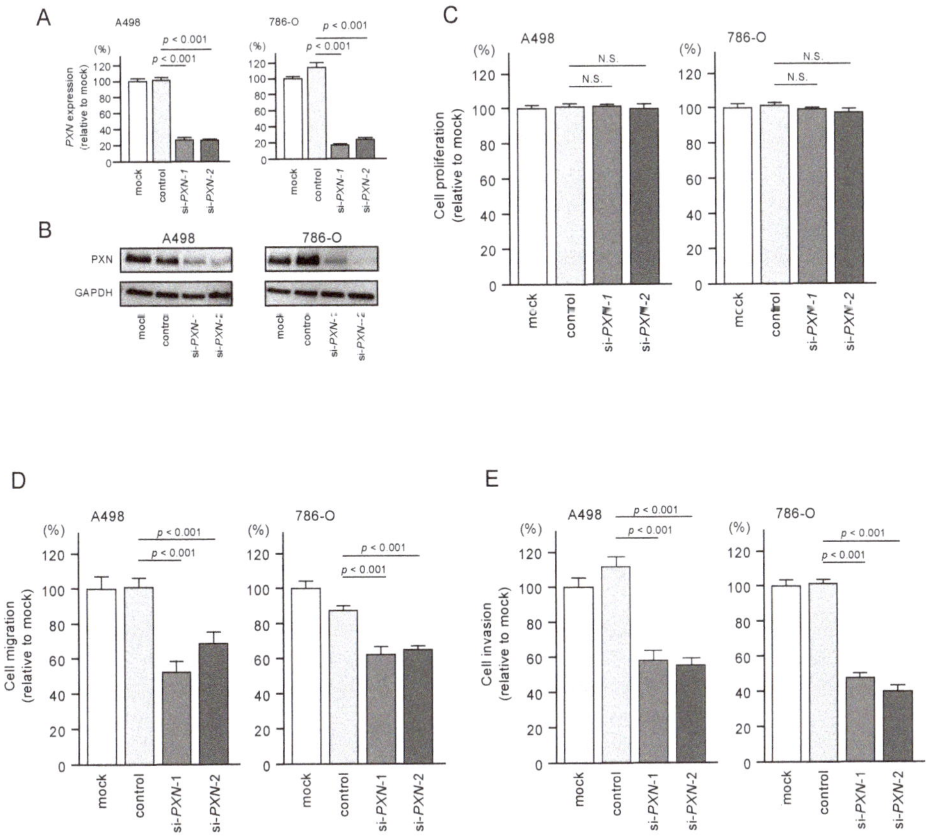

Figure 7. Effects of *PXN* knockdown on cell proliferation, migration, and invasion in RCC cells. (**A**,**B**) Expression of PXN was successfully reduced after siRNA transfection in RCC cells. (**C**) Cell proliferation was determined by XTT assays. N.S.: not significant. (**D**) Cell migration was determined by wound healing assays. (**E**) Cell invasion was determined by Matrigel invasion assays.

Next, functional assays were performed in RCC cells transfected with these siRNAs. Similar to *miR-139-3p* transfection, transfection of the siRNAs did not suppress cell proliferation (Figure 7C). However, cell migration and invasion were significantly suppressed by siRNA transfection in both cell lines (Figure 7D,E).

3.7. PXN-Mediated Pathways in RCC Cells

We performed gene set enrichment analysis to determine differentially expressed genes between the high and low *PXN* expression groups of the TCGA-KIRC cohort. In support of the functional assays using siRNAs, the most enriched signaling pathway in the high *PXN* expression group was "epithelial–mesenchymal transition" (Figure 8 and Table S3). Other significantly enriched pathways (FDR q-value < 0.05) were "hypoxia", "KRAS signaling", "myogenesis", "angiogenesis", and "apical junction", supporting the hypothesis that *PXN* is related to the metastatic ability of cancer cells (Figure 7 and Table S2).

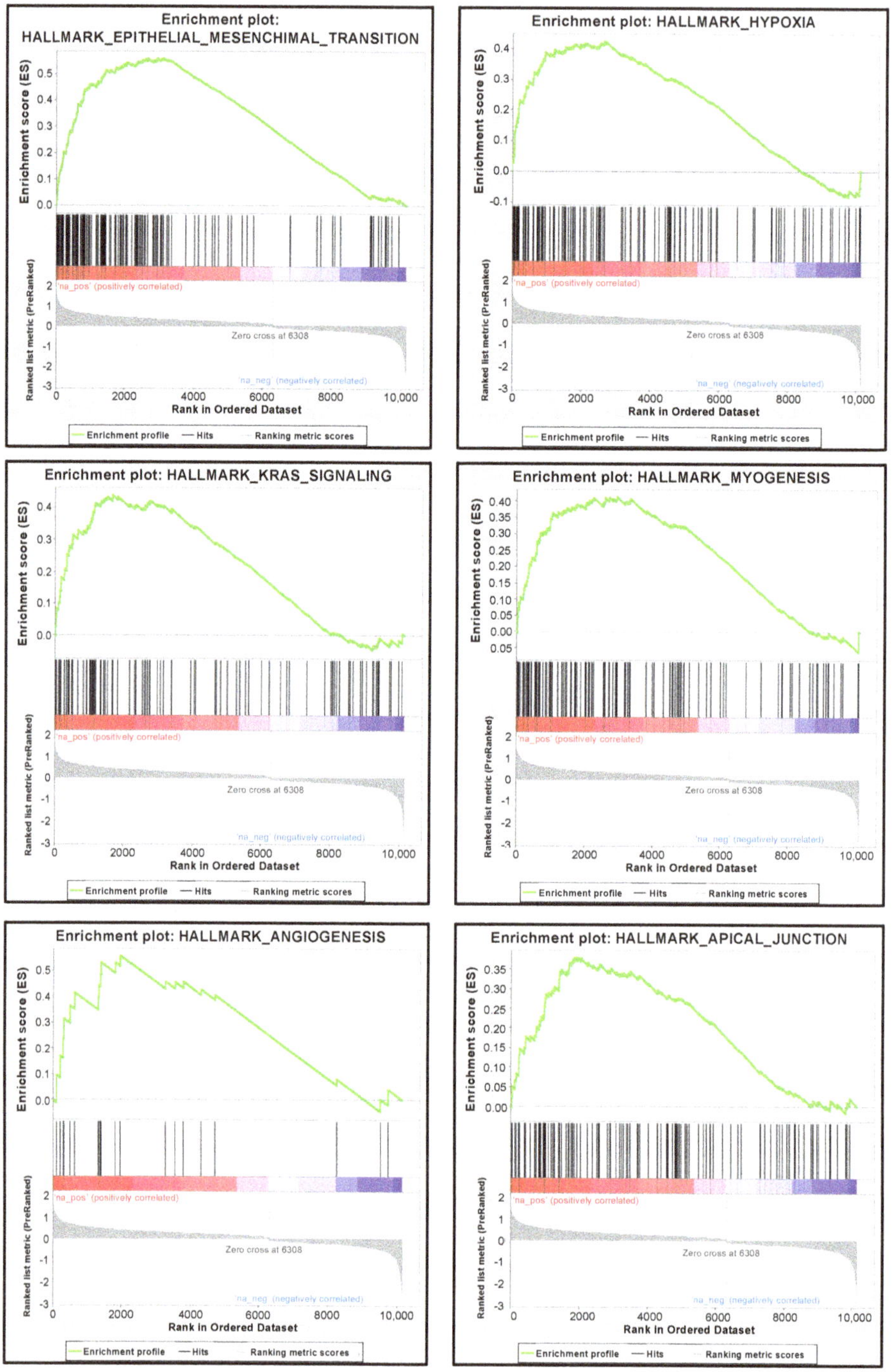

Figure 8. Pathways enriched among the differentially expressed genes in the high *PXN* expression group according to gene set enrichment analysis. The six significantly enriched pathways (FDR q-value < 0.05) are shown.

4. Discussion

Recent RNA sequencing techniques have accelerated the establishment of miRNA expression signatures. Our recent studies revealed that some passenger strands of miRNAs (e.g., *miR-30c-2-3p*, *miR-99a-3p*, *miR-101-5p*, *miR-143-5p*, *miR-145-3p*, and *miR-150-3p*) act as tumor-suppressive miRNAs by targeting several oncogenes in a wide range of cancers [10–16]. Involvement of the passenger strands of miRNAs in cancer pathogenesis is a new development in cancer research.

A recent in silico study (analysis of molecular profiles in more than 5200 patient samples from 14 different cancers) revealed that both strands of miRNAs are functional across different cancer types. For example, downregulation of both strands of *miR-30a* and *miR-145* was frequently observed in multiple cancers, and their downregulation enhanced aberrant expression of cell cycle-related genes. These malignant events were found to affect the prognosis of cancer patients [17]. Simultaneous analysis of both strands of miRNA duplexes will lead to elucidation of a new molecular mechanism in cancer cells.

In this study, we showed that both strands of the *miR-139* duplex acted as tumor-suppressive miRNAs in RCC cells. Accumulating evidence has shown a tumor-suppressive role of *miR-139-5p* in multiple types of cancers, including RCC [20–22]. Notably, *miR-139-5p* target genes are involved in cancer activation pathways, e.g., the PI3K/AKT/mTORC1, Wnt/β-catenin, and RTK/RAS/MAPK pathways [23–27]. Therefore, *miR-139-5p* plays a central role in controlling the malignant transformation of human cancers.

In contrast to *miR-139-5p*, few studies have evaluated the importance of *miR-139-3p* in cancer because it is a passenger strand. Recently, it was reported that *miR-139-3p* directly regulates *KDM5B* (lysine demethylase 5B), a key regulator of histone 3 lysine 4 demethylation in laryngeal squamous cell carcinoma [28]. Ectopic expression of *miR-139-3p* suppressed cancer cell malignant behavior by inhibiting the Wnt/β-catenin pathway [28]. In the HeLa cervical cancer cell line, *miR-139-3p* was reduced upon increasing expression of circular RNA hsa_circ_0031288 [29]. In turn, the decreased expression of *miR-139-3p* resulted in increased expression of B-cell CLL/lymphoma 6, as a target molecule. These regulatory effects promoted malignant transformation of HeLa cells. In bladder cells, both strands of the *miR-139* duplex were downregulated in cancer tissues and exhibited tumor-suppressive effects [30]. Interestingly, matrix metalloprotease 11 (*MMP11*) was directly regulated by both *miR-139* strands, and knockdown of *MMP11* attenuated the aggressive phenotype of bladder cancer cells [30]. Our study is the first to show that *miR-139-3p* (the passenger strand) is involved in RCC pathogenesis, and both strands of the *miR-139* duplex are pivotal players in RCC oncogenesis.

We are also interested in the presence of oncogenes controlled by tumor-suppressive *miR-139-5p* and *miR-139-3p* in RCC cells. A total of 19 and 22 genes were identified as putative targets of *miR-139-5p* and *miR-139-3p* regulation in RCC cells, respectively. Taking advantage of survival data from TCGA datasets, we performed multivariate analysis and found that, among the *miR-139* duplex target genes, *PXN*, *ELK1*, *ARHGEF19*, and *IKBKB* were independent prognostic factors for RCC patient survival. Further genomic analyses of these genes will contribute to elucidating the molecular pathogenesis of RCC.

ELK1 is a well-established transcription factor that is phosphorylated by MAPKs and induces transcription of the c-fos proto-oncogene [31,32]. *ARHGEF19* is a RhoGEF that reportedly activates the MAPK pathway and interacts with BRAF in lung cancer [33]. The role of *ARHGEF19* in RCC is not well characterized, but the interaction of *ARHGEF19* with *ELK1* via the MAPK pathway may play a role in RCC development. *IKBKB* encodes IKK-beta, which phosphorylates the inhibitor of NF-kB, resulting in activation of the NF-kB pathway [34,35]. NF-kB and its downstream inflammatory signaling pathway are closely related to RCC carcinogenesis and aggressiveness [36].

In this study, we focused on *PXN* (paxillin) because its expression was found to be directly controlled by *miR-139-3p* in RCC cells and strongly related to RCC molecular pathogenesis. PXN is a focal adhesion scaffold/adaptor protein that contains five LD domains (leucine-aspartate motifs) located at the N-terminus and four cysteine–histidine-rich LIM domains at the C-terminus [37,38]. The LD domains act as a docking site for focal adhesion-related proteins, e.g., Src (tyrosine-protein kinase), FAK (focal adhesion kinase), PAK (p21-activated kinase), and ILK (integrin-linked kinase) [37,38]. The LIM domains act as binding sites for protein–protein interactions [37,38]. The pathways activated by PXN enhance cancer cell malignant progression and metastasis in a wide range of cancers [39]. Moreover, two key proteins in focal adhesion complexes, PXN and integrin B4, directly bind to each other, and this complex enhanced cisplatin resistance in lung cancer cells [40]. Another study showed that PXN phosphorylation may contribute to cisplatin resistance via the ERK-mediated activation of Bcl-2 transcription in lung cancer [41]. PXN-mediated pathways may be therapeutic targets for attenuating drug-resistance in cancer cells.

In our GSEA analysis, *PXN*-high-expressed RCC specimen was enriched with epithelial-mesenchymal transition (EMT) signaling pathways. A previous report has shown the suppressive effect of *PXN* on EMT pathway in cell lines [42]. These findings suggest the speculation that inhibitions of cancer cell migration and invasion by *PXN* knockdown in our study are due to the regulation of epithelial-mesenchymal transition.

It has been reported that several miRNAs (e.g., *miR-132*, *miR-137*, *miR-145*, *miR-212*, and *miR-218*) directly control *PXN* expression in cancer cells [43–47]. Among these miRNAs, we previously reported the downregulation of *miR-145* and *miR-218* in RCC cells [48,49]. More recently, *DLX6-AS1* (a long noncoding RNA that adsorbs *miRNA-199b*) promoted epithelial–mesenchymal transition and cisplatin resistance via the *miR-199b-5p/PXN* axis in breast cancer cells [50]. This is the first report that *PXN* is directly regulated by tumor-suppressive *miR-139-3p* in RCC cells. Non-coding RNA-mediated epigenetic regulation of *PXN* expression will be assessed in the future.

5. Conclusions

Both strands of the *miR-139* duplex are closely involved in RCC oncogenesis. This is the first report to reveal that *miR-139-3p* (the passenger strand) acts as a tumor-suppressive miRNA in RCC. Several genes are controlled by *miR-139* and contribute to RCC molecular pathogenesis. Notably, the expression of PXN was directly regulated by *miR-139-3p*, and its overexpression enhanced RCC malignant transformation. Analyses of tumor-suppressive miRNAs (including the passenger strands) will contribute to the elucidation of new molecular networks in RCC.

Supplementary Materials: The following are available online at http://www.mdpi.com/2227-9059/8/12/599/s1, Figure S1: Schematic of the strategy used to narrow down the *miR-139-5p/miR-139-3p* target genes, Figure S2: Western blot images, Table S1: Reagents used in this study, Table S2: The number of cases per group for multivariate analysis, Table S3: Gene set enrichment analysis of genes enriched among the differentially expressed genes in the TCGA-KIRC high *PXN* expression group.

Author Contributions: Conceptualization, N.S.; methodology, N.S.; validation, R.O., S.M., and S.A.; formal analysis, R.O., Y.G., and Y.Y.; investigation, R.O. and Y.Y.; resources, Y.G., Y.Y., M.K., and T.I.; data curation, R.O., Y.G., Y.Y., and S.A.; writing—original draft preparation, R.O., Y.G., and N.S.; writing—review and editing, R.O., Y.G., and S.M.; visualization, R.O., Y.Y., and N.S.; supervision, N.S.; project administration, N.S.; funding acquisition, Y.G., M.K., and N.S. All authors have read and agreed to the published version of the manuscript.

Funding: This research was funded by JSPS KAKENHI, grant numbers; 20K22970, 18K16724, 18K09338.

Conflicts of Interest: The authors declare no conflict of interest.

References

1. Bray, F.; Ferlay, J.; Soerjomataram, I.; Siegel, R.L.; Torre, L.A.; Jemal, A. Global cancer statistics 2018: GLOBOCAN estimates of incidence and mortality worldwide for 36 cancers in 185 countries. *CA Cancer J. Clin.* **2018**, *68*, 394–424. [CrossRef] [PubMed]
2. Pierorazio, P.M.; Johnson, M.H.; Patel, H.D.; Sozio, S.M.; Sharma, R.; Iyoha, E.; Bass, E.B.; Allaf, M.E. Management of renal masses and localized renal cancer: Systematic review and meta-analysis. *J. Urol.* **2016**, *196*, 989–999. [CrossRef] [PubMed]
3. Cairns, P. Renal cell carcinoma. *Cancer Biomark* **2010**, *9*, 461–473. [CrossRef]
4. Cohen, H.T.; McGovern, F.J. Renal-cell carcinoma. *N. Engl. J. Med.* **2005**, *353*, 2477–2490. [CrossRef] [PubMed]

5. Anfossi, S.; Babayan, A.; Pantel, K.; Calin, G.A. Clinical utility of circulating non-coding RNAs-an update. *Nat. Rev. Clin. Oncol.* **2018**, *15*, 541–563. [CrossRef] [PubMed]
6. Ha, M.; Kim, V.N. Regulation of microRNA biogenesis. *Nat. Rev. Mol. Cell Biol.* **2014**, *15*, 509–524. [CrossRef]
7. Gebert, L.F.R.; MacRae, I.J. Regulation of microRNA function in animals. *Nat. Rev. Mol. Cell Biol.* **2019**, *20*, 21–37. [CrossRef]
8. Lin, S.; Gregory, R.I. MicroRNA biogenesis pathways in cancer. *Nat. Rev. Cancer* **2015**, *15*, 321–333. [CrossRef]
9. Rupaimoole, R.; Slack, F.J. MicroRNA therapeutics: Towards a new era for the management of cancer and other diseases. *Nat. Rev. Drug Discov.* **2017**, *16*, 203–222. [CrossRef]
10. Goto, Y.; Kurozumi, A.; Arai, T.; Nohata, N.; Kojima, S.; Okato, A.; Kato, M.; Yamazaki, K.; Ishida, Y.; Naya, Y.; et al. Impact of novel miR-145-3p regulatory networks on survival in patients with castration-resistant prostate cancer. *Br. J. Cancer* **2017**, *117*, 409–420. [CrossRef]
11. Toda, H.; Seki, N.; Kurozumi, S.; Shinden, Y.; Yamada, Y.; Nohata, N.; Moriya, S.; Idichi, T.; Maemura, K.; Fujii, T.; et al. RNA-sequence-based microRNA expression signature in breast cancer: Tumor-suppressive miR-101-5p regulates molecular pathogenesis. *Mol. Oncol.* **2020**, *14*, 426–446. [CrossRef] [PubMed]
12. Wada, M.; Goto, Y.; Tanaka, T.; Okada, R.; Moriya, S.; Idichi, T.; Noda, M.; Sasaki, K.; Kita, Y.; Kurahara, H.; et al. RNA sequencing-based microRNA expression signature in esophageal squamous cell carcinoma: Oncogenic targets by antitumor miR-143-5p and miR-143-3p regulation. *J. Hum. Genet.* **2020**, *65*, 1019–1034. [CrossRef] [PubMed]
13. Tanaka, T.; Okada, R.; Hozaka, Y.; Wada, M.; Moriya, S.; Satake, S.; Idichi, T.; Kurahara, H.; Ohtsuka, T.; Seki, N. Molecular pathogenesis of pancreatic ductal adenocarcinoma: Impact of miR-30c-5p and miR-30c-2-3p regulation on oncogenic genes. *Cancers (Basel)* **2020**, *12*, 2731. [CrossRef] [PubMed]
14. Mizuno, K.; Tanigawa, K.; Nohata, N.; Misono, S.; Okada, R.; Asai, S.; Moriya, S.; Suetsugu, T.; Inoue, H.; Seki, N. FAM64A: A novel oncogenic target of lung adenocarcinoma regulated by both strands of miR-99a (miR-99a-5p and miR-99a-3p). *Cells* **2020**, *9*, 2083. [CrossRef] [PubMed]
15. Yamada, Y.; Nohata, N.; Uchida, A.; Kato, M.; Arai, T.; Moriya, S.; Mizuno, K.; Kojima, S.; Yamazaki, K.; Naya, Y.; et al. Replisome genes regulation by antitumor miR-101-5p in clear cell renal cell carcinoma. *Cancer Sci.* **2020**, *111*, 1392–1406. [CrossRef] [PubMed]
16. Misono, S.; Seki, N.; Mizuno, K.; Yamada, Y.; Uchida, A.; Sanada, H.; Moriya, S.; Kikkawa, N.; Kumamoto, T.; Suetsugu, T.; et al. Molecular pathogenesis of gene regulation by the miR-150 duplex: miR-150-3p regulates TNS4 in lung adenocarcinoma. *Cancers (Basel)* **2019**, *11*, 601. [CrossRef] [PubMed]
17. Mitra, R.; Adams, C.M.; Jiang, W.; Greenawalt, E.; Eischen, C.M. Pan-cancer analysis reveals cooperativity of both strands of microRNA that regulate tumorigenesis and patient survival. *Nat. Commun.* **2020**, *11*, 968. [CrossRef]
18. Alpha, K.M.; Xu, W.; Turner, C.E. Paxillin family of focal adhesion adaptor proteins and regulation of cancer cell invasion. *Int. Rev. Cell Mol. Biol.* **2020**, *355*, 1–52. [CrossRef]
19. Noh, K.; Bach, D.H.; Choi, H.J.; Kim, M.S.; Wu, S.Y.; Pradeep, S.; Ivan, C.; Cho, M.S.; Bayraktar, E.; Rodriguez-Aguayo, C.; et al. The hidden role of paxillin: Localization to nucleus promotes tumor angiogenesis. *Oncogene* **2020**. [CrossRef]
20. Khalili, N.; Nouri-Vaskeh, M.; Hasanpour Segherlou, Z.; Baghbanzadeh, A.; Halimi, M.; Rezaee, H.; Baradaran, B. Diagnostic, prognostic, and therapeutic significance of miR-139-5p in cancers. *Life Sci.* **2020**, *256*, 117865. [CrossRef]
21. Wu, X.; Weng, L.; Li, X.; Guo, C.; Pal, S.K.; Jin, J.M.; Li, Y.; Nelson, R.A.; Mu, B.; Onami, S.H.; et al. Identification of a 4-microRNA signature for clear cell renal cell carcinoma metastasis and prognosis. *PLoS ONE* **2012**, *7*, e35661. [CrossRef] [PubMed]
22. Fridman, E.; Dotan, Z.; Barshack, I.; David, M.B.; Dov, A.; Tabak, S.; Zion, O.; Benjamin, S.; Benjamin, H.; Kuker, H.; et al. Accurate molecular classification of renal tumors using microRNA expression. *J. Mol. Diagn* **2010**, *12*, 687–696. [CrossRef] [PubMed]
23. Catanzaro, G.; Besharat, Z.M.; Miele, E.; Chiacchiarini, M.; Po, A.; Carai, A.; Marras, C.E.; Antonelli, M.; Badiali, M.; Raso, A.; et al. The miR-139-5p regulates proliferation of supratentorial paediatric low-grade gliomas by targeting the PI3K/AKT/mTORC1 signalling. *Neuropathol. Appl. Neurobiol.* **2018**, *44*, 687–706. [CrossRef] [PubMed]

24. Noorolyai, S.; Mokhtarzadeh, A.; Baghbani, E.; Asadi, M.; Baghbanzadeh Kojabad, A.; Mogaddam, M.M.; Baradaran, B. The role of microRNAs involved in PI3-kinase signaling pathway in colorectal cancer. *J. Cell Physiol.* **2019**, *234*, 5664–5673. [CrossRef] [PubMed]
25. Ji, X.; Guo, H.; Yin, S.; Du, H. miR-139-5p functions as a tumor suppressor in cervical cancer by targeting TCF4 and inhibiting Wnt/β-catenin signaling. *Onco Targets Ther.* **2019**, *12*, 7739–7748. [CrossRef] [PubMed]
26. Shirjang, S.; Mansoori, B.; Asghari, S.; Duijf, P.H.G.; Mohammadi, A.; Gjerstorff, M.; Baradaran, B. MicroRNAs in cancer cell death pathways: Apoptosis and necroptosis. *Free Radic. Biol. Med.* **2019**, *139*, 1–15. [CrossRef]
27. Niveditha, D.; Jasoria, M.; Narayan, J.; Majumder, S.; Mukherjee, S.; Chowdhury, R.; Chowdhury, S. Common and unique microRNAs in multiple carcinomas regulate similar network of pathways to mediate cancer progression. *Sci. Rep.* **2020**, *10*, 2331. [CrossRef]
28. Ma, Y.; Chen, Z.; Yu, G. microRNA-139-3p inhibits malignant behaviors of laryngeal cancer cells via the KDM5B/SOX2 axis and the Wnt/β-catenin pathway. *Cancer Manag. Res.* **2020**, *12*, 9197–9209. [CrossRef]
29. Xu, Y.J.; Yu, H.; Liu, G.X. Hsa_circ_0031288/hsa-miR-139-3p/Bcl-6 regulatory feedback circuit influences the invasion and migration of cervical cancer HeLa cells. *J. Cell Biochem.* **2020**, *121*, 4251–4260. [CrossRef]
30. Yonemori, M.; Seki, N.; Yoshino, H.; Matsushita, R.; Miyamoto, K.; Nakagawa, M.; Enokida, H. Dual tumor-suppressors miR-139-5p and miR-139-3p targeting matrix metalloprotease 11 in bladder cancer. *Cancer Sci.* **2016**, *107*, 1233–1242. [CrossRef]
31. Hipskind, R.A.; Rao, V.N.; Mueller, C.G.; Reddy, E.S.; Nordheim, A. Ets-related protein Elk-1 is homologous to the c-fos regulatory factor p62TCF. *Nature* **1991**, *354*, 531–534. [CrossRef] [PubMed]
32. Gille, H.; Kortenjann, M.; Thomae, O.; Moomaw, C.; Slaughter, C.; Cobb, M.H.; Shaw, P.E. ERK phosphorylation potentiates Elk-1-mediated ternary complex formation and transactivation. *EMBO J.* **1995**, *14*, 951–962. [CrossRef] [PubMed]
33. Li, Y.; Ye, Z.; Chen, S.; Pan, Z.; Zhou, Q.; Li, Y.Z.; Shuai, W.D.; Kuang, C.M.; Peng, Q.H.; Shi, W.; et al. ARHGEF19 interacts with BRAF to activate MAPK signaling during the tumorigenesis of non-small cell lung cancer. *Int. J. Cancer* **2018**, *142*, 1379–1391. [CrossRef] [PubMed]
34. DiDonato, J.A.; Hayakawa, M.; Rothwarf, D.M.; Zandi, E.; Karin, M. A cytokine-responsive IkappaB kinase that activates the transcription factor NF-kappaB. *Nature* **1997**, *388*, 548–554. [CrossRef] [PubMed]
35. Mitchell, S.; Vargas, J.; Hoffmann, A. Signaling via the NFκB system. *Wiley. Interdiscip. Rev. Syst. Biol. Med.* **2016**, *8*, 227–241. [CrossRef]
36. Oya, M.; Takayanagi, A.; Horiguchi, A.; Mizuno, R.; Ohtsubo, M.; Marumo, K.; Shimizu, N.; Murai, M. Increased nuclear factor-kappa B activation is related to the tumor development of renal cell carcinoma. *Carcinogenesis* **2003**, *24*, 377–384. [CrossRef] [PubMed]
37. Yang, W.J.; Zhong, J.; Yu, J.G.; Zhao, F.; Xiang, Y. The structure and functions of paxillin and its roles in neovascularization. *Eur. Rev. Med. Pharmacol. Sci.* **2017**, *21*, 1768–1773.
38. López-Colomé, A.M.; Lee-Rivera, I.; Benavides-Hidalgo, R.; López, E. Paxillin: A crossroad in pathological cell migration. *J. Hematol. Oncol.* **2017**, *10*, 50. [CrossRef]
39. Deakin, N.O.; Pignatelli, J.; Turner, C.E. Diverse roles for the paxillin family of proteins in cancer. *Genes Cancer* **2012**, *3*, 362–370. [CrossRef]
40. Mohanty, A.; Nam, A.; Pozhitkov, A.; Yang, L.; Srivastava, S.; Nathan, A.; Wu, X.; Mambetsariev, I.; Nelson, M.; Subbalakshmi, A.R.; et al. A non-genetic mechanism involving the integrin β4/paxillin axis contributes to chemoresistance in lung cancer. *iScience* **2020**, *23*, 101496. [CrossRef]
41. Wu, D.W.; Wu, T.C.; Wu, J.Y.; Cheng, Y.W.; Chen, Y.C.; Lee, M.C.; Chen, C.Y.; Lee, H. Phosphorylation of paxillin confers cisplatin resistance in non-small cell lung cancer via activating ERK-mediated Bcl-2 expression. *Oncogene* **2014**, *33*, 4385–4395. [CrossRef] [PubMed]
42. Wen, L.; Zhang, X.; Zhang, L.; Chen, S.; Ma, Y.; Hu, J.; Yue, T.; Wang, J.; Zhu, J.; Wu, T.; et al. Paxillin knockdown suppresses metastasis and epithelial-mesenchymal transition in colorectal cancer via the ERK signalling pathway. *Oncol. Rep.* **2020**, *44*, 1105–1115. [CrossRef] [PubMed]
43. Li, D.; Li, Z.; Xiong, J.; Gong, B.; Zhang, G.; Cao, C.; Jie, Z.; Liu, Y.; Cao, Y.; Yan, Y.; et al. MicroRNA-212 functions as an epigenetic-silenced tumor suppressor involving in tumor metastasis and invasion of gastric cancer through down-regulating PXN expression. *Am. J. Cancer Res.* **2015**, *5*, 2980–2997. [PubMed]
44. Qin, J.; Wang, F.; Jiang, H.; Xu, J.; Jiang, Y.; Wang, Z. MicroRNA-145 suppresses cell migration and invasion by targeting paxillin in human colorectal cancer cells. *Int. J. Clin. Exp. Pathol.* **2015**, *8*, 1328–1340.

45. Qin, J.; Ke, J.; Xu, J.; Wang, F.; Zhou, Y.; Jiang, Y.; Wang, Z. Downregulation of microRNA-132 by DNA hypermethylation is associated with cell invasion in colorectal cancer. *Onco. Targets Ther.* **2015**, *8*, 3639–3648. [CrossRef]
46. Bi, Y.; Han, Y.; Bi, H.; Gao, F.; Wang, X. miR-137 impairs the proliferative and migratory capacity of human non-small cell lung cancer cells by targeting paxillin. *Hum. Cell* **2014**, *27*, 95–102. [CrossRef]
47. Wu, D.W.; Cheng, Y.W.; Wang, J.; Chen, C.Y.; Lee, H. Paxillin predicts survival and relapse in non-small cell lung cancer by microRNA-218 targeting. *Cancer Res.* **2010**, *70*, 10392–10401. [CrossRef]
48. Yoshino, H.; Enokida, H.; Itesako, T.; Kojima, S.; Kinoshita, T.; Tatarano, S.; Chiyomaru, T.; Nakagawa, M.; Seki, N. Tumor-suppressive microRNA-143/145 cluster targets hexokinase-2 in renal cell carcinoma. *Cancer Sci.* **2013**, *104*, 1567–1574. [CrossRef]
49. Yamasaki, T.; Seki, N.; Yoshino, H.; Itesako, T.; Hidaka, H.; Yamada, Y.; Tatarano, S.; Yonezawa, T.; Kinoshita, T.; Nakagawa, M.; et al. MicroRNA-218 inhibits cell migration and invasion in renal cell carcinoma through targeting caveolin-2 involved in focal adhesion pathway. *J. Urol.* **2013**, *190*, 1059–1068. [CrossRef]
50. Du, C.; Wang, Y.; Zhang, Y.; Zhang, J.; Zhang, L.; Li, J. LncRNA DLX6-AS1 contributes to epithelial-mesenchymal transition and cisplatin resistance in triple-negative breast cancer via modulating Mir-199b-5p/paxillin axis. *Cell Transpl.* **2020**, *29*. [CrossRef]

Publisher's Note: MDPI stays neutral with regard to jurisdictional claims in published maps and institutional affiliations.

biomedicines MDPI

Article

Identification of miR-29c-3p as a Robust Normalizer for Urine microRNA Studies in Bladder Cancer

Julia Oto [1,†], Emma Plana [1,2,†], Álvaro Fernández-Pardo [1], Fernando Cana [1], Manuel Martínez-Sarmiento [3], César D. Vera-Donoso [3], Francisco España [1] and Pilar Medina [1,*]

1 Haemostasis, Thrombosis, Arteriosclerosis and Vascular Biology Research Group, Medical Research Institute Hospital La Fe, 46026 Valencia, Spain; juliaotomartinez@gmail.com (J.O.); plana_emm@gva.es (E.P.); alvarofernandezpardo@gmail.com (Á.F.-P.); fernandocana1998@gmail.com (F.C.); espanya_fra@gva.es (F.E.)
2 Angiology and Vascular Surgery Service, La Fe University and Polytechnic Hospital, 46026 Valencia, Spain
3 Department of Urology, La Fe University and Polytechnic Hospital, 46026 Valencia, Spain; mmarsar@gmail.com (M.M.-S.); cdveradonoso@gmail.com (C.D.V.-D.)
* Correspondence: medina_pil@gva.es; Tel.: +34-961246636
† The authors contributed equally to this work.

Received: 17 September 2020; Accepted: 19 October 2020; Published: 22 October 2020

Abstract: Bladder cancer (BC) is among the most frequent malignancies worldwide, being the most expensive cancer to treat and monitor and the most lethal urological cancer. Urine microRNAs (miRNAs) have been proposed as novel non-invasive biomarkers to early diagnose and monitor BC patients in order to avoid the performance of current aggressive diagnostic techniques. However, huge discrepancies arise among studies mainly due to the lack of standardization in the normalization, a crucial step in all miRNA studies. Our aim was to identify the best miRNA normalizer for miRNA studies in urine of BC patients. We evaluated the performance of 110 candidate miRNAs in urine of 35 BC patients and 15 healthy controls by Real Time quantitative Polymerase Chain Reaction (RT-qPCR) followed by a stability analysis with RefFinder. In this screening stage, miR-29c-3p arose as the most stably expressed miRNA in BC and controls, with a good expression level. Stability of miR-29c-3p expression was validated in an independent cohort of 153 BC patients and 57 controls. Finally, we evaluated the robustness of miR-29c-3p as normalizer in the expression study of miR-200c-3p, a potential diagnostic marker for BC. We propose miR-29c-3p as a normalizer for miRNA studies in BC urine. This is the first study that characterizes a reliable normalizer that may allow the comparison of future urine miRNA studies as non-invasive biomarkers for BC diagnosis and monitoring.

Keywords: bladder cancer; urine; miRNA; reference miRNA; normalizer; RT-qPCR

1. Introduction

Bladder cancer (BC) is among the most frequent malignancies worldwide. Indeed, BC accounts for 3% of all malignant tumors in adults and is the most lethal urological tumor type. According to the GLOBOCAN 2018 study, BC accounts for more than 540,000 new cases worldwide and almost 200,000 deaths annually [1,2]. Moreover, BC largely increases health expenses since it is the most expensive cancer to treat and the one that accounts for the highest monitoring costs. In fact, the cost of the muscle-invasive subtype reaches $150,000 per capita [3,4] and, by the end of the decade, BC is expected to account for >3% of all cancer-related medical expenses [3].

Bladder ultrasound, computerized tomography (CT) scan, cystoscopy or cytology are presently the gold standard techniques for BC diagnosis. However, cystoscopy is a highly invasive procedure that causes a high discomfort in patients and urine cytology, while being non-invasive, is unable to detect low-grade bladder tumors. Furthermore, none of both achieve a high sensitivity and specificity [5]. These last two techniques are routinely used as follow-up methods. As a consequence, novel non-invasive biomarkers are being explored to early and accurately diagnose and monitor BC patients in order to avoid the performance of this aggressive technique while reaching or even exceeding the sensitivity and specificity of cystoscopy [6]. Urine represents the most accessible source of biological markers for the analysis of BC and other urological tumors since it is rapid and easily obtained by the patient, avoids patient discomfort and potential complications from an invasive procedure, is available in copious amounts and re-sampling is easily achievable [7]. Thus, urine represents the ideal sample to develop novel tools to diagnose and monitor the prognosis of BC patients.

microRNAs (miRNAs) are small non-coding RNAs that regulate protein expression. They have been found to play a role in the different processes of tumor development [8,9] and have been proposed as regulatory molecules and biomarkers in virtually all cancer types [10–12]. miRNAs are known to be released from tissues into biological fluids such as urine and have been proposed as biomarkers for many disorders. Indeed, several profiles of miRNAs have been proposed as diagnostic and prognostic tools for BC [13–24]. However, strikingly only a small number of miRNAs are shared by these studies, what is probably due to the lack of standardization in the protocols used among laboratories. Particularly results are often not reproducible among publications because of different criteria used for selecting patients and controls, the use of different procedures for sample processing, RNA isolation and miRNA quantification but, mainly, it may be caused by different normalization strategies used, what certainly represents a crucial step in any miRNA study. Several RNA species have been proposed as normalizers for miRNAs studies: small nuclear RNAs (snRNAs), nucleolar RNAs (snoRNAs), ribosomal RNAs (rRNAs), miRNAs or even exogenous synthetic RNAs. However, some of these proposed normalizers differ from miRNAs in length (150 nt for snRNAs and 60–200 nt for snoRNAs, compared to 20–24 nt for miRNAs) and also in structure [25]. This disparity in lengths and structure can affect the isolation efficiency, reverse transcription and amplification. Consequently, the use of stably expressed miRNAs for the normalization of miRNAs studies is the best option rather than the use of other small RNA species [26]. Likewise, the addition of exogenous synthetic RNAs is intended to track the isolation and reverse transcription efficiency, not to normalize for amplification. To date, no consensus exists on internal reference miRNAs for BC studies performed in urine samples by Real Time quantitative Polymerase Chain Reaction (RT-qPCR)

In the present study, we aimed for the first time to ascertain the best miRNA normalizer for miRNA studies in BC performed in urine samples. The discovery of a good and reproducible internal miRNA normalizer will eliminate the current inconsistency among studies and will finally allow the comparison of results obtained in urine of BC patients. This is an inexorable requirement in order to apply this technique to clinical practice.

2. Experimental Section

2.1. Study Subjects

BC can be subdivided in two types, superficial or non-muscle-invasive bladder cancer (NMIBC) (70% of total), which comprehends Ta and T1 lesions, and muscle-invasive bladder cancer (MIBC) (30% of total), which comprehends T2, T3 and T4 lesions. Additionally, grading indicates the degree of cellular differentiation, being G1 well differentiated and less likely to spread, G2 moderately differentiated and G3 poorly differentiated and more likely to spread. This can be better envisaged in Figure 1, where the study workflow is also described.

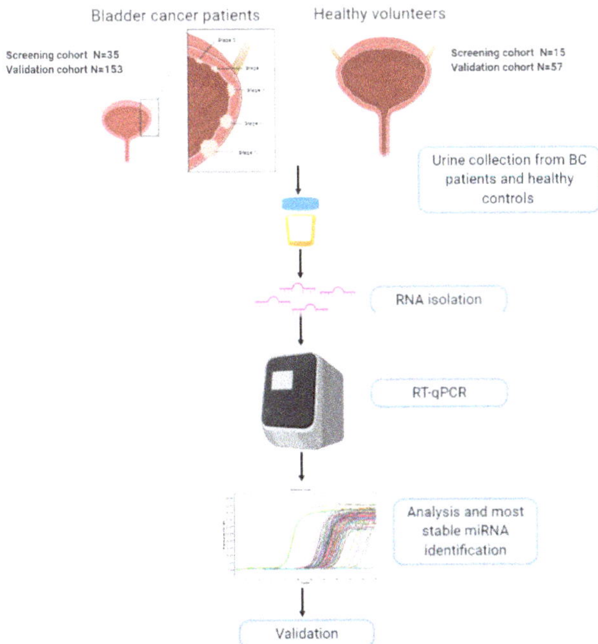

Figure 1. Graphical description of BC subtypes and study workflow.

For the initial screening stage, 35 BC patients (10 TaG1, 8 TaG3, 5 T1G3 and 12 T2G3) were recruited at La Fe University and Polytechnic Hospital (Valencia, Spain). Fifteen healthy volunteers (control group) with similar age and sex who underwent an ultrasound scan to rule out the presence of urological malignancies or other alterations were also recruited. For the validation stage, an independent cohort of 153 BC patients (33 TaG1, 13 TaG3, 29 T1G3, 54 TaG2, 9 T1G2, 4 T2G2, 10 T2G3 and 1 T3G3) and 57 healthy controls were additionally recruited.

Pre-operative clinical staging was performed through physical examination, urine cytology and CT scans of the chest, abdomen and pelvis. The tumor histological classification was done according to both the 1973 and 2004/2016 WHO classifications. Demographic and clinical data were collected.

The exclusion criteria were lack of informed consent, absence of histological confirmation and presence of other malignancies.

Informed consent was obtained from all participants according to protocols approved by the ethics review board at La Fe University and Polytechnic Hospital (2014/0314 and 2017/0474). The study was performed according to the declaration of Helsinki, as amended in Edinburgh in 2000.

2.2. Urine Collection

A first morning urine sample of 25 to 50 mL was collected in sterile containers from all participants and kept at 4 °C until processing. Urine was centrifuged at 805× g for 5 min at 4 °C to remove cellular debris and supernatant was aliquoted and frozen at −80 °C until analyzed. The concentration of creatinine in urine was measured by clinical laboratory standardized methods.

2.3. RNA Isolation and cDNA Synthesis from Urine

Total urine RNA (including miRNAs) was isolated from 600 µL of urine using the miRNeasy Mini Kit (Qiagen, Hilden, Germany) following manufacturer's instructions with several modifications optimized by our group [27]. Briefly, 200 µL of cell-free urine were transferred to a tube with 1 mL

Qiazol (Qiagen) and 1 µL carrier (1 µg/µL yeast RNA, Invitrogen, ThermoFisher Scientific, Waltham, MA, USA). This step was done in three independent tubes for each sample. In one of the three tubes 1 µL spike-in mix (UniSp2/4/5, Exiqon, Vedbaek, Denmark) was also added and each tube was gently mixed. After a 5 min incubation at room temperature, 200 µL chloroform were added to each tube, and centrifuged at 12,000× *g*, 15 min at 4 °C to allow phase separation. Ethanol in a proportion of 1.5:1 (volume:volume) was added to the liquid phase. The three tubes containing the urine sample from the same individual were pooled in one single column in order to increase the final RNA yield. Then, 4 cleaning steps with the buffers supplied in the kit were performed. Total RNA was finally eluted in 50 µL of DNase/RNase-free sterile distilled water.

The concentration and purity of the RNA was assessed by spectrophotometric quantification with the NanoDrop ND-1000 (Thermo Fisher Scientific). RNA was stored at −80 °C until used.

In the initial screening stage where predesigned panels were used, cDNA was obtained from 5 µL of urine RNA with the miRCURY LNA RT Kit (Qiagen) according to the supplied protocol in a final reaction volume of 25 µL. Due to the addition of a RNA carrier during the isolation, the final RNA yield includes the RNA isolated from urine plus the carrier RNA. Therefore, urine RNA retrotranscription was based on volume (µL) rather than RNA quantity (ng), according to the suppliers' recommendations. In the validation stage, where the expression level of selected miRNAs was conducted, cDNA was obtained from 2 µL of urine RNA using the same technology (final reaction volume 10 µL). In all cases, the reaction mix containing RNA, enzyme, buffer, RNAse-free water and UniSp6 RNA spike-in template, was incubated 60 min at 42 °C followed by 5 min at 95 °C for reverse transcriptase inactivation. Reactions were carried out in a thermocycler TC-412 (Techne, Minneapolis, MN, USA).

2.4. miRNAs Quantification

In the screening stage, 35 urine cDNA samples from BC patients and 15 from healthy controls were analyzed. In them, a total of 179 miRNAs were quantified using the commercially predesigned Serum/plasma miRNA PCR Panel V5 (Qiagen). This panel contains 179 miRNAs commonly found in human plasma and serum according to the manufacturer's in-house analyses of miRNA expression in blood, serum and plasma samples, as well as on the limited number of peer-reviewed published papers available. The list of all the quantified miRNAs is detailed in Table S1. miRCURY LNA SYBR Green Master Mix (Qiagen) was used as a fluorophore, according to manufacturer's indications. Briefly, cDNA (dilution 1/40), water and PCR master Mix (which includes SYBR Green) were added to a 384-well PCR plate supplied that includes the LNA^{TM} primer sets in a final reaction volume of 10 µL. Furthermore, each panel included the following internal controls: 5 synthetic RNAs of the RNA Spike-in kit aimed to monitor the RNA isolation and cDNA synthesis, and an inter-plate calibrator in triplicate and a negative control to evaluate qPCR performance. qPCR reactions were performed as follows: a polymerase activation/denaturation cycle of 2 min at 95 °C followed by 55 cycles of 10 s at 95 °C and 1 min at 56 °C with a ramp-rate of 2.2 °C/s. All RT-qPCR reactions were conducted in a LightCycler 480 II (Roche, Mannheim, Germany). In the validation stage, selected miRNAs were quantified using specific LNA PCR primer sets (Qiagen) in a total of 153 urine samples from BC patients and 57 healthy controls.

2.5. Selection of Candidate miRNA Normalizers and Analysis of Their Stability

To normalize the expression level of each miRNA, the best candidate with the highest stability and the lowest biological variance over the entire range of samples being investigated (BC and controls) was selected. To that aim, all miRNAs with a mean Ct < 35 in BC and controls were scrutinized. To select the best normalizer, the comprehensive tool RefFinder was employed which integrates the computational programs Genorm [28], BestKeeper [29], the comparative Delta Ct method [30] and NormFinder [31] (https://www.heartcure.com.au/for-researchers/) [32].

2.6. Statistical Analysis

Continuous variables were presented as median and interquartile range, and categorical variables as count and percentage. The analysis of variance (ANOVA) with the Tuckey Post Hoc test and unpaired t-test were used to identify significant differences in miRNA expression levels between BC patients and healthy controls. The statistical analysis was performed using the GraphPad Prism software v.8.0.1 (GraphPad software Inc., La Jolla, CA, USA). The Venn diagram was performed using the online tool available on http://bioinformatics.psb.ugent.be/webtools/Venn/. *p*-Values < 0.05 were considered statistically significant.

3. Results

3.1. Clinical Characteristics of the Study Subjects

A total of 188 BC patients were prospectively recruited together with 72 healthy volunteers (control group) with similar age and sex. The clinical characteristics of the study subjects are depicted in Table 1. The patients studied in the screening stage (n = 35) were: 10 patients (28.57%) with a TaG1 BC, 8 patients (22.86%) with TaG3, 5 patients (14.29%) with T1G3 and 12 patients (34.29%) with T2G3. The patients studied in the validation stage (n = 153) were: 33 patients (21.57%) with TaG1, 13 patients (8.5%) with TaG3, 29 patients (18.95%) with T1G3 and 15 patients (9.80%) with T2G2/T2G3/T3G3. Additionally, in order to validate the proposed normalizer in the whole spectrum of BC patients, 2 more groups of patients were included in the validation stage: TaG2 (54 patients, 35.29%) and T1G2 (9 patients, 5.88%).

Table 1. Clinical characteristics of the BC patients and healthy controls studied.

	BC Patients		Controls	
	Screening (N = 35)	**Validation (N = 153)**	**Screening (N = 15)**	**Validation (N = 57)**
Age, *y*	67 (61–74)	69 (63–75)	64 (51–76)	64 (56–68)
Male sex, *N* (%)	32 (91.43%)	129 (84.31%)	12 (80.00%)	43 (75.44%)
Urine creatinine, *mg/dL*	76.5 (37.4–123.3)	77.4 (51.9–118.9)	78.5 (49.0–100.2)	98.2 (69.4–161.6)
Tumor Stage and Grade, *N* (%)				
TaG1	10 (28.57%)	33 (21.57%)	-	-
TaG3	8 (22.86%)	13 (8.50%)	-	-
T1G3	5 (14.29%)	29 (18.95%)	-	-
TaG2	-	54 (35.29%)	-	-
T1G2	-	9 (5.88%)	-	-
T2G2/T2G3/T3G3	12 (34.29%)	15 (9.80%)	-	-

Continuous variables are presented as median and interquartile range and categorical variables are presented as count and percentage.

3.2. Quality Internal Control with Synthetic Spike-in RNAs

To ensure that miRNA quantification was not influenced by technical and interpersonal variability, synthetic non-human spike-in RNAs are frequently used. We assessed the RNA isolation step by adding the synthetic spike-in 2 and spike-in 4 RNAs during all RNA isolations, and the retrotranscription efficiency by adding the spike-in 6 RNA in all retrotranscription reactions. No differences were observed in any spike-in studied among the study groups (Figure 2), thus indicating a proper performance of isolation and retrotranscription steps. To evaluate the qPCR performance of all reactions, the inter-plate calibrator spike-in 3 RNA in triplicate and a negative control were included in each panel. No differences were observed in any comparison made.

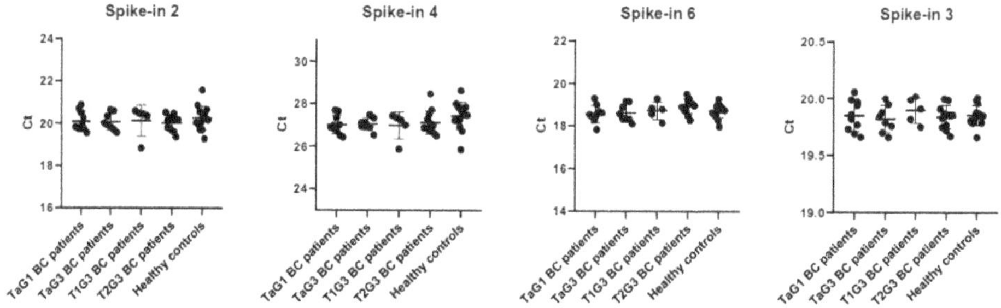

Figure 2. Differences in expression levels of three synthetic spike-in RNAs among BC patients with different stages and grades and healthy controls. Spike-in 2 and spike-in 4 monitor the RNA isolation step, spike-in 6 monitors the retrotranscription efficiency and spike-in 3 functions as inter-plate calibrator. Expression levels are represented as Ct values and error bars represent standard deviation.

3.3. Selection of Candidate miRNA Normalizers and Analysis of Their Stability

Of the 179 miRNAs quantified in each sample of the screening stage, we obtained high quality signals in 110 miRNAs (mean of the Ct < 35) both in BC patients and controls, thus they were included in the analysis with RefFinder. This tool comprehends the computational algorithms Genorm, BestKeeper, Delta Ct and NormFinder. Figure 3 shows the best 10 reference miRNAs selected by each algorithm. The stability analysis conducted with Genorm revealed that the greatest stability was reached by the combination of let-7e-5p and let-7a-5p (Figure 3a). BestKeeper revealed that the most stable miRNA was miR-2110 (Figure 3b). The Delta Ct method and NormFinder agreed with the most stable miRNA being miR-29c-3p (Figure 3c,d). Finally, the recommended comprehensive ranking that integrates all the previous analyses, rendered miR-29c-3p as the most stable miRNA (Figure 3e), being aligned with the results of the Delta Ct method and NormFinder. Next, we represented in a Venn diagram the overlap among the best 10 reference miRNAs selected by each algorithm (Figure 4). Three miRNAs were shared by the Delta Ct method, Genorm and NormFinder: miR-29c-3p, miR-26a-5p and miR-361-5p. Likewise, 7 miRNAs were shared by Delta Ct Method and NormFinder. In contrast, none of the 10 most stable miRNAs rendered by BestKeeper were selected by any other algorithm, thus none of them were included in the comprehensive ranking provided by RefFinder and consequently discarded as potential normalizers in our study.

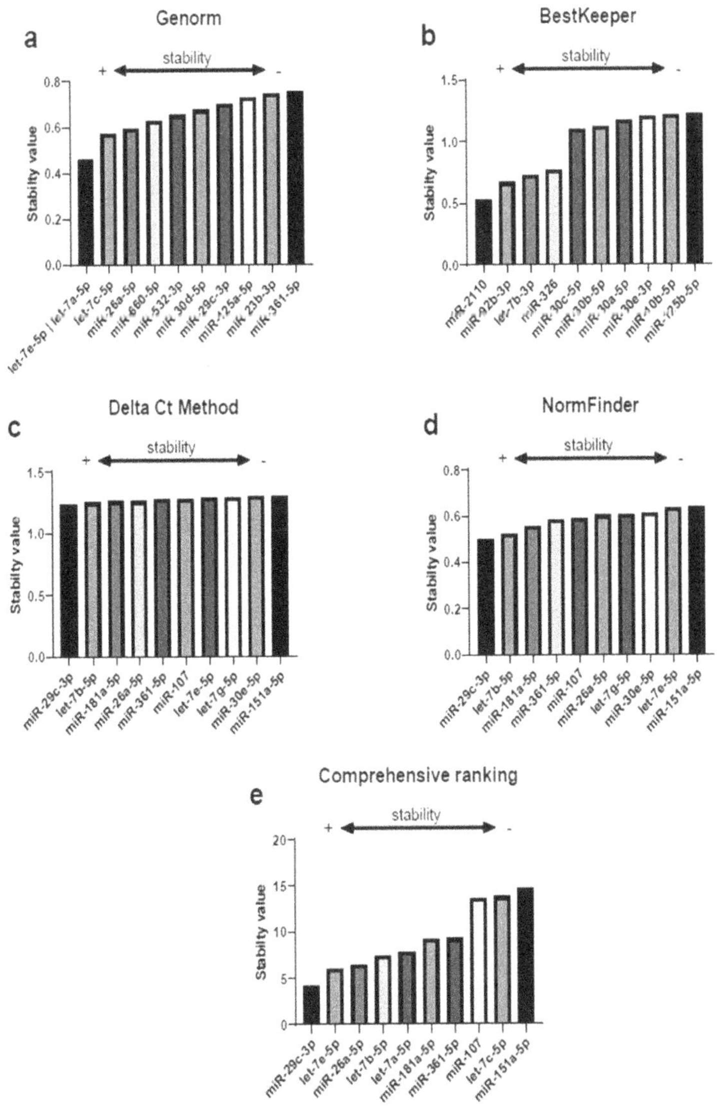

Figure 3. Selection of candidate miRNA normalizers and analysis of their stability conducted with the comprehensive tool RefFinder. Each graph represents the best 10 reference miRNAs selected by each algorithm: (**a**) Genorm, (**b**) BestKeeper, (**c**) Delta Ct method, (**d**) NormFinder and (**e**) Comprehensive ranking. The lower the stability value, the higher the stability of each miRNA.

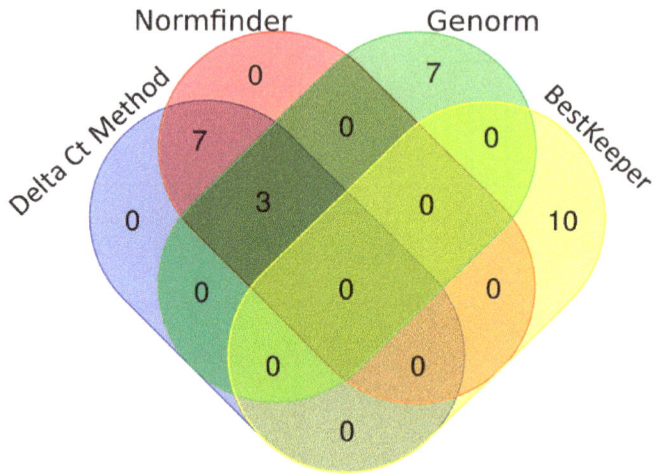

Figure 4. Venn diagram presenting the overlap among the best 10 reference miRNAs selected by each algorithm. Delta Ct method, NormFinder, Genorm and BestKeeper.

3.4. Differences in Expression Levels of Candidate miRNA Normalizers between BC Patients and Controls

A crucial characteristic of a good normalizer in miRNAs studies is the stable expression among the samples analyzed. Thus, we compared the mean Ct values of the best 10 reference miRNAs selected by the comprehensive ranking of RefFinder between BC patients and controls. No significant differences were observed in the expression of miR-29c-3p, let-7e-5p, miR-26a-5p, let-7a-5p and let-7c-5p (unpaired t-test, $p > 0.05$). In contrast, we found significant differences in the expression levels of let-7b-5p ($p = 0.029$), miR-181-5p ($p = 0.026$), miR-361-5p ($p = 0.025$), miR-107 ($p = 0.012$) and miR-151-5p ($p = 0.015$) (Figure 5). As expected, those miRNAs with the highest stability value had similar expression levels between BC and controls.

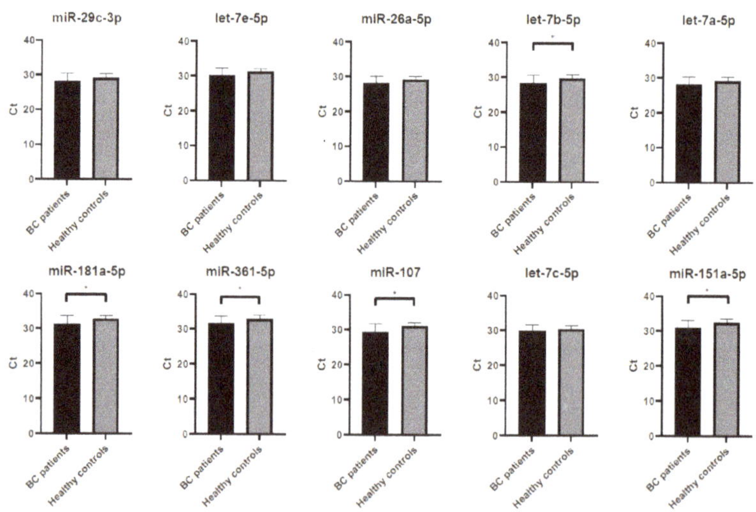

Figure 5. Differences in expression levels of the candidate miRNA normalizers selected by the comprehensive ranking of RefFinder between BC patients and controls. Expression levels are represented as Ct values and error bars represent standard deviation. Unpaired T-test: * $p < 0.05$.

3.5. Effect of Different Normalization Strategies on the Relative Quantification of a miRNA Closely Related with BC

miR-200c-3p has been previously proposed as urinary diagnostic biomarker for BC [24,33]. Thus, we evaluated the performance of the 10 most stable miRNAs, selected by the comprehensive ranking of RefFinder, as normalizers for miR-200c-3p quantification in the screening cohort. As seen in Figure 6, significant differences among the BC groups studied and controls were observed when any of the normalizers were used, although the results obtained were not always comparable in magnitude and trend.

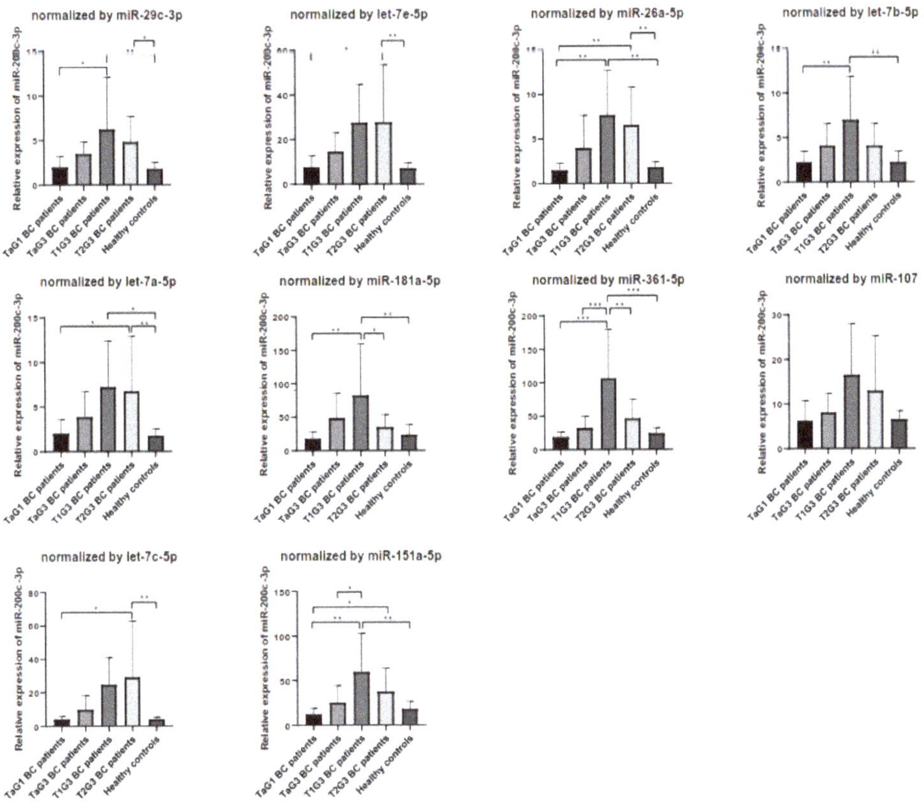

Figure 6. Relative expression of miR-200c-3p normalized by each candidate miRNA selected by the comprehensive ranking of RefFinder. Normalization was performed by the $2^{-\Delta\Delta Ct}$ method. Error bars represent the standard error of the mean. ANOVA with the Tuckey Post Hoc test: * $p < 0.05$; ** $p < 0.01$; *** $p < 0.001$.

In consideration of the aforementioned results, we selected miR-29c-3p as the best normalizer for the following reasons: (1) It was proposed as the most stable miRNA in the comprehensive analysis of RefFinder, and in two of the main algorithms independently tested (Figure 2); (2) Its expression level was optimal for a proper quantification (mean Ct = 28.4) (data not shown); and (3) Significant differences in the expression of miR-200c-3p, a miRNA related to BC, were observed when it was used as normalizer (Figure 6).

3.6. Validation of the Performance of miR-29c-3p as Normalizer in an Independent Cohort of BC Patients and Controls

We verified the stability of miR-29c-3p in an independent cohort of BC patients and healthy controls. As in the screening cohort, no significant differences were observed in the expression of miR-29c-3p between BC patients (mean Ct = 27.65) and healthy controls (mean Ct = 28.13) ($p = 0.37$). Next we validated the robustness and efficacy of miR-29c-3p as endogenous control for BC in the validation cohort using the $2^{-\Delta\Delta Ct}$ method by analyzing the expression level of the BC-related miR-200c-3p in every group of BC patients and controls. As occurred in the samples studied in the screening stage, we observed significant differences in the expression of miR-200c-3p among the different clinical groups studied in the validation cohort ($p < 0.001$), with an increase of miR-200c-3p with the increase in BC stage (Figure 7). As occurred in the screening cohort, the biggest differences were found in the T1G3 BC group. As the patients with mildest stage of BC seem to have a lower expression level of miR-200c-3p, we grouped all the patients with the Ta stage and repeated the analysis. Figure S1a confirms that the expression of miR-200c-3p significantly increases with the severity of NMIBC, being T1G3 the BC type with the highest expression. Next, we grouped all NMIBC patients (TaG1+TaG2+TaG3+T1G2+T1G3) and compared the expression of miR-200c-3p to that of to MIBC patients (T2G2+T2G3+T3G3) and healthy controls. While NMIBC patients still showed the highest expression level of miR-200c-3p, no significant differences were observed when compared to MIBC or healthy controls (Figure S1b), probably because the highest difference occurs within the NMIBC group.

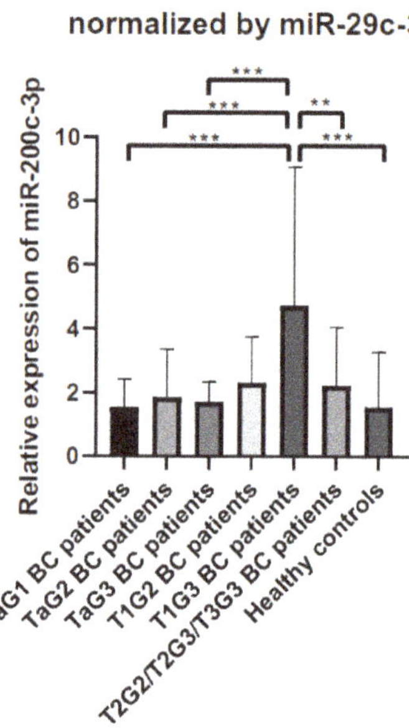

Figure 7. Relative expression of miR-200c-3p normalized by miR-29c-3p in the validation cohort. Normalization was performed by the $2^{-\Delta\Delta Ct}$ method. Error bars represent the standard error of the mean. ANOVA with the Tuckey Post Hoc test: ** $p < 0.01$; *** $p < 0.001$.

4. Discussion

New non-invasive markers are presently being under study to circumvent several drawbacks in BC diagnosis, monitoring and prognosis. Urine miRNAs are non-invasive promising biomarkers that have been previously proposed for BC diagnosis [13–17,19–21,23,24]. However, huge discrepancies arise among miRNA studies, to a great extent due to nonexistence of standardized procedures. Although the populations studied, sample processing, and RNA isolation and miRNA quantification methods are partly responsible for these inconsistencies; the normalization strategy used may represent the main hurdle. In fact, to minimize the effect occasioned by methodology-related factors on miRNA expression levels, an accurate data analysis ought to be performed using appropriate normalizers for external and internal variation correction [26]. These normalizers should be chosen from a selection of candidates that are expected to be stably expressed over the entire range of samples being investigated, since miRNAs can be affected by the condition under study. As an alternative, the mean expression value of all commonly expressed microRNAs in a given sample has been proposed for normalization [34]. Although this strategy presents good and robust results it implies that a large number of miRNAs have to be always profiled, which may not be possible or cost-effective in all studies [26,35]. Thus, as a general guideline suggested by the manufacturer, the use of the global mean for normalization is limited to the use of PCR panels that contain a larger number of microRNA assays, and it cannot be used for studies analyzing less than 20–50 different miRNAs.

Different RNA species have been proposed as normalizers (snRNAs, snoRNAs, rRNAs, miRNAs or exogenous synthetic RNAs); however, substantial differences in length and structure to that of miRNAs generate a high variability in the results [26]. Regarding the use of exogenous synthetic RNAs as normalizers, these are meant to track isolation and reverse transcription efficiency in order to eliminate deviations in the experimental process and make the results more reliable. However, it is important to remark that their use would never correct for deviations in sampling, in the quality of samples or in the amplification process. In fact, age, sample collection, preparation or storing can modify miRNA expression levels, which may be caused by cell lysis or miRNA degradation [26].

Several molecules have been proposed as normalizers for miRNAs studies in different disorders, mainly in blood [36], plasma [37–39], cell cultures studies [25,40], tissue [41] and urine [35,42–44]. However, there is no consensus, even in the same sample type, regarding which is the best normalizer for miRNA studies. Indeed, no consensus exists on a robust normalizer for urine studies. A recent study proposed miRNA miR-193a and miR-448 as normalizers [35]. However, in this study, only urine from healthy donors was analyzed and these results could vary when urine from cancer patients is investigated. In prostate cancer, miR-191-5p showed the lowest degree of variation and the highest stability value [42]. In our study, miR-191-5p ranked 29 out of 110 according to the comprehensive ranking of RefFinder, thus it is not a reliable normalizer for BC studies in urine. miR-16 was identified as the most stable endogenous reference miRNA in chronic kidney disease, making it a suitable normalizer for urinary exosome-derived miRNA [43]. Conversely, in our data set, miR-16-5p ranked 106 out of 110 according to the comprehensive ranking of RefFinder, turning it an ineffectual normalizer for urine BC studies. Finally, U6 has been proposed as normalizer for miRNA studies in urinary sediment of IgA nephropathy [44]. Although U6 was widely used as normalizer in countless studies at the origins of miRNA investigation, it is a member of the larger small RNA species which have a different biogenesis pathway (originate from the nucleus), may not be secreted or protected in cell-free biofluids in the same way that microRNAs are and may also behave differently during RNA isolation. Thus, U6 is an unreliable normalizer for urine miRNA studies. Other studies have employed the combination of two stable miRNAs as normalizers however, this strategy presents several drawbacks: it may increase technical variability, it implies a higher economic cost since the number of miRNAs to be quantified by RT-qPCR increases and it is more time consuming. Altogether, the use of a combination of miRNAs as normalizers hampers the direct translation of miRNA studies to daily clinical practice with diagnostic/staging purposes.

In the present study, we set for the first time the aim to ascertain the best miRNA normalizer for miRNA studies in urine of BC patients in order to avoid future inconsistencies among studies. We evaluated the performance of 110 candidate miRNAs with the comprehensive tool RefFinder that integrates 4 programs (Genorm, BestKeeper, Delta Ct method and NormFinder) in 35 BC patients and 15 healthy controls. We selected miR-29c-3p as the best normalizer for miRNA studies in urine of BC. It was the most stable miRNA according to the comprehensive analysis of RefFinder among the 110 studied, and also according to the Delta Ct method and NormFinder. Moreover, it had a good expression level in urine (mean Ct value in BC patients and healthy controls = 28.4) and no differences were observed between BC patients and controls, both in the screening and the validation cohorts.

Urine miR-200c-3p has been previously related to BC and it has been proposed as diagnostic and staging marker [24,33]. miR-200c-3p appears to control the epithelial-to-mesenchymal transition process through BMI-1 in BC cells, and it inhibits their proliferation by down-regulating E2F3 [45]. Thus, we selected this miRNA to test the robustness of miR-29c-3p as endogenous normalizer. We found significant differences in miR-200c-3p among the different clinical groups studied both in the screening and validation cohorts, with a trend in the increase of miR-200c-3p with the severity of NMIBC. The evaluation of additional stably expressed miRNAs proposed by RefFinder may have rendered other potential normalizers for urine miRNA studies in the context of BC, what represents a limitation of our study. Nonetheless, our results confirm previous findings and reinforce the use of miR-29c-3p as normalizer.

5. Conclusions

In summary, our study is the first report characterizing a reliable normalizer for the analysis of urine miRNAs in BC patients. miR-29c-3p, being one of the most stably expressed miRNAs in urine of BC patients and healthy individuals, arises as an optimal reference miRNA that may allow the comparison of future urine miRNA studies as non-invasive biomarkers for BC diagnosis and monitoring.

Supplementary Materials: The following are available online at http://www.mdpi.com/2227-9059/8/11/447/s1, Table S1: miRNAs and internal controls included in the Serum/plasma miRNA PCR Panel V5 (Qiagen); Figure S1. Relative expression of miR-200c-3p normalized by miR-29c-3p in the validation cohort. (a) Comparison of the mildest stage of NMIBC patients (TaG1+TaG2+TaG3) and the other clinical groups. (b) Comparison of NMIBC patients (TaG1+TaG2+TaG3+T1G2+T1G3), MIBC (T2G2+T2G3+T3G3) and healthy controls. Normalization was performed by the $2^{-\Delta\Delta Ct}$ method. Error bars represent the standard error of the mean. ANOVA with the Tuckey Post Hoc test: * $p < 0.05$; ** $p < 0.01$; *** $p < 0.001$.

Author Contributions: J.O. processed samples, performed the research, analyzed the data and wrote the manuscript. E.P. analyzed the data and wrote the manuscript. Á.F.-P. performed the research and critically revised the manuscript. F.C. processed the samples and prepared the databases. M.M.-S. recruited the patients, revised the clinical records and critically revised the manuscript. C.D.V.-D. recruited the patients, revised the clinical records and critically revised the manuscript. F.E. designed and supervised the study and critically revised the manuscript. P.M. designed and supervised the study, analyzed the data and wrote the manuscript. All authors have read and agreed to the published version of the manuscript.

Funding: This research was funded by research grants from Instituto de Salud Carlos III (PI17/00495), FEDER una manera de hacer Europa, Generalitat Valenciana (ACIF/2017/138), Sociedad Española de Trombosis y Hemostasia and Sociedad Española de Angiología y Cirugía Vascular.

Acknowledgments: We would like to thank María José Solmoirago for her technical support.

Conflicts of Interest: The authors declare no conflict of interest. The funders had no role in the design of the study; in the collection, analyses, or interpretation of data; in the writing of the manuscript, or in the decision to publish the results.

References

1. Bray, F.; Me, J.F.; Soerjomataram, I.; Siegel, R.L.; Torre, L.A.; Jemal, A. Global cancer statistics 2018: GLOBOCAN estimates of incidence and mortality worldwide for 36 cancers in 185 countries. *CA Cancer J. Clin.* **2018**, *68*, 394–424. [CrossRef]
2. Siegel, R.L.; Miller, K.D.; Jemal, A. Cancer statistics, 2019. *CA Cancer J. Clin.* **2019**, *69*, 7–34. [CrossRef]
3. Svatek, R.S.; Hollenbeck, B.K.; Holmäng, S.; Lee, R.; Kim, S.P.; Stenzl, A.; Lotan, Y. The Economics of Bladder Cancer: Costs and Considerations of Caring for This Disease. *Eur. Urol.* **2014**, *66*, 253–262. [CrossRef]
4. Cumberbatch, M.G.; Noon, A.P. Epidemiology, aetiology and screening of bladder cancer. *Transl. Androl. Urol.* **2019**, *8*, 5–11. [CrossRef]
5. Zhu, C.-Z.; Ting, H.-N.; Ng, K.-H.; Ong, T.-A. A review on the accuracy of bladder cancer detection methods. *J. Cancer* **2019**, *10*, 4038–4044. [CrossRef]

6. Amuran, G.G.; Peker, I.; Tinay, I.; Akkiprik, M. New Insights in Bladder Cancer Diagnosis: Urinary miRNAs and Proteins. *Med. Sci.* **2018**, *6*, 113. [CrossRef]
7. Urquidi, V.; Rosser, C.J.; Goodison, S. Molecular diagnostic trends in urological cancer: Biomarkers for non-invasive diagnosis. *Curr. Med. Chem.* **2012**, *19*, 3653–3663. [CrossRef]
8. Chen, P.-S.; Hung, M.-C.; Hung, M.-C. Dysregulation of MicroRNAs in cancer. *J. Biomed. Sci.* **2012**, *19*, 90. [CrossRef]
9. Chan, S.-H.; Wang, L.-H. Regulation of cancer metastasis by microRNAs. *J. Biomed. Sci.* **2015**, *22*, 9. [CrossRef]
10. Bartel, D.P. MicroRNAs: Target Recognition and Regulatory Functions. *Cell* **2009**, *136*, 215–233. [CrossRef]
11. Tam, W. The emergent role of microRNAs in molecular diagnostics of cancer. *J. Mol. Diagn.* **2008**, *10*, 411–414. [CrossRef]
12. Søkilde, R.; Vincent, M.; Møller, A.K.; Hansen, A.; Høiby, P.E.; Blondal, T.; Nielsen, B.S.; Daugaard, G.; Møller, S.; Litman, T. Efficient Identification of miRNAs for Classification of Tumor Origin. *J. Mol. Diagn.* **2014**, *16*, 106–115. [CrossRef] [PubMed]
13. Hanke, M.; Hoefig, K.; Merz, H.; Feller, A.C.; Kausch, I.; Jocham, D.; Warnecke, J.M.; Sczakiel, G. A robust methodology to study urine microRNA as tumor marker: microRNA-126 and microRNA-182 are related to urinary bladder cancer. *Urol. Oncol.* **2010**, *28*, 655–661. [CrossRef]
14. Miah, S.; Dudziec, E.; Drayton, R.M.; Zlotta, A.R.; Morgan, S.L.; Rosario, D.J.; Hamdy, F.C.; Catto, J.W. An evaluation of urinary microRNA reveals a high sensitivity for bladder cancer. *Br. J. Cancer* **2012**, *107*, 123–128. [CrossRef]
15. Mengual, L.; Lozano, J.J.; Ingelmo-Torres, M.; Gazquez, C.; Ribal, M.J.; Alcaraz, A. Using microRNA profiling in urine samples to develop a non-invasive test for bladder cancer. *Int. J. Cancer* **2013**, *133*, 2631–2641. [PubMed]
16. Ratert, N.; Meyer, H.-A.; Jung, M.; Lioudmer, P.; Mollenkopf, H.J.; Wagner, I.; Miller, K.; Kilic, E.; Erbersdobler, A.; Weikert, S.; et al. miRNA Profiling Identifies Candidate miRNAs for Bladder Cancer Diagnosis and Clinical Outcome. *J. Mol. Diagn.* **2013**, *15*, 695–705. [CrossRef] [PubMed]
17. Zhang, D.-Z.; Lau, K.-M.; Chan, E.S.Y.; Wang, G.; Szeto, C.; Wong, K.; Choy, R.K.W.; Ng, C.-F. Cell-Free Urinary MicroRNA-99a and MicroRNA-125b Are Diagnostic Markers for the Non-Invasive Screening of Bladder Cancer. *PLoS ONE* **2014**, *9*, e100793. [CrossRef]
18. Du, M.; Shi, D.; Yuan, L.; Li, P.; Chu, H.; Qin, C.; Yin, C.; Zhang, Z.; Wang, M. Circulating miR-497 and miR-663b in plasma are potential novel biomarkers for bladder cancer. *Sci. Rep.* **2015**, *5*, 10437. [CrossRef]
19. Sasaki, H.; Yoshiike, M.; Nozawa, S.; Usuba, W.; Katsuoka, Y.; Aida, K.; Kitajima, K.; Kudo, H.; Hoshikawa, M.; Yoshioka, Y.; et al. Expression Level of Urinary MicroRNA-146a-5p Is Increased in Patients With Bladder Cancer and Decreased in Those After Transurethral Resection. *Clin. Genitourin. Cancer* **2016**, *14*, e493–e499. [CrossRef]
20. Urquidi, V.; Netherton, M.; Gomes-Giacoia, E.; Serie, D.J.; Eckel-Passow, J.; Rosser, C.J.; Goodison, S. A microRNA biomarker panel for the non-invasive detection of bladder cancer. *Oncotarget* **2016**, *7*, 86290–86299. [CrossRef]
21. Homami, A.; Ghazi, F. MicroRNAs as biomarkers associated with bladder cancer. *Med. J. Islam. Repub. Iran* **2016**, *30*, 475. [PubMed]
22. Xie, Y.; Ma, X.; Chen, L.; Li, H.; Gu, L.; Gao, Y.; Zhang, Y.; Li, X.; Fan, Y.; Chen, J.; et al. MicroRNAs with prognostic significance in bladder cancer: A systematic review and meta-analysis. *Sci. Rep.* **2017**, *7*, 5619. [CrossRef] [PubMed]
23. Hofbauer, S.L.; De Martino, M.; Lucca, I.; Haitel, A.; Susani, M.; Shariat, S.F.; Klatte, T. A urinary microRNA (miR) signature for diagnosis of bladder cancer. *Urol. Oncol. Semin. Orig. Investig.* **2018**, *36*, 531–e1. [CrossRef] [PubMed]
24. Pardini, B.; Cordero, F.; Naccarati, A.; Viberti, C.; Birolo, G.; Oderda, M.; Di Gaetano, C.; Arigoni, M.; Martina, F.; Calogero, R.A.; et al. microRNA profiles in urine by next-generation sequencing can stratify bladder cancer subtypes. *Oncotarget* **2018**, *9*, 20658–20669. [CrossRef]
25. Drobna, M.; Szarzynska-Zawadzka, B.; Daca-Roszak, P.; Kosmalska, M.; Jaksik, R.; Witt, M.; Dawidowska, M. Identification of endogenous control mirnas for rt-qpcr in t-cell acute lymphoblastic leukemia. *Int. J. Mol. Sci.* **2018**, *19*, 2858. [CrossRef]

26. Schwarzenbach, H.; Da Silva, A.M.; Calin, G.A.; Pantel, K. Data Normalization Strategies for MicroRNA Quantification. *Clin. Chem.* **2015**, *61*, 1333–1342. [CrossRef]
27. Ramon-Nunez, L.A.; Martos, L.; Fernandez-Pardo, A.; Oto, J.; Medina, P.; Espana, F.; Navarro, S. Comparison of protocols and rna carriers for plasma mirna isolation. Unraveling rna carrier influence on mirna isolation. *PLoS ONE* **2017**, *12*, e0187005. [CrossRef]
28. Vandesompele, J.; De Preter, K.; Pattyn, F.; Poppe, B.; Van Roy, N.; De Paepe, A.; Speleman, F. Accurate normalization of real-time quantitative RT-PCR data by geometric averaging of multiple internal control genes. *Genome Biol.* **2002**, *3*, RESEARCH0034. [CrossRef]
29. Pfaffl, M.W.; Tichopad, A.; Prgomet, C.; Neuvians, T.P. Determination of stable housekeeping genes, differentially regulated target genes and sample integrity: BestKeeper – Excel-based tool using pair-wise correlations. *Biotechnol. Lett.* **2004**, *26*, 509–515. [CrossRef]
30. Silver, N.; Best, S.; Jiang, J.; Thein, S.L. Selection of housekeeping genes for gene expression studies in human reticulocytes using real-time PCR. *BMC Mol. Biol.* **2006**, *7*, 33. [CrossRef]
31. Andersen, C.L.; Jensen, J.L.; Ørntoft, T.F. Normalization of Real-Time Quantitative Reverse Transcription-PCR Data: A Model-Based Variance Estimation Approach to Identify Genes Suited for Normalization, Applied to Bladder and Colon Cancer Data Sets. *Cancer Res.* **2004**, *64*, 5245–5250. [CrossRef]
32. Xie, F.; Xiao, P.; Chen, D.; Xu, L.; Zhang, B. miRDeepFinder: A miRNA analysis tool for deep sequencing of plant small RNAs. *Plant Mol. Biol.* **2012**, *80*, 75–84. [CrossRef] [PubMed]
33. Armstrong, D.A.; Green, B.B.; Seigne, J.D.; Schned, A.R.; Marsit, C.J. MicroRNA molecular profiling from matched tumor and bio-fluids in bladder cancer. *Mol. Cancer* **2015**, *14*, 194. [CrossRef] [PubMed]
34. Mestdagh, P.; Van Vlierberghe, P.; De Weer, A.; Muth, D.; Westermann, F.; Speleman, F.; Vandesompele, J. A novel and universal method for microrna rt-qpcr data normalization. *Genome Biol.* **2009**, *10*, R64. [CrossRef] [PubMed]
35. Martinez-Fernandez, M.; Paramio, J.M.; Duenas, M. Rna detection in urine: From rna extraction to good normalizer molecules. *J. Mol. Diagn.* **2016**, *18*, 15–22. [CrossRef] [PubMed]
36. Serafin, A.; Foco, L.; Blankenburg, H.; Picard, A.; Zanigni, S.; Zanon, A.; Pramstaller, P.P.; Hicks, A.A.; Schwienbacher, C. Identification of a set of endogenous reference genes for miRNA expression studies in Parkinson's disease blood samples. *BMC Res. Notes* **2014**, *7*, 715. [CrossRef]
37. Solayman, M.H.; Langaee, T.; Patel, A.; El-Wakeel, L.; El-Hamamsy, M.; Badary, O.; Johnson, J.A. Identification of suitable endogenous normalizers for qrt-pcr analysis of plasma microrna expression in essential hypertension. *Mol. Biotechnol.* **2016**, *58*, 179–187. [CrossRef]
38. Santamaria-Martos, F.; Benítez, I.; Zapater, A.; Girón, C.; Pinilla, L.; Fernandez-Real, J.M.; Barbé, F.; Ortega, F.J.; Sanchez-De-La-Torre, M. Identification and validation of circulating miRNAs as endogenous controls in obstructive sleep apnea. *PLoS ONE* **2019**, *14*, e0213622. [CrossRef]
39. Prado, M.S.G.; De Goes, T.C.; De Jesus, M.L.; Mendonça, L.S.O.; Nascimento, J.S.; Kaneto, C. Identification of miR-328-3p as an endogenous reference gene for the normalization of miRNA expression data from patients with Diabetic Retinopathy. *Sci. Rep.* **2019**, *9*, 19677. [CrossRef]
40. Morata-Tarifa, C.; Picon-Ruiz, M.; Grinan-Lison, C.; Boulaiz, H.; Peran, M.; Garcia, M.A.; Marchal, J.A. Validation of suitable normalizers for mir expression patterns analysis covering tumour heterogeneity. *Sci. Rep.* **2017**, *7*, 39782. [CrossRef]
41. Schaefer, A.; Jung, M.; Miller, K.; Lein, M.; Kristiansen, G.; Erbersdobler, A.; Jung, K. Suitable reference genes for relative quantification of miRNA expression in prostate cancer. *Exp. Mol. Med.* **2010**, *42*, 749–758. [CrossRef] [PubMed]
42. Egidi, M.G.; Cochetti, G.; Guelfi, G.; Zampini, D.; Diverio, S.; Poli, G.; Mearini, E. Stability Assessment of Candidate Reference Genes in Urine Sediment of Prostate Cancer Patients for miRNA Applications. *Dis. Markers* **2015**, *2015*, 973597. [CrossRef] [PubMed]
43. Lange, T.; Stracke, S.; Rettig, R.; Lendeckel, U.; Kuhn, J.; Schluter, R.; Rippe, V.; Endlich, K.; Endlich, N. Identification of mir-16 as an endogenous reference gene for the normalization of urinary exosomal mirna expression data from ckd patients. *PLoS ONE* **2017**, *12*, e0183435. [CrossRef] [PubMed]

44. Duan, Z.-Y.; Cai, G.-Y.; Li, J.-J.; Bu, R.; Wang, N.; Yin, P.; Chen, X.-M. U6 can be used as a housekeeping gene for urinary sediment miRNA studies of IgA nephropathy. *Sci. Rep.* **2018**, *8*, 10875. [CrossRef] [PubMed]
45. Liu, L.; Qiu, M.; Tan, G.; Liang, Z.; Qin, Y.; Chen, L.; Chen, H.; Liu, J. miR-200c inhibits invasion, migration and proliferation of bladder cancer cells through down-regulation of BMI-1 and E2F3. *J. Transl. Med.* **2014**, *12*, 305. [CrossRef] [PubMed]

Publisher's Note: MDPI stays neutral with regard to jurisdictional claims in published maps and institutional affiliations.

biomedicines

MDPI

Review

MicroRNAs Regulating Hippo-YAP Signaling in Liver Cancer

Na-Hyun Lee [1,2], So Jung Kim [1,2] and Jeongeun Hyun [1,2,3,*]

1 Institute of Tissue Regeneration Engineering (ITREN), Dankook University, Cheonan 31116, Korea; nhlee0609@dankook.ac.kr (N.-H.L.); r00by8340@dankook.ac.kr (S.J.K.)
2 Department of Nanobiomedical Science and BK21 PLUS NBM Global Research Center for Regenerative Medicine, Dankook University, Cheonan 31116, Korea
3 Department of Regenerative Dental Medicine, College of Dentistry, Dankook University, Cheonan 31116, Korea
* Correspondence: j.hyun@dankook.ac.kr; Tel.: +82-41-550-3082

Abstract: Liver cancer is one of the most common cancers worldwide, and its prevalence and mortality rate are increasing due to the lack of biomarkers and effective treatments. The Hippo signaling pathway has long been known to control liver size, and genetic depletion of Hippo kinases leads to liver cancer in mice through activation of the downstream effectors yes-associated protein (YAP) and transcriptional coactivator with PDZ-binding motif (TAZ). Both YAP and TAZ not only reprogram tumor cells but also alter the tumor microenvironment to exert carcinogenic effects. Therefore, understanding the mechanisms of YAP/TAZ-mediated liver tumorigenesis will help overcome liver cancer. For decades, small noncoding RNAs, microRNAs (miRNAs), have been reported to play critical roles in the pathogenesis of many cancers, including liver cancer. However, the interactions between miRNAs and Hippo-YAP/TAZ signaling in the liver are still largely unknown. Here, we review miRNAs that influence the proliferation, migration and apoptosis of tumor cells by modulating Hippo-YAP/TAZ signaling during hepatic tumorigenesis. Previous findings suggest that these miRNAs are potential biomarkers and therapeutic targets for the diagnosis, prognosis, and treatment of liver cancer.

Keywords: microRNAs; hepatocellular carcinoma; Hippo kinase; Yes-associated protein; liver disease; diagnosis; prognosis; cancer therapy

Citation: Lee, N.-H.; Kim, S.J.; Hyun, J. MicroRNAs Regulating Hippo-YAP Signaling in Liver Cancer. *Biomedicines* **2021**, *9*, 347. https://doi.org/10.3390/biomedicines9040347

Academic Editor: Francesca Lovat

Received: 9 March 2021
Accepted: 25 March 2021
Published: 30 March 2021

1. Introduction

Liver cancer is one of the leading causes of cancer-related death worldwide, with a 5-year survival rate as low as 30–40% [1,2]. Hepatocellular carcinoma (HCC) is the most common type of primary liver cancer and occurs in patients with chronic liver disease [3]. Despite remarkable advances in research to overcome viral hepatitis, which is the biggest cause of HCC, the incidence and mortality of liver cancer are still increasing along with the increased prevalence of nonviral steatohepatitis. However, the complex pathophysiology of liver cancer has limited the development of effective diagnosis and therapeutic intervention, prompting a comprehensive understanding of liver carcinogenesis.

Many studies have revealed that several developmental pathways, such as Wnt/β-catenin, Hedgehog (Hh) and Hippo/Yes-associated protein (YAP) signaling, contribute to hepatic carcinogenesis [4]. For example, Hh signaling is inactive in the normal liver of adult mice and humans but reactivated in chronic liver diseases and liver cancers, promoting liver fibrosis and hyperplasia and constructing a tumor-favorable microenvironment [5]. Likewise, overexpression of YAP in mice massively increases the size of the liver due to hyperproliferation of liver cells and ultimately promotes hepatic tumorigenesis [6]. The hepatocyte-specific induction of YAP nuclear localization dedifferentiates hepatocytes into ductal-like progenitor cells that are highly proliferative [7]. Similar results have been found after genetic ablation of one of the key components of the Hippo signaling pathway, including Neurofibromin 2 (Nf2), Mammalian STE20-like 1/2 (Mst1/2), and Large Tumor

Suppressor 1/2 (Lats1/2), which regulate the activity of YAP and transcriptional coactivator with PDZ-binding motif (TAZ), another downstream effector of Hippo signaling, by phosphorylation and subsequent proteasomal degradation [8,9]. For example, Nf2 conditional null mice show hyperproliferation of hepatic progenitor cells, known as a ductular reaction, and develop both HCC and intrahepatic cholangiocarcinoma (ICC) [10,11]. Notably, codeletion of Yap suppresses liver overgrowth, progenitor expansion and tumor development in Nf2 knockout mice, demonstrating that Yap drives hepatic tumorigenesis caused by inactivated Hippo signaling [10]. Accordingly, it has been observed that the expression of NF2 negatively correlates with the expression of YAP in liver tissues of patients with HCC or ICC [12,13]. Except for NF2, for which missense mutations have been discovered in 1.9% and 5.3% of human HCC and ICC cases, respectively [12], somatic or germline mutations in neither MST1/2 nor LATS1/2 genes have been reported in common cancers [14]. This suggests that post-transcriptional regulation or epigenetic silencing, rather than DNA mutations, of the Hippo signaling pathway may play central roles in the aberrant inactivation of Hippo signaling and hyperactivation of YAP/TAZ transcriptional coactivators in liver cancers. Nevertheless, whether and how the activity of the Hippo and YAP/TAZ signaling pathways are regulated post-transcriptionally has not been well investigated.

MicroRNAs (miRNAs) are a group of small noncoding RNAs that are approximately 21–25 nucleotides in length [15]. miRNAs regulate the expression of approximately 30% of protein-coding genes at the post-transcriptional level by binding directly to target messenger RNA (mRNA), which results in translational suppression or degradation of target mRNAs [16]. Many studies have demonstrated that miRNAs play important roles in diverse biological processes, including cell proliferation, differentiation, and death; thus, abnormal regulation of miRNAs can lead to various pathological conditions, including liver cancers [17]. miRNAs have become promising candidates for biomarkers and therapeutic targets in many cancers due to their regulatory functions and detectable properties in various biological fluids, including blood, saliva, and urine.

Recently, several miRNAs have been reported to be associated with Hippo-YAP/TAZ signaling in the control of liver cancer cell behaviors [18,19]. Here, we introduce recent findings of the interaction between miRNAs and Hippo-YAP/TAZ signaling components in the development and progression of liver cancers. We also discuss what remains to be addressed in future studies to improve our knowledge of the underlying mechanisms whereby dysregulation of miRNA and Hippo-YAP/TAZ signaling contributes to hepatic malignancies. Finally, we propose the potential of these miRNAs in clinical applications for the diagnosis, prognosis, and treatment of liver cancers.

2. Hippo-YAP/TAZ Signaling Pathway

Hippo signaling, which was first discovered in *Drosophila*, is evolutionarily conserved in vertebrates as a key regulator of organ growth [20,21]. In mammals, two serine/threonine kinases, the STE20-like protein kinase MST1/2 (Hpo in *Drosophila*) and the NDR family protein kinase LATS1/2 (Wts in *Drosophila*), consist of a core kinase cascade of the Hippo signaling pathway [8,22] (Figure 1). The MST1/2 kinases form a complex with Salvador 1 (SAV1, Sav in *Drosophila*) to phosphorylate and activate LATS1/2 kinases, which in turn phosphorylate and sequester the two major downstream effectors of the Hippo pathway, YAP and TAZ (Yki in *Drosophila*), by promoting the association of YAP/TAZ with 14-3-3 proteins in the cytoplasm where ubiquitin-mediated degradation of YAP and TAZ occurs [8,23–25]. MST1/2 also interact with Mps one binder kinase activator-like 1A/B (MOBKL1A/B, Mats in *Drosophila*), which enhances the activity of LATS1/2 kinases [26,27]. When Hippo signaling activity is suppressed, YAP and TAZ are dephosphorylated, translocate into the nucleus, and function as transcriptional coactivators via interactions with various transcription factors, such as p73, TEA domain family member (TEAD, Sd in *Drosophila*), SMAD, and Runt-related transcription factor (RUNX), to induce the expression of genes promoting cell proliferation and inhibiting apoptosis [25]. For example, *AREG* [28], *BIRC5* [29], *CCNE1* [30], *CTGF* [31,32], *CYR61* [30,32], and *GLI2* [33] have been identified as direct target genes of both YAP and TAZ. The upstream regulators of the Hippo kinase cascade include FAT1-4 (Fat in *Drosophila*), Merlin (encoded by the *NF2* gene; Mer in *Drosophila*), KIBRA (Kibra in *Drosophila*), RASSF, and Ajuba [25]. MST1/2 kinases are activated by FAT1-4 through the apical protein FRMD6/Willin (Ex in *Drosophila*) that forms a complex with two other apically localized proteins, Merlin and KIBRA [34,35]. In contrast, both RASSF and Ajuba inhibit the Hippo signaling pathway by competing with SAV1 for binding with MST1/2 [36]. In addition, YAP and TAZ sense extracellular mechanical stimuli, such as extracellular matrix (ECM) stiffness and cell geometry, and integrate and convert them into intracellular molecular signals, resulting in changes in cellular behaviors, including cell proliferation, migration, and transdifferentiation [37].

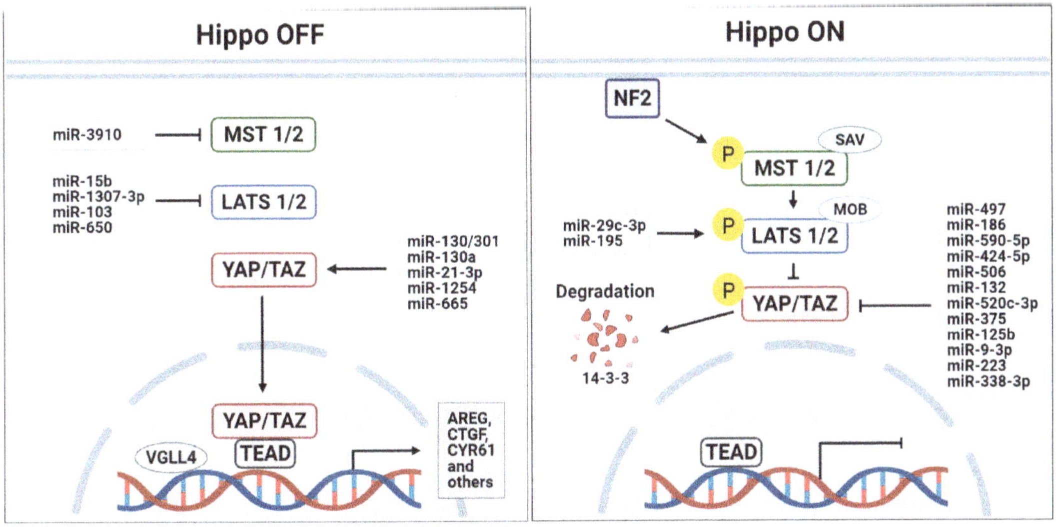

Figure 1. MicroRNAs that regulate the core components of Hippo signaling pathway. Several microRNAs are involved in kinase cascade (MST1/2 and LATS1/2) and downstream effectors (YAP/TAZ) of the Hippo signaling pathway in liver cancer as either oncogenes or tumor suppressors. Abbreviations: NF2, Neurofibromin 2; MST1/2, Mammalian STE20-like 1/2; SAV, Salvador; LATS1/2, Large Tumor Suppressor 1/2; MOB, Mps one binder kinase activator; YAP/TAZ, Yes-associated protein and transcriptional coactivator with PDZ-binding motif; TEAD, TEA domain family member; VGLL4, Vestigial like family member 4; AREG, amphiregulin; CTGF, connective tissue growth factor; CYR61, Cysteine-rich angiogenic inducer 61.

3. The Roles of the Hippo-Yap/Taz Signaling Pathway in Hepatic Tumorigenesis

Hippo-YAP/TAZ signaling is well known to control organ size during development and to mediate the expansion of tissue-specific progenitor cells during tissue regeneration and normal cell proliferation [38]. Accumulated evidence shows aberrant expression of Hippo kinases, YAP/TAZ and their partners in many human cancers, including liver cancers [39–41]. In normal livers, Hippo kinases act as tumor suppressors by inhibiting hepatocyte proliferation and maintaining the differentiated state of hepatocytes [42]. In contrast, loss of Hippo kinase activities, as in mice with a genetic deletion of Nf2 [10,43], Mst1/2 [43,44], Lats1/2 [45,46], or Sav1 [47], causes hepatomegaly and liver cancers, including HCC, ICC, and/or the HCC/ICC mixed form. YAP, on the other hand, functions as an oncogene [48]. Overexpression of YAP phenocopies Hippo signaling deficiency in mice, as shown by liver overgrowth, which is mediated by an increase in hepatocyte proliferation coordinated with a decrease in hepatocyte death [49]. Moreover, removal of Yap in mice with Hippo signaling components knocked out prevented hepatomegaly and hepatic tumor development [10], indicating that YAP is required for hyperplastic cell proliferation and oncogenic transformation of liver cells. It has also been revealed that YAP induces the epithelial-to-mesenchymal transition (EMT), suppression of apoptosis, growth factor-independent proliferation, and anchorage-independent growth of cancer cells, which are attributes of cancer stem cells that are responsible for the major causes of cancer mortality, such as chemoresistance, metastasis, and recurrence [40].

Recent studies have demonstrated that activation of YAP and TAZ in tumor cells also fosters a tumor-favorable microenvironment by communicating with neighboring stromal cells [50,51]. For example, activated YAP/TAZ in hepatocyte-specific Mst1/2 knockout mice creates an inflammatory tumor microenvironment by increasing the production of inflammatory cytokines to suppress immune clearance of transformed hepatocytes by recruiting tumor-associated macrophages and to promote liver cancer development [38,41]. In addition, both tumor cells and cancer-associated fibroblasts increase the stiffness of extracellular matrices, which activates YAP/TAZ mechanosensors to mediate the metabolic crosstalk between tumor cells and cancer-associated fibroblasts, providing them with sufficient nutrients for tumor growth and maintenance of a cancer-prone microenvironment [52–54]. Hence, Hippo-YAP/TAZ signaling has been targeted for developing anti-cancer therapeutics [21].

Most Hippo-YAP/TAZ signaling genes are rarely mutated in liver cancers, which strongly suggests that molecular events, such as epigenetic or post-transcriptional regulation in response to mechanical stresses from the tumor microenvironment, other than DNA mutations, may cause dysregulation of the Hippo-YAP/TAZ signaling pathway in in liver cancers [19,40].

4. MiRNAs Interacting with the Hippo-Yap/Taz Signaling Pathway in Liver Cancer

Dysregulation of biogenesis and expression of miRNAs affect the incidence and progression of liver cancers [55]. miRNAs can function as either tumor suppressors or oncogenes depending on their target mRNAs and resultant phenotypic changes in the cells [56]. In this section, we discuss the interplay between miRNAs and the components of the Hippo-YAP/TAZ signaling pathway that play important roles in hepatic cancer biology (Figure 1).

4.1. MiRNAs as Tumor Suppressors

4.1.1. MiRNAs Targeting YAP or TAZ

Bioinformatic algorithms performed by Liu et al. [57] identified that miR-375 binds directly to the 3′ untranslated region of YAP mRNA. In PLC/PRF/5 and MHCC-97L HCC cell lines, ectopic expression of miR-375 downregulated the expression level of YAP protein and CTGF mRNA, which can be transcribed by activated YAP [57]. Recently, Dinh and Jewell et al. [58] found that miR-375 is the most downregulated miRNA in primary fibrolamellar carcinoma (pFLC), a rare liver cancer that primarily affects adolescents and young adults, compared with nonmalignant human livers. The loss of miR-375 was induced by the presence of the DNAJB1-PRKACA fusion gene, a hallmark of FLC, although it is yet unknown how DNAJB1-PRKACA inhibits miR-375 expression [58]. As in HCC cells, the overexpression of miR-375 in FLC cells inhibited YAP and CTGF, mitigating the proliferative and migratory ability of tumor cells [58]. Another miRNA, miR-186 [59], also decreases the expression of YAP and CTGF by directly disrupting YAP mRNA, inhibiting the proliferation, migration, and invasion of HepG2, Hep3B and SNU398 HCC cell lines [59]. CTGF plays a central role in tissue remodeling and liver fibrosis, which increases the risk for cancer development and progression [60]. Moreover, CTGF is associated with tumor progression by forcing crosstalk between cancer cells and hepatic stellate cells to form a tumor-favorable microenvironment [60]. Thus, the miRNAs discussed above might have tumor suppressive effects not only by killing cancer cells but also by targeting the tumor microenvironment (Table 1).

Both miR-506 and miR-132 also have binding sites in the 3′ untranslated region of YAP mRNA [61,62]. It was shown that the expression of miR-506 is significantly reduced in human HCC tissues and inversely correlates with the expression of the YAP gene [61]. It was experimentally revealed that miR-506 inhibits the growth of HepG2 and H7402

HCC cell lines by downregulating YAP expression [61]. miR-132 promotes apoptosis and suppresses proliferation and invasion of Huh-7 and HepG2 HCC cell lines through direct inhibition of YAP [62]. When miR-132 was cotransfected with miR-520-3p, another tumor suppressive miRNA modulating GPC3, into the Huh-7 HCC cell line, the antiproliferative and proapoptotic functions of both miRNAs on the HCC cell line were shown to be enhanced by minimizing the level of YAP [63]. Similarly, Piao et al. [64] reported that miR-424-5p decreases YAP expression, attenuates proliferation and induces the apoptosis of ICC cell lines, including CCLP-1, RBE and HuCCT-1 [64]. Zhang et al. [65] showed the in vitro anticancer effect of miR-497 by suppressing its functional target, YAP. Overexpression of miR-497 inhibits the growth and survival of HCC cell lines such as HepG2 and Huh-7, whereas silencing of miR-497 has the opposite effect [65].

Chemoresistance is one of the major issues in HCC treatment [66]. Chen et al. [66] found that the level of miR-590-5p negatively correlates with YAP expression in HCC, which is resistant to adriamycin (also known as doxorubicin), the most common first-line chemotherapeutic agent for transarterial chemoembolization [66]. Furthermore, they found that miR-590-5p directly targets YAP and that dysregulation of the miR-590-5p/YAP axis leads to chemoresistance in HCC. The potential anticancer role of miR-509-3p through the suppression of YAP in cholangiocarcinoma (CCA) has been reported by Jung et al. [67]. They found that anticancer drugs such as gemcitabine show synergistic effects when treated in combination with a novel inhibitor of histone deacetylase, CG200745, in CCA cell lines and that CG200745 effectively suppresses tumor growth in xenograft mouse models of gemcitabine-resistant CCA [67]. By performing a microarray and a miRNA array, they further found that CG200745 inhibits the expression of YAP-TEAD4 signaling and increases the expression of miR-509-3p, which downregulates the level of YAP protein [67], possibly by directly targeting YAP mRNA, as in ovarian cancer cells [68].

miRNAs that target TAZ, a paralog of YAP, in HCC include miR-338-3p, miR-9-3p, and miR-125b [69–71]. miR-338-3p has also been reported to be suppressed by the hepatitis B virus (HBV)-encoded protein preS2, upregulating the expression of TAZ in HBV-associated HCC [69]. Both miR-9-3p and miR-125b are known to be downregulated in human HCC tissues and hepatic tumor cells, while the expression of TAZ is upregulated [70,71]. In particular, miR-9-3p is involved in the noninvasive proliferation of tumor cells via signaling pathways, including AKT, ERK1/2, and β-catenin [70]. miR-125b is capable of inhibiting cell invasion and migration through the regulation of TAZ expression [71]. However, the roles of miR-9-3p and miR-125b in human cancer are controversial, and they could function as either tumor suppressors or oncogenes depending on the type of cancer [70,71]. For example, miR-9-3p has been reported as a tumor suppressor in liver, breast, oral, and gastric cancers, whereas it is known as an oncogene primarily in brain cancer [70]. miR-223 plays a critical role in the progression of nonalcoholic steatohepatitis (NASH) to HCC by targeting Taz and the inflammatory gene Cxcl10 in mouse hepatocytes [72]. He et al. [72] found that chronic high-fat diet treatment in miR-223 knockout mice increases the prevalence of liver cancer by activating oncogenic and inflammatory pathways through TAZ and CXCL10 compared to wild-type mice. These findings are consistent with other reports that the level of miR-223 is substantially downregulated in human HCC [72] and that chronic inflammatory injury promotes hepatobiliary carcinogenesis [72].

Table 1. List of tumor suppressive microRNAs inactivating YAP/TAZ and their effect(s) on liver cancer.

MiRNA Name	Direct Target(s) † (Activator or Suppressor)	Effect(s) ‡	Disease Type	Ref.
miR-29c-3p	DNMT3B suppressor	LATS1 methylation ↓	HCC	[73]
miR-195	LATS2 activator	Apoptosis ↑	HCC	[74]
miR-497	YAP1 suppressor	Proliferation ↓, apoptosis ↑	HCC	[65]
miR-186	YAP1 suppressor	Proliferation, migration ↓	HCC	[59]
miR-590-5p	YAP1 suppressor	Chemoresistance ↓	HCC	[66]
miR-424-5p	YAP1 suppressor	Proliferation ↑, apoptosis ↓	HCC	[64]
miR-506	YAP1 suppressor	Proliferation ↓	HCC	[61]
miR-132	YAP1 suppressor	Apoptosis ↑	HCC	[62]
miR-375	YAP1 and CTGF suppressor	Growth, invasion ↓	FLC	[58]
	YAP1/2 suppressor	Proliferation, invasion ↓	HCC	[57]
miR-125b	TAZ suppressor	Migration, invasion ↓	HCC	[71]
miR-9-3p	TAZ suppressor	Proliferation ↓	HCC	[70]
miR-223	TAZ suppressor	Neutrophil activation(proinflammatory mediators) ↓	NASH, HCC	[72]
miR-338-3p	TAZ suppressor	preS2 expression ↓	HCC (HBV derived)	[69]

Abbreviation: miR, microRNA; HCC, hepatocellular carcinoma; FLC, fibrolamellar carcinoma; NASH, nonalcoholic steatohepatitis; HBV, hepatitis B virus. † All direct targets were confirmed by luciferase reporter assay. ‡ ↑ indicates the promotion, ↓ indicates the suppression.

4.1.2. MiRNAs Interacting with LATS1/2

Yang et al. [75] reported that miR-195 is one of a few miRNAs expressed differently between the drug-resistant HCC cell line BEL-7402/5-FU and its parental cell line BEL-7402. miR-195 was shown to be downregulated in HCC cells that acquired drug resistance [75]. Overexpression of miR-195 upregulates the expression of LATS2 and downregulates BCL-w, an anti-apoptotic protein, which sensitizes BEL-7402/5-FU cells to anticancer drugs by suppressing proliferation and inducing apoptosis of the cells [74,75].

Some miRNAs play a tumor suppressive role by interacting with other epigenetic regulators [76]. For example, Wu et al. [73] found that miR-29c-3p promotes DNA demethylation of the LATS1 gene through the direct inhibition of DNA methyltransferase 3B (DNMT3B), resulting in upregulated LATS1 expression (and thus increased Hippo kinase activity) to suppress oncogenic YAP activation. Consistent with the findings that the expression of DNMT3B is increased in many malignancies [73], Wu et al. [73] observed that the level of DNMT3B is higher in HCC tumor tissues than in adjacent nontumor tissues, while miR-29c-3p and LATS1 are expressed at lower levels in HCC tumors than in nontumors. Moreover, patients with either high DNMT3B, low LATS1 or low miR-29c-3p/LATS1 with high DNMT3B have worse outcomes [73].

4.2. *MiRNAs as Oncogenes*

4.2.1. MiRNAs Targeting MST1 and LATS1/2

Cheng et al. [77] showed that miR-3910 is highly expressed in human HCC tissues and various HCC cell lines compared to its level in nontumor liver tissues and normal liver cells. They found that miR-3910 directly inhibits the expression of MST1, promoting the growth and migration of HCC cells in vitro and tumor formation in vivo through the activation of oncogenic YAP (Table 2).

Other studies have shown that miRNAs, including miR-103 [78] and miR-650 [79], directly bind to and inhibit LATS2 mRNA, leading to an increase in YAP, which promotes EMT, metastasis, and invasion of cancer cells, particularly in the liver. In addition, miR-1307-3p is induced by MEIS2, a homeobox protein that promotes HCC development and downregulates LATS1 [80]. Thus, the oncogenic function of MEIS2 is accomplished by the miR-1307-3p/LATS1 axis promoting YAP nuclear translocation in addition to its association with Wnt/β-catenin signaling [80]. Interestingly, a recent study reported that miR-15b in extracellular vesicles (EVs) derived from macrophages after exposure to arsenite, a carcinogen, is transferred to HCC cells and inactivates Hippo signaling by directly targeting LATS1 to promote the proliferation, migration, and invasion of HCC cells [81]. This study suggested the role of miR-15b as a messenger between tumor-associated macrophages and tumor cells in the progression of HCC and that targeting miR-15b could be a strategy for the treatment of liver cancer.

4.2.2. MiRNAs Interacting with YAP/TAZ

Fibrosis is known as a precancerous condition that raises the risk for cancers, including liver cancer [82], and ECM stiffness increases in liver fibrosis and cirrhosis [83]. The mechanosensitive miRNAs of the miR-130/301 family have been shown to regulate fibrosis-related pathways, and their expression depends on the activation of YAP/TAZ, which is promoted by and enforces ECM stiffening [84]. miR-130a also plays a central role in positive feedback regulation of YAP expression by directly inhibiting VGLL4, an antagonist of YAP for TEAD binding, after its expression is induced by YAP [85]. Consequently, endogenous miR-130a strengthens the YAP-TEAD complex, which is a major player in hepatic carcinogenesis [85].

In addition, Hu et al. [86] reported that miR-665 is highly expressed in HCC, suppresses LATS1 activity and enhances activated YAP by negatively regulating the tyrosine phosphatase receptor type B (PTPRB) gene, the protein of which phosphorylates the LATS1 Hippo kinase. As a result, overexpression of miR-665 markedly enhances EMT, cell cycle progression, migration, and invasion of HepG2 cells in vitro [86]. Furthermore, Hu et al. [86] demonstrated that miR-665 promotes tumor growth and metastasis in HepG2 xenograft mouse models in vivo. Recently, Lu et al. [87] reported that miR-1254 is upregulated in human HCC tissues compared with adjacent nontumor tissues and various HCC cell lines, including Hep3B and Huh-7. They also demonstrated that miR-1254 promotes the proliferation, migration, and invasion of HCC cells in vitro and enhances tumor size, vascular invasion, histological grade of HCC assessed by the Edmondson-Steiner scoring system, and lung metastasis in xenografts in vivo [87]. Notably, they found that paired box gene 5 (PAX5) is directly regulated by miR-1254 and that miR-1254 inhibits the phosphorylation of LATS1 and YAP, which seems to be dependent on the availability of PAX5 [87]. Consequently, overexpression of miR-1254 promotes HCC progression [87].

TGF-β signaling is closely associated with YAP/TAZ signaling, and both signaling pathways are known to involve many cellular processes, including proliferation, differentiation and tumor formation and progression [88,89]. Hong et al. [90] reported that miR-21-3p is enriched in human HCC tissues and exerts oncogenic effects on hepatocarcinogenesis and HCC progression by targeting SMAD7, one of the negative regulators of the TGF-β signaling pathway [90], to upregulate oncogenic YAP.

Table 2. List of oncogenic microRNAs activating YAP/TAZ and their effect(s) on liver cancer.

MiRNA Name	Direct Target(s) † (Activator or Suppressor)	Effect(s) ‡	Disease Type	Ref
miR-3910	MST1 suppressor	YAP-TEAD ↑	HCC	[77]
miR-15b	LATS1 suppressor	M2 polarization (MΦ) ↑, proliferation, migration, invasion (tumor) ↑	HCC	[81]
miR-1307-3p	LATS1 suppressor	YAP-Wnt/β-catenin signaling ↑	HCC	[80]
miR-103	LATS2 suppressor	Metastasis, EMT ↑	HCC	[78]
miR-650	LATS2 suppressor	Metastasis, EMT ↑	HCC	[79]
miR-130/301	YAP/TAZ activator	ECM remodeling, fibrosis ↑	NASH	[84]
miR-130a	YAP-TEAD activator	VGLL4 (YAP antagonist) ↓	HCC	[85]
miR-21-3p	SMAD7 suppressor	TGF-β, YAP1 ↑	HCC	[90]
miR-1254	PAX5 suppressor	Hippo pathway ↓	HCC	[87]
miR-665	PTPRB suppressor	Hippo pathway ↓, EMT ↑	HCC	[86]

Abbreviation: miR, microRNA; MΦ, macrophage; HCC, hepatocellular carcinoma; NASH, nonalcoholic steatohepatitis. † All direct targets were confirmed by luciferase reporter assay. ‡ ↑ indicates the promotion, ↓ indicates the suppression.

5. Potential for Clinical Use of MiRNAs Interacting with the Hippo-YAP/TAZ Signaling Pathway in Liver Cancer

Liver cancer is often asymptomatic in the early stages and is usually diagnosed after the advent of metastasis and advanced stages, limiting the chances for surgical treatment with an optimistic prognosis [17]. Although liver biopsy is currently the gold standard for diagnosing liver cancer, it can cause severe complications due to its invasiveness and is restricted to some subset of patients [91]. Various blood-based tests are used clinically, but the currently available biomarkers show insufficient specificity and sensitivity [92]. Hence, there is an unmet need for biomarkers that can diagnose liver cancer early and differentiate the stage of liver cancer. Over the past several years, many studies have focused on identifying circulating miRNAs that have the potential to be used as biomarkers of liver cancer, since they can be detected in cell culture media and different biological fluids, such as serum, plasma, saliva, tears, urine, and breast milk [93–95]. Unlike cellular miRNAs or other RNAs that are degraded within a few seconds, circulating miRNAs are relatively stable, viable for a long time and resistant to endogenous RNase activity in the extracellular environment [96]. Chen et al. [97] reported that circulating miRNAs remain stable under harsh conditions, such as boiling, high or low pH, prolonged storage time, and multiple freeze–thaw cycles. Furthermore, circulating miRNAs in serum maintain their expression patterns after incubation for 24 h at room temperature and are still detectable in serum after a maximum of 10 freeze–thaw cycles [97].

However, no reports have determined whether the miRNAs involved in hepatic carcinogenesis that regulate Hippo-YAP/TAZ signaling can be released into the circulation under either normal or pathological conditions. However, previous findings that some of the miRNAs discussed above (e.g., miR-15b [81,98], miR-9-3p [70], miR-223 [72]) are also detected in the serum of liver cancer patients give rise to the intriguing question of whether they could predict the risk or progression of liver cancer. In lung cancer, for example, upregulation of miR-328-3p targeting NF2 to inactivate Hippo kinase activities is promoted by hypoxic bone marrow mesenchymal stem cells, which deliver EVs containing miR-328-3p to the lung tissue [99]. miR-328-3p is detected not only in cancer tissues but also in the serum of lung cancer patients, suggesting the potential of miR-328-3p as both a biomarker and a therapeutic target of lung cancer [99]. Moreover, miRNAs can monitor tissue responses to therapeutic interventions [100]. Bie et al. [101] have shown that the miRNA expression profiles changes in BEL-7402 HCC cells after treatment with baicalein, an anticancer drug. Notably, the putative target genes for the differentially expressed miRNAs after baicalein treatment are enriched in pathways involved in cell proliferation, including the Hippo signaling pathway [101], indicating that the miRNAs interacting with Hippo-YAP/TAZ signaling represent the status of liver cancer.

Small molecules, such as pazopanib, dasatinib, and statins, which are under investigation in clinical trials or used currently for the treatment of liver cancer, have been reported to activate the Hippo signaling pathway, thereby reducing cancer cell viability and sensitizing tumor cells to chemotherapeutics [102,103]. Verteporfin, a YAP inhibitor, disrupts the YAP-TEAD interaction by promoting the degradation of YAP [104]. Thus, an anticancer approach using verteporfin has been suggested for liver cancers with YAP overexpression and chemoresistance [105,106]. Nevertheless, adverse side effects of these small molecules have been reported, and toxicity, short life, and unintended outcomes also limit the use of these inhibitors [107]. miRNAs may become the alternative since miRNA-based therapy is effective and biologically safe [108,109]. The efficacy of mimics of tumor suppressive miRNAs or inhibitors of oncogenic miRNAs in the prevention and treatment of liver cancer has been evaluated in preclinical models of HCC. To improve the efficiency of miRNA delivery and targeting to specific organs, nanoparticles such as liposomes have been used as delivery vehicles [110,111]. The first-in-human phase I study of miRNA-based therapy was recently completed and used a liposomal miR-34 mimic (known as MRX34, Mirna Therapeutics, Inc.) in solid tumors, including HCC [112]. Although MRX34 was shown to regulate its target genes dose-dependently, unexpected immune-related adverse events occurred in a small subset of patients, the reason for which needs to be elucidated [112]. Hence, further studies are strongly encouraged to understand the mechanism of action of miRNA-based cancer therapeutics to develop more therapeutic candidates that can be used in clinical trials.

6. Conclusions

Genetic ablation of Hippo signaling and overactivation of YAP cause liver cancer in mice [41,113], but the Hippo-YAP/TAZ signaling pathway is dysregulated in human liver cancer primarily by molecular events other than mutations [114]. Recently, it has been demonstrated that alternative RNA splicing of Hippo signaling regulators, including NF2 and CSNK1D, is important for their activities, and certain exon skipping of their mRNAs promotes hepatocyte proliferation and loss of mature hepatocyte functions, suggesting a novel post-transcriptional regulation of the Hippo-YAP/TAZ signaling pathway [115,116]. As well-known post-transcriptional regulators, several miRNAs have been proposed as regulators of the expression of Hippo-YAP/TAZ signaling components, and dysregulation of those miRNAs can lead to hepatic tumorigenesis [117].

In this review, we summarized the miRNAs that contribute to the development and progression of liver cancer by directly binding to the mRNAs of Hippo-YAP/TAZ signaling components or indirectly through interactions with related signaling pathways. Accumulating evidence indicates that these miRNAs could be used as biomarkers for the early detection, prognosis and monitoring of liver cancer and therapeutic targets against liver carcinogenesis. However, most of the studies have been conducted using liver cancer cell lines and xenograft models or on a relatively small number of human liver cancer specimens. Therefore, an increasing number of studies are required to investigate the functions of miRNAs in liver cancers with various etiologies and at different stages during carcinogenesis using relevant models of liver cancer and large-scale cohorts.

In conclusion, miRNAs that interact with the Hippo-YAP/TAZ signaling pathway are promising therapeutic targets for liver cancer. They play pivotal roles in hepatic tumorigenesis by affecting oncogenic transformation, proliferation and migration of tumor cells and modulating the cancer microenvironment.

Author Contributions: Conceptualization, J.H.; investigation, N.-H.L., S.J.K. and J.H.; writing—original draft preparation, N.-H.L. and S.J.K.; writing—review and editing, J.H.; supervision, J.H. All authors have read and agreed to the published version of the manuscript.

Funding: This research was funded by National Research Foundation of Korea, grant number 2020R1F1A1073695 (J.H.) and 2021R1C1C1003904 (J.H.).

Institutional Review Board Statement: Not applicable.

Informed Consent Statement: Not applicable.

Data Availability Statement: Not applicable.

Conflicts of Interest: The authors declare no conflict of interest.

References

1. Mattiuzzi, C.; Lippi, G. Current Cancer Epidemiology. *J. Epidemiol. Glob. Health* **2019**, *9*, 217–222. [CrossRef] [PubMed]
2. Craig, A.J.; von Felden, J.; Garcia-Lezana, T.; Sarcognato, S.; Villanueva, A. Tumour evolution in hepatocellular carcinoma. *Nat. Rev. Gastroenterol. Hepatol.* **2020**, *17*, 139–152. [CrossRef] [PubMed]
3. Yang, J.D.; Hainaut, P.; Gores, G.J.; Amadou, A.; Plymoth, A.; Roberts, L.R. A global view of hepatocellular carcinoma: Trends, risk, prevention and management. *Nat. Rev. Gastroenterol. Hepatol.* **2019**, *16*, 589–604. [CrossRef] [PubMed]
4. Kolluri, A.; Ho, M. The Role of Glypican-3 in Regulating Wnt, YAP, and Hedgehog in Liver Cancer. *Front. Oncol.* **2019**, *9*, 708. [CrossRef]
5. Omenetti, A.; Choi, S.; Michelotti, G.; Diehl, A.M. Hedgehog signaling in the liver. *J. Hepatol.* **2011**, *54*, 366–373. [CrossRef] [PubMed]
6. Kowalik, M.A.; Saliba, C.; Pibiri, M.; Perra, A.; Ledda-Columbano, G.M.; Sarotto, I.; Ghiso, E.; Giordano, S.; Columbano, A. Yes-associated protein regulation of adaptive liver enlargement and hepatocellular carcinoma development in mice. *Hepatology* **2011**, *53*, 2086–2096. [CrossRef] [PubMed]
7. Yimlamai, D.; Christodoulou, C.; Galli, G.G.; Yanger, K.; Pepe-Mooney, B.; Gurung, B.; Shrestha, K.; Cahan, P.; Stanger, B.Z.; Camargo, F.D. Hippo pathway activity influences liver cell fate. *Cell* **2014**, *157*, 1324–1338. [CrossRef]
8. Mo, J.S.; Park, H.W.; Guan, K.L. The Hippo signaling pathway in stem cell biology and cancer. *EMBO. Rep.* **2014**, *15*, 642–656. [CrossRef]
9. Totaro, A.; Panciera, T.; Piccolo, S. YAP/TAZ upstream signals and downstream responses. *Nat. Cell Biol.* **2018**, *20*, 888–899. [CrossRef]
10. Zhang, N.; Bai, H.; David, K.K.; Dong, J.; Zheng, Y.; Cai, J.; Giovannini, M.; Liu, P.; Anders, R.A.; Pan, D. The Merlin/NF2 tumor suppressor functions through the YAP oncoprotein to regulate tissue homeostasis in mammals. *Dev. Cell* **2010**, *19*, 27–38. [CrossRef]
11. Yi, C.; Kissil, J.L. Merlin in organ size control and tumorigenesis: Hippo versus EGFR? *Genes Dev.* **2010**, *24*, 1673–1679. [CrossRef]
12. Zhang, N.; Zhao, Z.; Long, J.; Li, H.; Zhang, B.; Chen, G.; Li, X.; Lv, T.; Zhang, W.; Ou, X.; et al. Molecular alterations of the NF2 gene in hepatocellular carcinoma and intrahepatic cholangiocarcinoma. *Oncol. Rep.* **2017**, *38*, 3650–3658. [CrossRef]
13. Park, J.; Kim, J.S.; Nahm, J.H.; Kim, S.K.; Lee, D.H.; Lim, D.S. WWC1 and NF2 Prevent the Development of Intrahepatic Cholangiocarcinoma by Regulating YAP/TAZ Activity through LATS in Mice. *Mol. Cells* **2020**, *43*, 491–499. [CrossRef] [PubMed]
14. Harvey, K.F.; Zhang, X.; Thomas, D.M. The Hippo pathway and human cancer. *Nat. Rev. Cancer* **2013**, *13*, 246–257. [CrossRef] [PubMed]
15. Calin, G.A.; Croce, C.M. MicroRNA signatures in human cancers. *Nat. Rev. Cancer* **2006**, *6*, 857–866. [CrossRef] [PubMed]
16. Li, M.; Marin-Muller, C.; Bharadwaj, U.; Chow, K.H.; Yao, Q.; Chen, C. MicroRNAs: Control and loss of control in human physiology and disease. *World. J. Surg.* **2009**, *33*, 667–684. [CrossRef] [PubMed]
17. Shen, S.; Lin, Y.; Yuan, X.; Shen, L.; Chen, J.; Chen, L.; Qin, L.; Shen, B. Biomarker MicroRNAs for Diagnosis, Prognosis and Treatment of Hepatocellular Carcinoma: A Functional Survey and Comparison. *Sci. Rep.* **2016**, *6*, 38311. [CrossRef]
18. Shen, S.; Luo, X.; Gao, K.; Sun, Y.; Yao, D.; Zhu, L. Identification and integrative analysis of microRNAs and mRNAs involved in proliferation and invasion of pressure-treated human liver cancer cell lines. *Mol. Med. Rep.* **2019**, *20*, 375–387. [CrossRef]
19. Wang, Y.; Xu, X.; Maglic, D.; Dill, M.T.; Mojumdar, K.; Ng, P.K.; Jeong, K.J.; Tsang, Y.H.; Moreno, D.; Bhavana, V.H.; et al. Comprehensive Molecular Characterization of the Hippo Signaling Pathway in Cancer. *Cell Rep.* **2018**, *25*, 1304–1317.e5. [CrossRef]
20. Moya, I.M.; Halder, G. Hippo-YAP/TAZ signalling in organ regeneration and regenerative medicine. *Nat. Rev. Mol. Cell Biol.* **2019**, *20*, 211–226. [CrossRef]
21. Ma, Y.; Yang, Y.; Wang, F.; Wei, Q.; Qin, H. Hippo-YAP signaling pathway: A new paradigm for cancer therapy. *Int. J. Cancer* **2015**, *137*, 2275–2286. [CrossRef]
22. Driskill, J.H.; Pan, D. The Hippo Pathway in Liver Homeostasis and Pathophysiology. *Annu. Rev. Pathol.* **2021**, *16*, 299–322. [CrossRef]
23. Zeng, Q.; Hong, W. The emerging role of the hippo pathway in cell contact inhibition, organ size control, and cancer development in mammals. *Cancer Cell* **2008**, *13*, 188–192. [CrossRef] [PubMed]
24. Bennett, F.C.; Harvey, K.F. Fat cadherin modulates organ size in Drosophila via the Salvador/Warts/Hippo signaling pathway. *Curr. Biol.* **2006**, *16*, 2101–2110. [CrossRef] [PubMed]
25. Yu, F.X.; Guan, K.L. The Hippo pathway: Regulators and regulations. *Genes Dev.* **2013**, *27*, 355–371. [CrossRef] [PubMed]

26. Wei, X.; Shimizu, T.; Lai, Z.-C. Mob as tumor suppressor is activated by Hippo kinase for growth inhibition in Drosophila. *EMBO J.* **2007**. [CrossRef]
27. Praskova, M.; Xia, F.; Avruch, J. MOBKL1A/MOBKL1B phosphorylation by MST1 and MST2 inhibits cell proliferation. *Curr. Biol.* **2008**, *18*, 311–321. [CrossRef]
28. Pan, D. The hippo signaling pathway in development and cancer. *Dev. Cell* **2010**, *19*, 491–505. [CrossRef] [PubMed]
29. Hong, W.; Guan, K.L. The YAP and TAZ transcription co-activators: Key downstream effectors of the mammalian Hippo pathway. *Semin. Cell. Dev. Biol.* **2012**, *23*, 785–793. [CrossRef]
30. LaQuaglia, M.J.; Grijalva, J.L.; Mueller, K.A.; Perez-Atayde, A.R.; Kim, H.B.; Sadri-Vakili, G.; Vakili, K. YAP Subcellular Localization and Hippo Pathway Transcriptome Analysis in Pediatric Hepatocellular Carcinoma. *Sci. Rep.* **2016**, *6*, 30238. [CrossRef] [PubMed]
31. Zhou, Y.; Huang, T.; Cheng, A.S.; Yu, J.; Kang, W.; To, K.F. The TEAD Family and Its Oncogenic Role in Promoting Tumorigenesis. *Int. J. Mol. Sci.* **2016**, *17*, 138. [CrossRef]
32. Lai, D.; Ho, K.C.; Hao, Y.; Yang, X. Taxol resistance in breast cancer cells is mediated by the hippo pathway component TAZ and its downstream transcriptional targets Cyr61 and CTGF. *Cancer Res.* **2011**, *71*, 2728–2738. [CrossRef]
33. Xu, S.; Zhang, H.; Chong, Y.; Guan, B.; Guo, P. YAP Promotes VEGFA Expression and Tumor Angiogenesis Though Gli2 in Human Renal Cell Carcinoma. *Arch. Med. Res.* **2019**, *50*, 225–233. [CrossRef] [PubMed]
34. Angus, L.; Moleirinho, S.; Herron, L.; Sinha, A.; Zhang, X.; Niestrata, M.; Dholakia, K.; Prystowsky, M.B.; Harvey, K.F.; Reynolds, P.A.; et al. Willin/FRMD6 expression activates the Hippo signaling pathway kinases in mammals and antagonizes oncogenic YAP. *Oncogene* **2012**, *31*, 238–250. [CrossRef]
35. Xiao, L.; Chen, Y.; Ji, M.; Dong, J. KIBRA regulates Hippo signaling activity via interactions with large tumor suppressor kinases. *J. Biol. Chem.* **2011**, *286*, 7788–7796. [CrossRef]
36. Polesello, C.; Huelsmann, S.; Brown, N.H.; Tapon, N. The Drosophila RASSF homolog antagonizes the hippo pathway. *Curr. Biol.* **2006**, *16*, 2459–2465. [CrossRef] [PubMed]
37. d'Angelo, M.; Benedetti, E.; Tupone, M.G.; Catanesi, M.; Castelli, V.; Antonosante, A.; Cimini, A. The Role of Stiffness in Cell Reprogramming: A Potential Role for Biomaterials in Inducing Tissue Regeneration. *Cells* **2019**, *8*, 1036. [CrossRef] [PubMed]
38. Song, H.; Mak, K.K.; Topol, L.; Yun, K.; Hu, J.; Garrett, L.; Chen, Y.; Park, O.; Chang, J.; Simpson, R.M.; et al. Mammalian Mst1 and Mst2 kinases play essential roles in organ size control and tumor suppression. *Proc. Natl. Acad. Sci. USA* **2010**, *107*, 1431–1436. [CrossRef]
39. Zhang, S.; Zhou, D. Role of the transcriptional coactivators YAP/TAZ in liver cancer. *Curr. Opin. Cell. Biol.* **2019**, *61*, 64–71. [CrossRef]
40. Zanconato, F.; Cordenonsi, M.; Piccolo, S. YAP/TAZ at the Roots of Cancer. *Cancer Cell* **2016**, *29*, 783–803. [CrossRef]
41. Liu, Y.; Wang, X.; Yang, Y. Hepatic Hippo signaling inhibits development of hepatocellular carcinoma. *Clin. Mol. Hepatol.* **2020**, *26*, 742–750. [CrossRef]
42. Michalopoulos, G.K. Hepatostat: Liver regeneration and normal liver tissue maintenance. *Hepatology* **2017**, *65*, 1384–1392. [CrossRef] [PubMed]
43. Hong, L.; Cai, Y.; Jiang, M.; Zhou, D.; Chen, L. The Hippo signaling pathway in liver regeneration and tumorigenesis. *Acta. Biochim. Biophys. Sin.* **2015**, *47*, 46–52. [CrossRef] [PubMed]
44. Patel, S.H.; Camargo, F.D.; Yimlamai, D. Hippo Signaling in the Liver Regulates Organ Size, Cell Fate, and Carcinogenesis. *Gastroenterology* **2017**, *152*, 533–545. [CrossRef] [PubMed]
45. Li, H.; Wang, S.; Wang, G.; Zhang, Z.; Wu, X.; Zhang, T.; Fu, B.; Chen, G. Yes-associated protein expression is a predictive marker for recurrence of hepatocellular carcinoma after liver transplantation. *Dig. Surg.* **2014**, *31*, 468–478. [CrossRef] [PubMed]
46. Moroishi, T.; Hayashi, T.; Pan, W.W.; Fujita, Y.; Holt, M.V.; Qin, J.; Carson, D.A.; Guan, K.L. The Hippo Pathway Kinases LATS1/2 Suppress Cancer Immunity. *Cell* **2016**, *167*, 1525–1539.e17. [CrossRef]
47. Jeong, S.H.; Kim, H.B.; Kim, M.C.; Lee, J.M.; Lee, J.H.; Kim, J.H.; Kim, J.W.; Park, W.Y.; Kim, S.Y.; Kim, J.B.; et al. Hippo-mediated suppression of IRS2/AKT signaling prevents hepatic steatosis and liver cancer. *J. Clin. Investig.* **2018**, *128*, 1010–1025. [CrossRef] [PubMed]
48. Baia, G.S.; Caballero, O.L.; Orr, B.A.; Lal, A.; Ho, J.S.; Cowdrey, C.; Tihan, T.; Mawrin, C.; Riggins, G.J. Yes-associated protein 1 is activated and functions as an oncogene in meningiomas. *Mol. Cancer Res.* **2012**, *10*, 904–913. [CrossRef]
49. Huo, X.; Zhang, Q.; Liu, A.M.; Tang, C.; Gong, Y.; Bian, J.; Luk, J.M.; Xu, Z.; Chen, J. Overexpression of Yes-associated protein confers doxorubicin resistance in hepatocellullar carcinoma. *Oncol. Rep.* **2013**, *29*, 840–846. [CrossRef] [PubMed]
50. Ma, H.; Wang, J.; Zhao, X.; Wu, T.; Huang, Z.; Chen, D.; Liu, Y.; Ouyang, G. Periostin Promotes Colorectal Tumorigenesis through Integrin-FAK-Src Pathway-Mediated YAP/TAZ Activation. *Cell Rep.* **2020**, *30*, 793–806 e796. [CrossRef]
51. Tocci, P.; Blandino, G.; Bagnato, A. YAP and endothelin-1 signaling: An emerging alliance in cancer. *J. Exp. Clin. Cancer Res.* **2021**, *40*, 27. [CrossRef]
52. Kadel, D.; Zhang, Y.; Sun, H.R.; Zhao, Y.; Dong, Q.Z.; Qin, L.X. Current perspectives of cancer-associated fibroblast in therapeutic resistance: Potential mechanism and future strategy. *Cell Biol. Toxicol.* **2019**, *35*, 407–421. [CrossRef]
53. Shiga, K.; Hara, M.; Nagasaki, T.; Sato, T.; Takahashi, H.; Takeyama, H. Cancer-Associated Fibroblasts: Their Characteristics and Their Roles in Tumor Growth. *Cancers* **2015**, *7*, 2443–2458. [CrossRef]

54. Raghunathan, V.K.; Morgan, J.T.; Dreier, B.; Reilly, C.M.; Thomasy, S.M.; Wood, J.A.; Ly, I.; Tuyen, B.C.; Hughbanks, M.; Murphy, C.J.; et al. Role of substratum stiffness in modulating genes associated with extracellular matrix and mechanotransducers YAP and TAZ. *Investig. Ophthalmol. Vis. Sci.* **2013**, *54*, 378–386. [CrossRef]
55. Lin, S.; Gregory, R.I. MicroRNA biogenesis pathways in cancer. *Nat. Rev. Cancer* **2015**, *15*, 321–333. [CrossRef]
56. Giordano, S.; Columbano, A. MicroRNAs: New tools for diagnosis, prognosis, and therapy in hepatocellular carcinoma? *Hepatology* **2013**, *57*, 840–847. [CrossRef]
57. Liu, A.M.; Poon, R.T.; Luk, J.M. MicroRNA-375 targets Hippo-signaling effector YAP in liver cancer and inhibits tumor properties. *Biochem. Biophys. Res. Commun.* **2010**, *394*, 623–627. [CrossRef] [PubMed]
58. Dinh, T.A.; Jewell, M.L.; Kanke, M.; Francisco, A.; Sritharan, R.; Turnham, R.E.; Lee, S.; Kastenhuber, E.R.; Wauthier, E.; Guy, C.D.; et al. *MicroRNA*.-375 Suppresses the Growth and Invasion of Fibrolamellar Carcinoma. *Cell Mol. Gastroenterol. Hepatol.* **2019**, *7*, 803–817. [CrossRef]
59. Ruan, T.; He, X.; Yu, J.; Hang, Z. MicroRNA-186 targets Yes-associated protein 1 to inhibit Hippo signaling and tumorigenesis in hepatocellular carcinoma. *Oncol. Lett.* **2016**, *11*, 2941–2945. [CrossRef] [PubMed]
60. Makino, Y.; Hikita, H.; Kodama, T.; Shigekawa, M.; Yamada, R.; Sakamori, R.; Eguchi, H.; Morii, E.; Yokoi, H.; Mukoyama, M.; et al. CTGF Mediates Tumor-Stroma Interactions between Hepatoma Cells and Hepatic Stellate Cells to Accelerate HCC Progression. *Cancer Res.* **2018**, *78*, 4902–4914. [CrossRef] [PubMed]
61. Wang, Y.; Cui, M.; Sun, B.D.; Liu, F.B.; Zhang, X.D.; Ye, L.H. MiR-506 suppresses proliferation of hepatoma cells through targeting YAP mRNA 3′UTR. *Acta. Pharmacol. Sin.* **2014**, *35*, 1207–1214. [CrossRef]
62. Lei, C.J.; Li, L.; Gao, X.; Zhang, J.; Pan, Q.Y.; Long, H.C.; Chen, C.Z.; Ren, D.F.; Zheng, G. Hsa-miR-132 inhibits proliferation of hepatic carcinoma cells by targeting YAP. *Cell Biochem. Funct.* **2015**, *33*, 326–333. [CrossRef] [PubMed]
63. Lei, C.J.; Yao, C.; Li, D.K.; Long, Z.X.; Li, Y.; Tao, D.; Liou, Y.P.; Zhang, J.Z.; Liu, N. Effect of co-transfection of miR-520c-3p and miR-132 on proliferation and apoptosis of hepatocellular carcinoma Huh7. *Asian. Pac. J. Trop. Med.* **2016**, *9*, 898–902. [CrossRef]
64. Piao, L.; Wang, F.; Wang, Y.; Yang, Z.; Li, Q.; Cui, L.; Yu, Q. miR-424-5p Regulates Hepatoma Cell Proliferation and Apoptosis. *Cancer Biother. Radiopharm.* **2019**, *34*, 196–202. [CrossRef]
65. Zhang, L.; Yu, Z.; Xian, Y.; Lin, X. microRNA-497 inhibits cell proliferation and induces apoptosis by targeting YAP1 in human hepatocellular carcinoma. *FEBS Open Bio* **2016**, *6*, 155–164. [CrossRef]
66. Chen, M.; Wu, L.; Tu, J.; Zhao, Z.; Fan, X.; Mao, J.; Weng, Q.; Wu, X.; Huang, L.; Xu, M.; et al. miR-590-5p suppresses hepatocellular carcinoma chemoresistance by targeting YAP1 expression. *EBioMedicine* **2018**, *35*, 142–154. [CrossRef] [PubMed]
67. Jung, D.E.; Park, S.B.; Kim, K.; Kim, C.; Song, S.Y. CG200745, an HDAC inhibitor, induces anti-tumour effects in cholangiocarcinoma cell lines via miRNAs targeting the Hippo pathway. *Sci. Rep.* **2017**, *7*, 10921. [CrossRef]
68. Pan, Y.; Robertson, G.; Pedersen, L.; Lim, E.; Hernandez-Herrera, A.; Rowat, A.C.; Patil, S.L.; Chan, C.K.; Wen, Y.; Zhang, X.; et al. miR-509-3p is clinically significant and strongly attenuates cellular migration and multi-cellular spheroids in ovarian cancer. *Oncotarget* **2016**, *7*, 25930–25948. [CrossRef]
69. Liu, P.; Zhang, H.; Liang, X.; Ma, H.; Luan, F.; Wang, B.; Bai, F.; Gao, L.; Ma, C. HBV preS2 promotes the expression of TAZ via miRNA-338-3p to enhance the tumorigenesis of hepatocellular carcinoma. *Oncotarget* **2015**, *6*, 2904859. [CrossRef] [PubMed]
70. Higashi, T.; Hayashi, H.; Ishimoto, T.; Takeyama, H.; Kaida, T.; Arima, K.; Taki, K.; Sakamoto, K.; Kuroki, H.; Okabe, H.; et al. miR-9-3p plays a tumour-suppressor role by targeting TAZ (WWTR1) in hepatocellular carcinoma cells. *Br. J. Cancer* **2015**, *113*, 252–258. [CrossRef]
71. Li, J.; Fang, L.; Yu, W.; Wang, Y. MicroRNA-125b suppresses the migration and invasion of hepatocellular carcinoma cells by targeting transcriptional coactivator with PDZ-binding motif. *Oncol. Lett.* **2015**, *9*, 1971–1975. [CrossRef] [PubMed]
72. He, Y.; Hwang, S.; Cai, Y.; Kim, S.J.; Xu, M.; Yang, D.; Guillot, A.; Feng, D.; Seo, W.; Hou, X.; et al. *MicroRNA*.-223 Ameliorates Nonalcoholic Steatohepatitis and Cancer by Targeting Multiple Inflammatory and Oncogenic Genes in Hepatocytes. *Hepatology* **2019**, *70*, 1150–1167. [CrossRef]
73. Wu, H.; Zhang, W.; Wu, Z.; Liu, Y.; Shi, Y.; Gong, J.; Shen, W.; Liu, C. miR-29c-3p regulates DNMT3B and LATS1 methylation to inhibit tumor progression in hepatocellular carcinoma. *Cell Death Dis.* **2019**, *10*, 48. [CrossRef]
74. Yang, X.; Yu, J.; Yin, J.; Xiang, Q.; Tang, H.; Lei, X. MiR-195 regulates cell apoptosis of human hepatocellular carcinoma cells by targeting LATS2. *Pharmazie* **2012**, *67*, 645–651. [PubMed]
75. Yang, X.; Yin, J.; Yu, J.; Xiang, Q.; Liu, Y.; Tang, S.; Liao, D.; Zhu, B.; Zu, X.; Tang, H.; et al. *miRNA*.-195 sensitizes human hepatocellular carcinoma cells to 5-FU by targeting BCL-w. *Oncol. Rep.* **2012**, *27*, 250–257. [CrossRef] [PubMed]
76. Memari, F.; Joneidi, Z.; Taheri, B.; Aval, S.F.; Roointan, A.; Zarghami, N. Epigenetics and Epi-miRNAs: Potential markers/therapeutics in leukemia. *Biomed. Pharmacother.* **2018**, *106*, 1668–1677. [CrossRef]
77. Cheng, L.; Wang, H.; Han, S. MiR-3910 Promotes the Growth and Migration of Cancer Cells in the Progression of Hepatocellular Carcinoma. *Dig. Dis. Sci.* **2017**, *62*, 2812–2820. [CrossRef]
78. Han, L.L.; Yin, X.R.; Zhang, S.Q. miR-103 promotes the metastasis and EMT of hepatocellular carcinoma by directly inhibiting LATS2. *Int. J. Oncol.* **2018**, *53*, 2433–2444. [CrossRef]
79. Han, L.L.; Yin, X.R.; Zhang, S.Q. miR-650 Promotes the Metastasis and Epithelial-Mesenchymal Transition of Hepatocellular Carcinoma by Directly Inhibiting LATS2 Expression. *Cell Physiol. Biochem.* **2018**, *51*, 1179–1192. [CrossRef]

80. Guan, L.; Li, T.; Ai, N.; Wang, W.; He, B.; Bai, Y.; Yu, Z.; Li, M.; Dong, S.; Zhu, Q.; et al. MEIS2C and MEIS2D promote tumor progression via Wnt/beta-catenin and hippo/YAP signaling in hepatocellular carcinoma. *J. Exp. Clin. Cancer Res.* **2019**, *38*, 417. [CrossRef]
81. Li, J.; Xue, J.; Ling, M.; Sun, J.; Xiao, T.; Dai, X.; Sun, Q.; Cheng, C.; Xia, H.; Wei, Y.; et al. MicroRNA-15b in extracellular vesicles from arsenite-treated macrophages promotes the progression of hepatocellular carcinomas by blocking the LATS1-mediated Hippo pathway. *Cancer Lett.* **2021**, *497*, 137–153. [CrossRef] [PubMed]
82. Higashi, T.; Friedman, S.L.; Hoshida, Y. Hepatic stellate cells as key target in liver fibrosis. *Adv. Drug Deliv. Rev.* **2017**, *121*, 27–42. [CrossRef]
83. Zhang, D.Y.; Friedman, S.L. Fibrosis-dependent mechanisms of hepatocarcinogenesis. *Hepatology* **2012**, *56*, 769–775. [CrossRef] [PubMed]
84. Bertero, T.; Cottrill, K.A.; Annis, S.; Bhat, B.; Gochuico, B.R.; Osorio, J.C.; Rosas, I.; Haley, K.J.; Corey, K.E.; Chung, R.T.; et al. A YAP/TAZ-miR-130/301 molecular circuit exerts systems-level control of fibrosis in a network of human diseases and physiologic conditions. *Sci. Rep.* **2015**, *5*, 18277. [CrossRef]
85. Shen, S.; Guo, X.; Yan, H.; Lu, Y.; Ji, X.; Li, L.; Liang, T.; Zhou, D.; Feng, X.H.; Zhao, J.C.; et al. *A. miR*.-130a-YAP positive feedback loop promotes organ size and tumorigenesis. *Cell Res.* **2015**, *25*, 997–1012. [CrossRef] [PubMed]
86. Hu, Y.; Yang, C.; Yang, S.; Cheng, F.; Rao, J.; Wang, X. miR-665 promotes hepatocellular carcinoma cell migration, invasion, and proliferation by decreasing Hippo signaling through targeting PTPRB. *Cell Death Dis.* **2018**, *9*, 954. [CrossRef]
87. Lu, X.; Yang, C.; Hu, Y.; Xu, J.; Shi, C.; Rao, J.; Yu, W.; Cheng, F. Upregulation of miR-1254 promotes Hepatocellular Carcinoma Cell Proliferation, Migration, and Invasion via Inactivation of the Hippo-YAP signaling pathway by decreasing PAX5. *J. Cancer* **2021**, *12*, 771–789. [CrossRef]
88. Qin, Z.; Xia, W.; Fisher, G.J.; Voorhees, J.J.; Quan, T. YAP/TAZ regulates TGF-beta/Smad3 signaling by induction of Smad7 via AP-1 in human skin dermal fibroblasts. *Cell Commun. Signal.* **2018**, *16*, 18. [CrossRef] [PubMed]
89. Noguchi, S.; Saito, A.; Nagase, T. YAP/TAZ Signaling as a Molecular Link between Fibrosis and Cancer. *Int. J. Mol. Sci.* **2018**, *19*, 3674. [CrossRef] [PubMed]
90. Hong, Y.; Ye, M.; Wang, D.; Wang, C.; Luo, J.; Liu, J.; Liu, J.; Zhao, Q.; Chang, Y. MiR-21-3p promotes hepatocellular carcinoma progression through regulating Smad7/Yap1. *Front. Oncol.* **2021**. [CrossRef]
91. Rockey, D.C.; Caldwell, S.H.; Goodman, Z.D.; Nelson, R.C.; Smith, A.D.; American Association for the Study of Liver Diseases. Liver biopsy. *Hepatology* **2009**, *49*, 1017–1044. [CrossRef] [PubMed]
92. Cheng, G. Circulating miRNAs: Roles in cancer diagnosis, prognosis and therapy. *Adv. Drug Deliv. Rev.* **2015**, *81*, 75–93. [CrossRef]
93. Weber, J.A.; Baxter, D.H.; Zhang, S.; Huang, D.Y.; Huang, K.H.; Lee, M.J.; Galas, D.J.; Wang, K. The microRNA spectrum in 12 body fluids. *Clin. Chem.* **2010**, *56*, 1733–1741. [CrossRef] [PubMed]
94. Hayes, C.N.; Chayama, K. MicroRNAs as Biomarkers for Liver Disease and Hepatocellular Carcinoma. *Int. J. Mol. Sci.* **2016**, *17*, 280. [CrossRef] [PubMed]
95. He, S.; Zhang, D.C.; Wei, C. MicroRNAs as biomarkers for hepatocellular carcinoma diagnosis and prognosis. *Clin. Res. Hepatol. Gastroenterol.* **2015**, *39*, 426–434. [CrossRef] [PubMed]
96. Kosaka, N.; Iguchi, H.; Ochiya, T. Circulating microRNA in body fluid: A new potential biomarker for cancer diagnosis and prognosis. *Cancer Sci.* **2010**, *101*, 2087–2092. [CrossRef]
97. Chen, X.; Ba, Y.; Ma, L.; Cai, X.; Yin, Y.; Wang, K.; Guo, J.; Zhang, Y.; Chen, J.; Guo, X.; et al. Characterization of microRNAs in serum: A novel class of biomarkers for diagnosis of cancer and other diseases. *Cell Res.* **2008**, *18*, 997–1006. [CrossRef]
98. Liu, A.M.; Yao, T.J.; Wang, W.; Wong, K.F.; Lee, N.P.; Fan, S.T.; Poon, R.T.; Gao, C.; Luk, J.M. Circulating miR-15b and miR-130b in serum as potential markers for detecting hepatocellular carcinoma: A retrospective cohort study. *BMJ Open* **2012**, *2*, e000825. [CrossRef]
99. Liu, X.; Jiang, F.; Wang, Z.; Tang, L.; Zou, B.; Xu, P.; Yu, T. Hypoxic bone marrow mesenchymal cell-extracellular vesicles containing miR-328-3p promote lung cancer progression via the NF2-mediated Hippo axis. *J. Cell Mol. Med.* **2021**, *25*, 96–109. [CrossRef]
100. Olivieri, F.; Capri, M.; Bonafe, M.; Morsiani, C.; Jung, H.J.; Spazzafumo, L.; Vina, J.; Suh, Y. Circulating miRNAs and miRNA shuttles as biomarkers: Perspective trajectories of healthy and unhealthy aging. *Mech. Ageing Dev.* **2017**, *165*, 162–170. [CrossRef]
101. Bie, B.; Sun, J.; Li, J.; Guo, Y.; Jiang, W.; Huang, C.; Yang, J.; Li, Z. Baicalein, a Natural Anti-Cancer Compound, Alters MicroRNA Expression Profiles in Bel-7402 Human Hepatocellular Carcinoma Cells. *Cell Physiol. Biochem.* **2017**, *41*, 1519–1531. [CrossRef]
102. Calses, P.C.; Crawford, J.J.; Lill, J.R.; Dey, A. Hippo Pathway in Cancer: Aberrant Regulation and Therapeutic Opportunities. *Trends. Cancer* **2019**, *5*, 297–307. [CrossRef] [PubMed]
103. Oku, Y.; Nishiya, N.; Shito, T.; Yamamoto, R.; Yamamoto, Y.; Oyama, C.; Uehara, Y. Small molecules inhibiting the nuclear localization of YAP/TAZ for chemotherapeutics and chemosensitizers against breast cancers. *FEBS Open Bio* **2015**, *5*, 542–549. [CrossRef] [PubMed]
104. Wang, C.; Zhu, X.; Feng, W.; Yu, Y. Verteporfin inhibits YAP function through up-regulating 14-3-3σ sequestering YAP in the cytoplasm. *Am. J. Cancer Res.* **2015**, *6*, 27–37, eCollection 2016. [PubMed]

105. Perra, A.; Kowalik, M.A.; Ghiso, E.; Ledda-Columbano, G.M.; Di Tommaso, L.; Angioni, M.M.; Raschioni, C.; Testore, E.; Roncalli, M.; Giordano, S.; et al. YAP activation is an early event and a potential therapeutic target in liver cancer development. *J. Hepatol.* **2014**, *61*, 1088–1096. [CrossRef] [PubMed]
106. Gavini, J.; Dommann, N.; Jakob, M.O.; Keogh, A.; Bouchez, L.C.; Karkampouna, S.; Julio, M.K.; Medova, M.; Zimmer, Y.; Schläfli, A.M.; et al. Verteporfin-induced lysosomal compartment dysregulation potentiates the effect of sorafenib in hepatocellular carcinoma. *Cell Death Dis.* **2019**, *10*, 749. [CrossRef]
107. Gurevich, E.V.; Gurevich, V.V. Therapeutic potential of small molecules and engineered proteins. *Handb. Exp. Pharmacol.* **2014**, *219*, 1–12. [CrossRef]
108. Ishida, M.; Selaru, F.M. miRNA-Based Therapeutic Strategies. *Curr. Anesthesiol. Rep.* **2013**, *1*, 63–70. [CrossRef]
109. Naidu, S.; Magee, P.; Garofalo, M. MiRNA-based therapeutic intervention of cancer. *J. Hematol. Oncol.* **2015**, *8*, 68. [CrossRef]
110. O'Neill, C.P.; Dwyer, R.M. Nanoparticle-Based Delivery of Tumor Suppressor microRNA for Cancer Therapy. *Cells* **2020**, *9*, 521. [CrossRef]
111. Toh, T.B.; Lim, J.J.; Chow, E.K. Epigenetics of hepatocellular carcinoma. *Clin. Transl. Med.* **2019**, *8*, 13. [CrossRef]
112. Hong, D.S.; Kang, Y.K.; Borad, M.; Sachdev, J.; Ejadi, S.; Lim, H.Y.; Brenner, A.J.; Park, K.; Lee, J.L.; Kim, T.Y.; et al. Phase 1 study of MRX34, a liposomal miR-34a mimic, in patients with advanced solid tumours. *Br. J. Cancer* **2020**, *122*, 1630–1637. [CrossRef]
113. Jie, L.; Fan, W.; Weiqi, D.; Yingqun, Z.; Ling, X.; Miao, S.; Ping, C.; Chuanyong, G. The hippo-yes association protein pathway in liver cancer. *Gastroenterol. Res. Pract.* **2013**, *2013*, 187070. [CrossRef]
114. Manmadhan, S.; Ehmer, U. Hippo Signaling in the Liver-A Long and Ever-Expanding Story. *Front. Cell Dev. Biol.* **2019**, *12*, 33. [CrossRef] [PubMed]
115. Bangru, S.; Arif, W.; Seimetz, J.; Bhate, A.; Chen, J.; Rashan, E.H.; Carstens, R.P.; Anakk, S.; Kalsotra, A. Alternative splicing rewires Hippo signaling pathway in hepatocytes to promote liver regeneration. *Nat. Struct. Mol. Biol.* **2018**, *25*, 928–939. [CrossRef] [PubMed]
116. Hyun, J.; Sun, Z.; Ahmadi, A.R.; Bangru, S.; Chembazhi, U.V.; Du, K.; Chen, T.; Tsukamoto, H.; Rusyn, I.; Kalsotra, A.; et al. Epithelial splicing regulatory protein 2-mediated alternative splicing reprograms hepatocytes in severe alcoholic hepatitis. *J. Clin. Investig.* **2020**, *130*, 2129–2145. [CrossRef] [PubMed]
117. Han, Y. Analysis of the role of the Hippo pathway in cancer. *J. Transl. Med.* **2019**, *17*, 116. [CrossRef] [PubMed]

biomedicines

MDPI

Review

Involvement of microRNA in Solid Cancer: Role and Regulatory Mechanisms

Ying-Chin Lin [1,2,†], Tso-Hsiao Chen [3,†], Yu-Min Huang [4,5], Po-Li Wei [4,6,7,8,9,*] and Jung-Chun Lin [10,11,12,*]

1 Department of Family Medicine, School of Medicine, College of Medicine, Taipei Medical University, Taipei 110, Taiwan
2 Department of Family Medicine, Wan Fang Hospital, Taipei Medical University, Taipei 116, Taiwan; greening1990@gmail.com
3 Division of Nephrology, Wan Fang Hospital, Taipei Medical University, Taipei 116, Taiwan; 88128@w.tmu.edu.tw
4 Department of Surgery, School of Medicine, College of Medicine, Taipei Medical University, Taipei 110, Taiwan
5 Division of Gastrointestinal Surgery, Department of Surgery, Taipei Medical University Hospital, Taipei Medical University, Taipei 110, Taiwan; yuminhuang26@gmail.com
6 Division of Colorectal Surgery, Department of Surgery, Taipei Medical University Hospital, Taipei Medical University, Taipei 110, Taiwan
7 Cancer Research Center, Taipei Medical University Hospital, Taipei Medical University, Taipei 110, Taiwan
8 Translational Laboratory, Department of Medical Research, Taipei Medical University Hospital, Taipei Medical University, Taipei 110, Taiwan
9 Graduate Institute of Cancer Biology and Drug Discovery, Taipei Medical University, Taipei 110, Taiwan
10 School of Medical Laboratory Science and Biotechnology, College of Medical Science and Technology, Taipei Medical University, Taipei 110, Taiwan
11 Program in Medical Biotechnology, College of Medical Science and Technology, Taipei Medical University, Taipei 110, Taiwan
12 Pulmonary Research Center, Wan Fang Hospital, Taipei Medical University, Taipei 110, Taiwan
* Correspondence: poliwei@tmu.edu.tw (P.-L.W.); lin2511@tmu.edu.tw (J.-C.L.); Tel.: +886-2-2736-1661 (ext. 3330) (J.-C.L.)
† Ying-Chin Lin and Tso-Hsiao Chen contributed equally to this work.

Citation: Lin, Y.-C.; Chen, T.-H.; Huang, Y.-M.; Wei, P.-L.; Lin, J.-C. Involvement of microRNA in Solid Cancer: Role and Regulatory Mechanisms. *Biomedicines* **2021**, 9, 343. https://doi.org/10.3390/biomedicines9040343

Academic Editor: Francesca Lovat

Received: 25 February 2021
Accepted: 24 March 2021
Published: 29 March 2021

Publisher's Note: MDPI stays neutral with regard to jurisdictional claims in published maps and institutional affiliations.

Abstract: MicroRNAs (miRNAs) function as the post-transcriptional factor that finetunes the gene expression by targeting to the specific candidate. Mis-regulated expression of miRNAs consequently disturbs gene expression profile, which serves as the pivotal mechanism involved in initiation or progression of human malignancy. Cancer-relevant miRNA is potentially considered the therapeutic target or biomarker toward the precise treatment of cancer. Nevertheless, the regulatory mechanism underlying the altered expression of miRNA in cancer is largely uncovered. Detailed knowledge regarding the influence of miRNAs on solid cancer is critical for exploring its potential of clinical application. Herein, we elucidate the regulatory mechanism regarding how miRNA expression is manipulated and its impact on the pathogenesis of distinct solid cancer.

Keywords: microRNA; solid cancer; post-transcriptional regulation

1. Introduction

Activation of tumorigenic processes mediate the neoplasia of normal cells, potentially resulting in initiation of malignancy. Exploration of regulatory mechanism relevant to the occurrence or progression of diverse cancers can be subjected to clinical application, such as early prevention, precise screening, or personal treatment. Among the regulatory factor involved in altered gene expression associated with carcinogenesis, the influence of microRNAs (miRNAs) has been widely pursued for decades [1]. MiRNA is a 22 nt-long noncoding RNA that functions as a post-transcriptional regulator for fine-tuning the coding efficiency of messenger RNA (mRNA) [2]. Targeting of the RNA-induced silencing complex (RISC) composed of single- strand miRNA, Argonaute (AGO), and

GW182 (also known as TNRC6A) mediates the translational repression or degradation of the. regulatory candidate [3]. miRNA-mediate regulation is relevant to diverse cellular or environmental stress, including starvation, oxidative stress, hypoxia, and DNA breakdown, thereby being implicated in malignant disease [4]. Accordingly, dysregulation of miRNA expression is documented to exhibit bidirectional impact toward oncogenesis or tumor suppression [5].

Disturbance in miRNA expression can be induced via multiple steps, including transcriptional regulation, epigenetic methylation of miRNA-containing loci, miRNA processing pathway, and sequestration with long non-coding RNA, which functions as the miRNA sponge (Figure 1) [6,7]. For instance, various p53-responsive miRNAs networks containing miR-34 or miR-27b is documented to mediate the quiescence of distinct cancer cells [8,9]. The presence of mutant p53 reversely diminishes the tumor suppressive influence of p53-regulated miRNAs on carcinogenic signature [10]. Recent studies disclose the correlation between miRNA expression and epigenetic control regarding the methylation of CpG island within promoter region in cancer [11]. The silence of miR-127, miR-124-1, or miR-129-2 is closely related to the hypermethylation of CpG island-containing promoter in various solid cancer [12–14]. In addition, the function or expression of miRNA processing machinery, such as Drosha or the DGCR8 protein, is frequently deregulated in diverse malignancies [15]. Even though the impact of Drosha or DGCR8 on carcinogenesis is controversial, the disturbance of miRNA processing machinery is closely related to the global change in the miRNA expression profile [16]. The mutations of *Dicer* gene lead to DICER1 syndrome, which is relevant to the incidence of multiple cancers in an individual [17]. The biogenesis of miRNAs and gene expression profiles is disrupted with the mutant Dicer protein [18]. Recent studies document that miRNA loci are frequently annotated within the chromosomal regions, which are susceptible to the cancer-associated copy number variation (CNV) [19]. Cancer-mediated genomic instability results in the amplification or deletion of miRNA loci, subsequently leading to variation in the miRNA copy number [20].

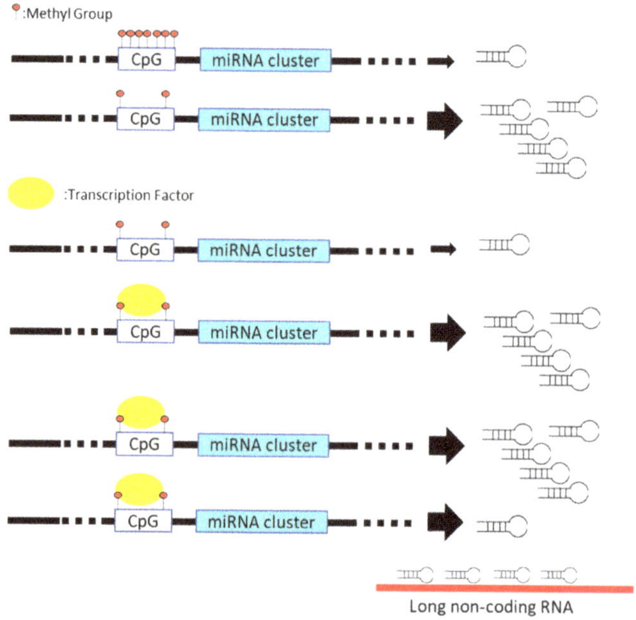

Figure 1. miRNA gene expression is manipulated via distinct mechanisms, including methylation (upper), transcription regulator (middle), and miRNA sponge (lower) in cancer cells.

According to the evidence gathered to date, we summarize the current knowledge regarding the impact of miRNA on the pathogenesis of frequent solid cancer, including colorectal cancer, lung cancer, breast cancer, and liver cancer. This review additionally evaluates the advantage or challenge associated with miRNA-based application in cancer treatment.

2. MiRNA and Solid Cancer

2.1. Colorectal Cancer

Colorectal cancer (CRC) is identified as the second leading cause of cancer-associated deaths worldwide with its high mortality and an increase in incidence [21]. In spite of progress in early screening, diagnosis, or prognostic prediction toward the incidence of CRC, miRNA is considered to be a potential biomarker for the assessment of CRC progression [22]. Decreases in the expression of miRNA clusters are frequently identified with the occurrence or development of CRC [23]. In the following section, the impact of several miRNA clusters on CRC are addressed.

2.2. Tumor-Suppressive miRNA in CRC

2.2.1. Clusters miR-1/133a and miR-206/133b

The miR-1-1/133a-2, miR-1-2/133a-1, and miR-206/133b clusters are transcribed from distinct host genes. The influence of these miRNAs on the development of striated muscle has been widely documented [24]. Hypermethylation of the promoter within the host gene results in the decrease in the expression of miR-1/133 cluster in CRC tissues [25]. Moreover, miR-133 and 206 are specifically sponged by the complementary long non-coding RNA (lncRNA) ABHD11-AS1, XIST, and multiple LINC RNAs [26–28]. With the inhibitory effect toward the carcinogenic signatures, miR-1, miR-133a, and miR-206 are considered to be the tumor suppressors targeting a wide range of specific candidates. The in vitro results indicate the repressive effect of miR-1 on proliferation, migration, motility, and metabolism of CRC cells by targeting vascular endothelial growth factor (VEGF), notch receptor 3 (NOTCH3), and hypoxia-inducible factor 1 subunit alpha (HIF1A) gene [29–31]. The presence of miR-133 is demonstrated to diminish the growth or motility of CRC cells by targeting multiple candidates, such as fascin actin-bundling protein 1 (FSCN1) and oncogenic SUMO-specific peptidase 1 (SENP1) gene [32]. Upregulated miR-206 is relevant to reduced migration, proliferation, and immortality of CRC cells via targeting formin-like 2 (FMNL2), NOTCH3, and BCL2 gene [33].

2.2.2. Clusters miR-15a/16-1 and miR-15b/16-2

miR-15 and miR-16 are encoded from two paralogues, miR-15a/16-1 and miR-15b/16-2 in the human genome. In CRC cells, an increase in the sirtuin 1 (SIRT1) protein mediates the down-regulated activity of promoter, which drives the transcription of miR-15b/16-2 cluster [34]. In addition, the mature miR-15/16 is sequestered by upregulated sponge LINC RNAs in distinct CRC cells [35,36]. Both in vivo and in vitro studies demonstrate the tumor-suppressive effect that is attributed to the presence of miR-15/16 clusters in CRC [37,38]. miR-15 and miR-16 are documented to target common candidates, including cyclin B1 and transcription factor AP-4, which majorly participated in the epithelial mesenchymal transition (EMT) [39]. Exogenous expression of miR-15 is reported to suppress the growth of CRC cells by targeting pro-survival BCL2 protein [40]. The presence of overexpressing miR-16 mediates the decrease in mortality and growth of CRC cells via suppressing the expression of KRAS proto-oncogene, GTPase (KRAS) protein both in vivo and in vitro [41]. Moreover, the reverse correlation between miR-16 level and VEGF receptor or MYB proto-oncogene is relevant to the prognosis of CRC patient [42].

2.2.3. Clusters miR-100/let-7a/miR-125 and miR-99/let-7c

miR-100/let-7a/miR-125 and miR-99/let-7c are separately transcribed from an independent host gene in human genome but categorized in the same family with the related function. Nevertheless, the regulatory mechanism contributes to the alteration of the abovementioned miRNA clusters in CRC cells or tissues not comprehensively disclosed. For example, the abundance of sponge lncRNA is reversely relevant to the level of let-7a in CRC model [43]. The expressions of miR-125a/125b are manipulated through a complex process, including transcriptional control, hypermethylation-regulated epigenetic regulation, or sequestration with lncRNAs [44,45]. Nevertheless, the decreases in these miRNAs are frequently identified in CRC patients with poor prognosis [46]. An increase in the let-7 members mediates the cell cycle arrest and diminished cell growth via targeting of PHD and ring finger domains 2, Rho effector rhotekin, insulin-like growth factor 1, or MYC genes [47–49]. Upregulation of let-7c or let-7e lessens the metastatic activity of CRC cells through targeting the candidates encoding matrix metallopeptidase 11, PBX homeobox 3, and double cortin-like kinase 1 protein [50,51]. Furthermore, the impact of the let-7 family member on the abovementioned candidates sensitizes the CRC cells toward the chemo- or radiotherapy [52]. The expression profiles of miR-99a/99b are relevant to the level of the mechanistic target of rapamycin kinase (MTOR) protein in CRC cell lines [53]. Overexpression of the miR-125 family member facilitates the apoptosis of CRC cells by targeting the related factor, such as BCL2, BCL2 family members like 12, and myeloid cell leukemia 1 (Mcl-1) gene [54]. Moreover, an increase in miR-125a level leads to the suppression of angiogenic or metastatic activity of CRC cells by targeting VEGFA, SMAD-specific E3 ubiquitin protein ligase 1, and cAMP-responsive element-binding protein 5 gene [55,56].

2.3. *Lung Cancer*

Lung cancer (LC) is the leading cause of cancer-associated deaths worldwide, resulting in over 1 million deaths annually [57]. It is classified into two subsets according to the pathological signature—small cell lung cancer and non-small cell lung cancer. The complex mechanism involved in tumor initiation or progression is not comprehensively elucidated, which impedes the application of gene-based screening. A growing body of evidence indicates the association between the occasion of LC and the altered miRNA expression, which can be divided into oncomiR and tumor-suppressive miRNAs.

2.3.1. Oncogenic miRNAs in LC Cells

MiRNA is widely considered a pivotal regulator in the control of cells growth [58]. With an increase in MYC expression, mis-regulated amplification of human miR-17-92 cluster, composed of six miRNAs, is noted in various solid tumor, including LC. An increase in the miR-17 cluster frequently leads to upregulated cell proliferation by targeting antiapoptotic factors, including transcription factor E2F1 or Phosphate and tensin homolog (PTEN) protein [59]. Immortality is a hallmark of cancerous cells, which is closely correlated with the decrease in p53 protein as previously illustrated [8,9]. In LC cells, the transcription of miR-34 family members is directly interfered with by the reduced p53 expression [60]. The cell cycle is subsequently disturbed by the augmentation of cyclin E2 and cyclin-dependent kinase, which is targeted by the miR-34 family in normal cells [60]. In contrast, the ectopic expression of miR-125b or miR-504 is both demonstrated to target p53 gene and in turn lessens the apoptotic sensitivity of LC cells toward environmental stress [61,62]. Metastasis of cancerous cells constitutes a predominant cause, which leads to the majority of cancer deaths [63]. Lung cancer is frequently diagnosed with the formation of metastases in the brain, bones, liver, and adrenal glands [64]. Activation of metastasis is associated with expression profiles of internal factors as well as external regulators involved in EMT process [65,66]. An increase in miR-10b level functions as a major factor attributed to active metastasis or enhance other oncogenic signatures of NSCLC cells [67,68].

2.3.2. Tumor-Suppressive miRNAs in LC

Angiogenesis functions as a critical process toward the initiation and growth of solid tumor [69]. Throughout the process, the expression of VEGF and related regulation in response to hypoxia, such as the Akt/eNOS pathway, constitutes a crucial mechanism [70].

The presence of overexpressing miR-128 is demonstrated to exert repressive influence toward the phosphorylation of PI3K and p38 MAPK signaling, in turn lessening the levels of VEGFA, VEGF receptor (VEGFR) 2, and VEGFR3 [71]. The suppressive effect of miR-206 on angiogenic activity of NSCLC is achieved by targeting STAT3, HIF-1, or VEGF pathway [72]. The suppressive impact of miR-135a on IGF-1 expression further mediates the decreases in angiogenesis-related factors, including VEGF, bFGF, and IL-8 protein in A549 cells [73]. The decreases in various miRNAs mediate the loss of the tumor-suppressive effect on metastasis of LC. For instance, the restoration of miR-126 generation consequently inhibits the metastatic activity of NSCLC cells via targeting the chemokine receptor 1 gene [74]. Targeting of the ectopically expressing miR-192-5p on the TRIM44 gene is highly relevant to the reduced metastasis of LC cells, which is associated with the inactivation of the AKT/mTOR signaling pathway [75]. A decrease in miR-7-5p lessens its effect on repressing expression of NOVA2, which acts as an important regulator participating in angiogenesis and growth of NSCLC [76]. miR-206 exerts its tumor-suppressive impact on NSCLC metastases via targeting the actin-binding protein coronin 1C gene, which mediates the growth and metastasis in other solid cancers as well [77]. Overexpression of the miR-335 leads to reduction of the TGFβ-mediated EMT process in NSCLC by downregulating the level of ROCK1 gene, which plays an activator role in the PI3K/AKT/FAK pathway [78]. The presence of the miR-98 interferes with the translational activity of the TGFβR1 gene, subsequently diminishing the proliferation, migration, and invasion in distinct LC cell lines [79]. In contrast, the let-7 family or miR-126 is demonstrated to exert a suppressive impact on the proliferation of LC cells [80]. Exogenous expression of let-7 family member is reported to diminish proliferation of the LC cells via targeting the ras gene [81]. In NSCLC cells, the presence of miR-126 is relevant to the downregulated activity of PTEN/PI3K/AKT signaling, which critically control the cell growth [82].

2.4. Breast Cancer

Breast cancer (BC) is classified as the major cause in terms of high morbidity and mortality in women worldwide [83]. Obesity is often related to an aberrantly high level of estrogen, estrogen receptor (ER), progesterone receptor (PR), or human epidermal growth factor receptor (HER2), which constitutes the predominant mechanism for the initiation and progression of BC. Accordingly, miRNA involved in the regulation of adipogenesis is considered to be a potential factor to BC development [84]. In addition to CRC, mis-regulated expression miRNA cluster is frequently identified in BC [85].

2.4.1. Oncogenic miRNAs in BC

An increase in this miR-183/96/182 cluster, which is located at 7q32.2, is identified in a variety of malignancies, including BC [86]. Bi-directional function of the miRNA cluster for oncogenesis or tumor suppressor is documented with the results of functional assays. Nevertheless, the alteration of this cluster is related to various cell process, such as apoptotic response, DNA repair, metabolism, or EMT process [87]. The transcription of the miR-183/96/182 cluster is activated with the increases in ZEB1 (zinc finger E-box binding homeobox 1) and HSF2 (heat shock transcription factor 2) protein in BC [86]. Moreover, activation of the HIF1 or PI3K/Akt pathway constitutes an additional mechanism attributed to the upregulation of the miR-183/96/182 cluster [88]. A growing body of study demonstrates that the presence of miR-183 manipulates the expression profiles composed of 45 genes in BC, including Programmed Cell Death 4, Early Growth Response1, Integrin Subunit Beta 1, p21, and p27, which is relevant to the abovementioned cell processes [89,90]. The presence of upregulated miR-96 facilitates the proliferation and migration of BC cells via targeting the protein tyrosine phosphatase, non-receptor type 9 gene [91]. Moreover, increases in the miR-96 and miR-182 synchronously interfere with the translation of FOXO1 protein, which maintains the homeostasis of glucose metabolism in normal lineages [92]. The members of the miR-183/96/182 cluster are incidentally noted in the exosome collected from the BC patients. Taken together, the miR-183/96/182 cluster mainly functions as the oncogenic factor toward the development of BC.

The miR-221/222 cluster is transcribed from Xp11.3 to encode two homologous miRNA members. The increases in miR-221/222 are highly relevant to the active cell proliferation, EMT process, and metastatic activity in BC cells via targeting a variety of tumor suppressors, including Transcriptional Repressor GATA Binding 1, Adiponectin Receptor 1, Suppressor of Cytokine Signaling 1, Cyclin Dependent Kinase Inhibitor 1B, ERα, p27, and TIMP Metallopeptidase Inhibitor 3 genes [93]. Upregulation of miR-221/222 leads to the transformation of ER-α positive tumors to ER-α negative BC by targeting the ER-α gene [94]. Moreover, the aberrant expression of the miR-221/222 cluster is advantageous to self-renewal of the BC stem cell self by targeting the PTEN-Akt signaling pathway [95]. Taken together, the miR-221/222 acts as a predominant oncogene in the origin of ER-negative BC with the aggressive signature.

2.4.2. Tumor-Suppressive miRNAs in BC

The miR-199a/214 cluster is annotated at chromosome1q24 to generate miR-199a-5p, miR-199a-3p, and miR-214 transcripts. The hedgehog-signaling Vitamin D Receptor and miR-214 are demonstrated to constitute a cross-talk axis involved in the downregulation of miR-199/214 expressions in BC cells [96]. The decrease in the miR-199/214 cluster results in the loss of its tumor-suppressive effect in triple negative BC (TNBC). For instance, depletion of the miR-199a/214 cluster induces the EMT-like phenotype in normal cell lineage [85]. Overexpression of the miR-199a/214 cluster reversely lessens the proliferative activity of TNBC cells by targeting hedgehog signaling, which is proposed as a potential therapeutic strategy toward BC [96]. The H3K27me3-related epigenetic silencing is associated with the levels of the Enhancer of zeste homolog 2 (Ezh2) protein, which induces neoplasia in various tissues. The inverse correlation between the miR-199/214 and Ezh2 or Ki-67 protein is identified by using an in vitro BC cell model [97]. The impact of miR-199/214 on interfering with the generation of Ezh2, β-catenin, or Ki-67 constitutes another mechanism for diminishing the proliferative activity of BC cells [98,99]. Accordingly, the tumor-suppressive influence of miR-199/214 cluster on inhibiting proliferation, migration, and invasion of BC cells is documented with majority of the present studies.

2.4.3. Oncogenic and Tumor-Suppressive of miR-23/27/24 Cluster

Two paralogs of this cluster, miR-23a/27a/24-2 and miR-23b/27b/24-1, are respectively annotated in chromosomes 19 and 9. The results of functional studies suggest that the transcribed member of the cluster executes both oncogenic and tumor-suppressive effects on the carcinogenic signature of BC cells. Upregulated transcription of miRNA-23/27/24 induces the progression of BC cells via targeting the Hypermethylated in Cancer 1 gene, which functions as a repressor to tumor growth [100]. Nevertheless, the impact of miR-23b and miR-27b targeting on tumor growth is controversial, based on the in vitro cultured or in vivo nude mice model [101]. Moreover, the members of this cluster share a variety of candidates, such as Sprouty RTK Signaling Antagonist 2, BCL2 Antagonist/Killer, PPARγ, and Nischarin gene [102]. This miRNA cluster also exerts its influence by collaborating with other miRNAs on BC development. For instance, FOXO1 expression is coordinatively regulated by the miRNA-23/27/24 cluster along with miR-27a, miR-96, or miR-182 in BC [92]. In contrast, targeting of miR-27a to the Transmembrane Protein 170B gene is demonstrated to suppress the Wnt/β-catenin pathway, in turn lessening the proliferative and migratory efficiency of BC cells [103]. Moreover, targeting of the overexpressing miR-23b-3p to PAK2 and phosphorylation of the myosin light chain II gene subsequently reduces the metastatic activity of BC cells in vitro [104]. Taken together, the comprehensive insight into the manipulation of miRNA-23/27/24 expression and its coordination with other miRNA is useful in controlling BC development.

2.5. Liver Cancer

Hepatocellular carcinoma (HCC) is identified as the third cause of cancer mortality in the Asia-Pacific region, which consequently accounts for over 80% of liver cancers [105]. In addition to viral infection, aflatoxin-contaminated food-, alcohol-, or obesity-mediated cirrhosis, the impact of miRNA is demonstrated in the context of HCC [106]. With its potential to manipulate gene expression profiles, miRNA is practicable as a novel therapeutic or emerging biomarker for the stratification of HCC patient.

2.5.1. Oncogenic miRNAs in HCC

miR-21 is a well-characterized oncomiR in a variety of solid tumors, including HCC [107]. An increase in miR-21 level with the incident of HCC is consistently revealed by using in vitro and in vivo experiments [108]. Aberrant expression of miRNA-21 with a concomitant decrease in tumor-suppressive PTEN protein leads to activation of focal adhesion kinase, Akt, and mTOR signaling, in turn resulting in active cell metastasis and proliferation [109]. Activation of a series kinase cascade is associated with the increase in MMP-2 and MMP-9, which further strengthens the pathogenesis and development of HCC [109]. An elevated level of miR-21 in plasma potentially provides high accuracy toward the screening or diagnosis of early-stage HCC [110].

miR-221 is identified as another oncomiR with an upregulated level in HCC [111]. By directly targeting the cell cycle inhibitors, including CDKN1B/p27 and CDKN1C/p57, the influence of miR-221 on active progression of HCC is consistently disclosed in multiple studies [112]. Direct targeting of miR-221 to the DNA damage-inducible transcript 4 gene relieves its tumor-suppressive activity in HCC model [111]. Moreover, the presence of miR-221 strengthens the anti-apoptosis activity of HCC cells by manipulating the level of the Bcl2 modifying factor protein (BMF), the pro-apoptotic member of the Bcl-2 family [113]. In contrast, inhibition of miR-221 increases the susceptibility of HCC to apoptotic stimuli with an elevated level of BMF and the downstream caspase pathway [113].

2.5.2. Tumor-Suppressive miRNAs in HCC

miR-29 family is comprised of four members, miRNA-29a, miRNA-29b-1, miRNA-29b-2, and miRNA-29c, which are transcribed from chromosomes 7q32.3 and 1q32.2. An miRNA profiling study first identified the downregulation of miR-29 in LC patients associated with poor prognosis and survival rate [114]. By using an in vitro cultured system, activation of TGF-β or nuclear factor kappa B (NF-κB) drives the downregulation of miRNA-29, in turn facilitating the expressions of extracellular matrix genes in hepatic stellate cells [115]. miR-29 family members are disclosed to share a variety of candidates, including CDC42, PIK3R1, Bcl-2, and Mcl-1 gene [116]. In line with these results, targeting of endogenous Bcl-2 and Mcl-1 gene by miR-29 family member sensitizes the HepG2 cells to apoptosis-induced chemotherapeutic treatment or serum starvation [116]. In addition to miR-29, the inverse correlation between Mcl-1 and miR-101 or miR-125b is also identified in liver tumor cells [117,118]. Similarly, respective targeting of miR-101 or miR-125b share the same effect on sensitizing the liver cancer cells with the presence of apoptotic treatment via targeting Mcl-1 gene. miR-122 is the most abundant miRNA, which fine-tunes a variety of cellular processes in hepatocyte [119]. The transcription of miRNA-122 is controlled by the expression of liver-specific factors, including CCAAT/enhancer-binding protein (C/EBP) and hepatocyte nuclear factor (HNF) family members [120]. A decrease in miRNA-122 level associated with hepatocarcinogenesis, active metastasis, and poor prognosis is identified in HCC tissue [121]. The presence of miRNA-122 exerts the tumor suppressive impact on HCC development via targeting cyclin G1, pyruvate kinase isoform M2, and Wnt family member 1 gene [112]. Th miR-122/cyclin G1 axis facilitates the stability of p53 and therefore promotes the sensitivity of HCC cells to doxorubicin-induced apoptosis [112].

2.6. Ovarian and Cervical Cancer

Ovarian cancer and cervical cancer is ranked as the first and fourth cause of cancer death in women worldwide [122,123], respectively. Around 65% of ovarian cancer is classified as surface epithelial according to the World Health Organization (WHO) classification [124]. Even though understanding the biological signature of ovarian cancer was gradually revealed with the recent progress and knowledge, the severity or mortality of ovarian cancer remains unchanged for the last 30 years. A promising biomarker, such as the cancer-specific miRNA, toward early screening or precise diagnosis of ovarian cancer is crucial. On the other hand, the majority of cervical cancer is initiated by the infection of certain subtypes of the human papilloma virus (HPV) [125]. The relevance of HPV infection with the chronic inflammation that subsequently mediates the initiation of cervical carcinogenesis remains controversial. Recent studies document that the aberrant miRNAs profile is identified throughout the initiation and development of cervical cancer [126]. Moreover, the altered miRNA level exhibits influence on manipulating the carcinogenic process of cervical cancer.

2.6.1. Oncogenic miRNA in Ovarian Cancer and Cervical Cancer

Metastasis of ovarian cancer is modulated via the interplay between miRNA and signaling factors involved in the EMT pathway [127]. An increase in miR-17-5p level is relevant to the active progression and EMT activity of ovarian cancer cells by targeting PTEN expression and downstream signaling [128]. In contrast, the administration of an miR-17-5p inhibitor interferes with the migration and invasion activity of ovarian cancer cells by using in vitro cultured assays [128]. The presence of miR-214 targets the PTEN expression, which in turn reduces the sensitivity of in vitro cultured ovarian cancer cells to cisplatin [129].

In cervical cancer, a growing body of studies demonstrate the oncogenic influence of miR-21 with the identification of diverse targets, including *PDCD4, PTEN, TIMP-3, TNF-α,* and *ANXA1* genes [130–132]. Therefore, upregulated miR-21 is highly related to the active inflammation and metastasis of cervical cancer cells. The upregulated level of miR-155 in peripheral blood and tissues collected from cervical cancer patients has been revealed in recent studies. The presence of miR-155 targets the expression of SOSC1, which in turn enhances progression and inflammation in cervical cancer [133].

2.6.2. Tumor-Suppressive miRNA in Ovarian Cancer and Cervical Cancer

A decrease in miR-150 is frequently identified in epithelial ovarian cancer cells [134]. The presence of overexpressing miR-150 lessens the invasive and metastatic activity of ovarian cancer cells by targeting the expression of Zinc Finger E-Box Binding Homeobox 1 (ZEB1) protein [135]. miR-150 is suggested as a potential therapeutic target for intervening the metastasis of ovarian cancer [135]. Similarly, the upregulation of miR-22, miR-183, or miR-31 level is reported to result in reduced migration or invasion of serous ovarian carcinoma by interfering with the expression of the TIAM1 protein [136]. Moreover, overexpressing miR-7 directly targets the expression of EGFR protein, which leads to reversion of the EMT signature in ovarian cancer through AKT and ERK1/2 pathways [137].

Multiple miRNAs exert a suppressive effect on the chronic inflammation, which is crucial for the development of cervical cancer [138]. Downregulation of miR-429 is relevant to the active inflammation in cervical cancer tissues with IL-6 and IFN-β production, which is driven through the NF-κB pathway [139]. IKKβ (the primary kinase toward NF-κB activation) is identified as a new target of miR-429 in cervical cancer cells [139]. miR-101 is documented to exert a tumor-suppressive impact on the proliferation, invasion, and inflammation of cervical cancer cells by directly targeting COX-2 protein [140]. The reverse association of the high mobility group box 1 (HMGB1) with miR-34a, miR-1284, and miR-142 is identified in the cervical cancer cells [141–143]. HMGB1 is a well-characterized oncogene that is involved in chronic inflammation, progressive tumorigenesis, active metastasis, and therapy resistance of cervical cancer tissues. Downregulation of miR-24, miR-451, let-7a, and miR-125a is noted as well in cervical cancer, which is relevant to the active inflammation. An increase in chitinase-3-like protein 1 with a concomitant decrease in miR-24 is proposed to facilitate the proliferation, metastasis, and inflammation in cervical cancer [144]. The tumor-suppressive impact of miR-451 on lessening the inflammation, invasion, angiogenesis, and proliferation of cervical cancer cells is demonstrated by directly targeting the expression of the IL-6 receptor [144]. The generation of HPV oncoprotein E6 mediates the decreases in let-7a and miR-125, in turn relieving the suppressive effect of let-7a and miR-125 on STAT3 expression [145,146]. The presence of STAT3 further facilitates expression of the HPV E6 protein through transcriptional regulation [145]. Taken together, STAT3 and HPV E6 constitute a feed-forward circuit that participates in the downregulation of let-7a and miR-125 throughout the development of cervical cancer. Reversely, the complementation of miR-125 leads to the decreases in STAT3, MMP-9, MMP-2, and N-cadherin levels and activities, subsequently diminishing the proliferation, metastasis, and inflammation of cervical cancer cells [146].

3. Role of Exosomal miRNA and Its Application

Exosomes are extracellular vesicles ranging from 30 to 150 nm in size that can be secreted by normal or cancer cells [147]. Secretion of exosome is demonstrated to deliver a messenger, including protein or miRNA, between normal and cancer cells [148]. It is documented that cancer cells secrete 10-fold more exosomes than that of normal cells, which is critical to the recruitment and development of carcinogenic environment [149]. Taking lung cancer as an example, miR-96-containing exosomes secreted from H1299 cells were demonstrated to exhibit oncogenic activity toward upregulated cell proliferation by directly targeting the production of the LIM-domain only protein 7 (LMO7) expression [150]. The presence of exosomal miR-23a secreted by lung cancer cells was documented to facilitate tumor angiogenesis under both normoxia and hypoxia conditions, suggesting that the genetic messenger was transmitted from lung cancer cells to distant endothelial cells [151]. Furthermore, the drug resistance of lung cancer cells is closely related to the existence of exosomal miRNA. A recent study reported that the gemcitabine-resistant A549 (A549-GR) cells assembled miR-222-3p-containing exosomes, which were transmitted into parental gemcitabine-sensitive cells and subsequently promoted their migration, invasion, and gemcitabine resistance by targeting the expression of the SOCS3 protein [152]. Additionally, the transfer of the miR-21-containing exosome assemble by gefitinib-resistant H827R cells to HCC827 cells activated AKT signaling and lead to gefitinib resistance of parental gefitinib-sensitive cells [153].

Taken together, exosomal miRNA is considered an ideal tool for diagnosis as well as therapeutic targets with its influence on the carcinogenic pathway or environment. As for lung cancer cells, five exosomal miRNAs, including miR-205, miR-19a/19b, miR-30b, and miR-20a, were considered the diagnostic markers of squamous cell lung carcinoma (SQCLC) with their decreases in the circulating levels after surgery [154]. Three SQCLC-related miRNAs, including miR-10b-5p, miR-15b-5p, and miR-320b, were demonstrated to be promising biomarkers with area under the ROC curve (AUC) values between 0.936 and 0.911 toward diagnosis of the disease [155]. In addition to diagnosis or early prediction, exosome is demonstrated as an ideal vehicle for drug delivery within recent studies [156]. For instance, the usage of engineered exosomes containing miR-21 sponge was reported to mediate a decrease in miR-21 in U87-MG glioma cell lines, in turn lessening its carcinogenic signature by relieving the miR-21-mediated suppression on the expression of PDCD4 and RECK protein [156]. Nevertheless, the problems of miRNA-mediated influence, such as off-target phenomenon, is not ignored with the promising reports.

4. Conclusions and Perspectives

MiRNAs are involved in diverse cell process to maintain the homeostasis of normal cells through complex network (Tables 1 and 2). This phenomenon makes aberrant miRNA profiles interesting biomarkers for initiation or progression of solid tumor as well as potential targets for precise treatment. Nevertheless, precise targeting or site-specific delivery of solid cancer-specific miRNA could be a major impediment in the use of miRNA-based therapy. In addition, a better knowledge regarding the off-target effect and comprehensive assessment of toxicity is another critical concern to be solved. In this review, we summarize the present knowledge of miRNA-mediated influence on the carcinogenic signature of frequent solid tumor, which may highlight a potential opportunity for clinical translation and potential application.

Table 1. The impact of classified oncomiR on the carcinogenic signature through the specific target.

miRNA	Disease	Manipulating Mechanism	Candidate	Physiological Influence	Reference
miR-17	LC; Ovarian cancer	CNVs; Transcriptional control (MYC)	E2F1, PTEN	Cell Growth, apoptosis, metastasis	[55,128]
miR-125b	LC	Epigenetic control	p53	Apoptosis	[60]
miR-504	LC	Transcriptional control (EGFR signaling)	p53	Apoptosis	[61]
miR-10b	LC	Epigenetic control	Homeobox D10,	Metastasis	[71,72]
miR-183	BC	Transcriptional control (ZEB1, HSF2)	PCD4, EGR1, p21, p27	Apoptosis, DNA repair, metabolism, EMT	[84–86]
miR-96	BC	miRNA sponge	FOXO1, PTPN9	Proliferation, migration, metabolism	[87]
miR-182	BC	miRNA sponge	FOXO1	Metabolism	[88]
miR-221/222	BC	Epigenetic control, miRNA sponge, transcriptional control (TGF-β)	Transcriptional Repressor GATA Binding 1, Adiponectin Receptor 1, Suppressor of Cytokine Signaling 1, Cyclin Dependent Kinase Inhibitor 1B, ERα, p27, and TIMP Metallopeptidase Inhibitor 3	Proliferation, EMT process, metastasis	[89]
miR-23/27/24	BC	Transcriptional control (HIC1)	Hypermethylated in Cancer 1 (HIC1), Sprouty RTK Signaling Antagonist 2, BCL2 Antagonist/Killer, PPARγ, Nischarin, Transmembrane Protein 170B, PAK2	Cell growth, migration	[88,99,100]
miR-21	HCC; Cervical Cancer	miRNA sponge	PTEN, MMP2, MMP9, PDCD4, PTEN, TIMP-3, TNF-α, ANXA1	Metastasis and proliferation	[105,130–132]
miR-221	HCC	Transcriptional control (NF-κB)	CDKN1B/p27, CDKN1C/p57 DNA damage-inducible transcript 4, BMF	Cell growth, apoptosis	[108,109]
miR-214	Ovarian Cancer	Transcriptional control (hedgehog signaling)	PTEN	Metastasis, chemoresistance	[129]
miR-155	Cervical Cancer	miRNA sponge, transcriptional control (c-MYB)	SOSC1	Inflammation	[133]

Table 2. The impact of classified tumor-suppressive miRs on carcinogenic the signature through the specific target.

miRNA	Disease	Manipulating Mechanism	Candidate	Physiological Influence	Reference
miR-1	CRC	Epigenetic control; miRNA sponge	VEGF, NOTCH3	Proliferation, migration, motility and metabolism	[25–27]
miR-133	CRC	Epigenetic control; miRNA sponge	FSCN1, SENP1	Growth or motility of CRC cells	[28]
miR-206	CRC, LC	Epigenetic control; miRNA sponge	FMNL2, NOTCH3, BCL2, STA3, HIF-1, Coronin 1C	Migration, proliferation, and immortality, metastasis	[29,65,76]
miR15/16	CRC	Transcriptional control (SIRT1)	cyclin B1, TFAP-4, Bcl-2,K-Ras, MYB	Epithelial mesenchymal transition (EMT), apoptosis	[35–38]
let-7 family	CRC; Cervical cancer	miRNA sponge	PHD, ring finger domains 2, RTKN, IGF-1, MYC, MMP11, PBX3, DCLK1, STAT3	Cell cycle arrest, metastasis	[43–47,145,146]
miR-125	CRC Cervical Cancer	miRNA sponge	Bcl-2, Mcl-1, SMURF1, VEGFA, CREB5, STAT3, MMP-9, MMP-2, N-cadherin	Apoptosis, angiogenic or metastatic activity, inflammaation	[50–52,146]
let-7 family	LC	miRNA sponge	Ras	Proliferation	[57]
miR-126	LC	miRNA sponge	PTEN, CX3CR1	Proliferation, metastasis	[58,73]
miR-34	LC; Cervical Cancer	Transcriptional control (p53)	Cyclin E2, HMGB1	Cell cycle arrest	[59,141]
miR-128	LC	miRNA sponge	VEGFA, VEGFR2, VEGFR3	Angiogenesis	[64]
miR-135a	LC	miRNA sponge	IGF-1	Angiogenesis	[66]
miR-192	LC	Transcriptional control (p53)	TRIM44	Metastasis	[74]
miR-7	LC; Ovarian cancer	miRNA sponge	Nova2, EGFR	Angiogenesis, EMT	[75,137]
miR-335	LC	miRNA sponge	ROCK1	EMT	[77]
miR-98	LC	miRNA sponge	TGF β R1	Proliferation, migration, and invasion	[78]
miR-199	BC	Transcriptional control (hedgehog signaling)	Ezh2, β-catenin, Ki-67	Proliferation, migration, and invasion	[93–95]
miR-214	BC	Transcriptional control (hedgehog signaling)	Ezh2, β-catenin, Ki-67	Proliferation, migration, and invasion	[93–95]
miR-29	HCC	Transcriptional control (NF-κB, TGF-β)	CDC42, PIK3R1, Bcl-2, Mcl-1	Cell cycle, apoptosis	[112]
miR-101	HCC; Cervical Cancer	miRNA sponge	Mcl-1, Cox-2	Apoptosis, inflammation, proliferation, invasion	[115,140]
miR-125b	HCC	Epigenetic, transcriptional control; miRNA sponge	Mcl-1	Apoptosis	[115]
miR-122	HCC	Transcriptional control (C/EBP, HNF)	Cyclin G1, PKM2, and Wnt family member 1	Cell cycle, apoptosis	[108]
miR-150	Ovarian cancer	miRNA sponge	ZEB1	Invasion, metastasis	[134,135]
miR-22, miR-183, miR-31	Ovarian cancer	Transcriptional control (Snail)	TIAM1	Invasion, migration	[136]
miR-429	Cervical Cancer	miRNA sponge	IL-6,IFN-β	Chronic inflammation	[139]
miR-142	Cervical Cancer	miRNA sponge	HMGB1	Chronic inflammation, progressive tumorigenesis, active metastasis	[141–143]
miR-24	Cervical Cancer	Uncertain	chitinase-3-like protein 1	Proliferation, metastasis, inflammation	[144]
miR-451	Cervical Cancer	Uncertain	IL-6 receptor	Inflammation, invasion, angiogenesis, proliferation	[144]

Author Contributions: This work was conceived by J.-C.L. and P.-L.W. The original draft preparation and writing: Y.-C.L.; T.-H.C.; Y.-M.H.; P.-L.W.; J.-C.L. Review and editing: J.-C.L. All authors have read and agreed to the published version of the manuscript.

Funding: This work was supported by a grant (MOST 109-2622-B-038-002-CC3) from the Ministry of Science and Technology, Taiwan.

Institutional Review Board Statement: Not applicable.

Informed Consent Statement: Not applicable.

Data Availability Statement: Not applicable.

Conflicts of Interest: The authors declare no conflict of interest.

References

1. Akgül, B.; Erdoğan, I. Intracytoplasmic Re-localization of miRISC Complexes. *Front. Genet.* **2018**, *9*, 403. [CrossRef]
2. O'Brien, J.; Hayder, H.; Zayed, Y.; Peng, C. Overview of MicroRNA Biogenesis, Mechanisms of Actions, and Circulation. *Front. Endocrinol.* **2018**, *9*, 402. [CrossRef]
3. Bartel, D.P. MicroRNAs: Target Recognition and Regulatory Functions. *Cell* **2009**, *136*, 215–233. [CrossRef]
4. Di Leva, G.; Garofalo, M.; Croce, C.M. MicroRNAs in Cancer. *Annu. Rev. Pathol. Mech. Dis.* **2014**, *9*, 287–314. [CrossRef]
5. Ventura, A.; Jacks, T. MicroRNAs and Cancer: Short RNAs Go a Long Way. *Cell* **2009**, *136*, 586–591. [CrossRef]
6. Menigatti, M.; Staiano, T.; Manser, C.N.; Bauerfeind, P.; Komljenovic, A.; Robinson, M.; Jiricny, J.; Buoli, F.; Marra, G. Epige-netic silencing of monoallelically methylated miRNA loci in precancerous colorectal lesions. *Oncogenesis* **2013**, *2*, e56. [CrossRef]
7. Cui, M.; Wang, H.; Yao, X.; Zhang, D.; Xie, Y.; Cui, R.; Zhang, X. Circulating MicroRNAs in Cancer: Potential and Challenge. *Front. Genet.* **2019**, *10*, 626. [CrossRef] [PubMed]
8. Wong, M.Y.M.; Walsh, W.R.; Yu, Y.; Yang, J.-L. microRNA-34 family and treatment of cancers with mutant or wild-type p53 (Review). *Int. J. Oncol.* **2011**, *38*, 1189–1195. [CrossRef] [PubMed]
9. Mu, W.; Hu, C.; Zhang, H.; Qu, Z.; Cen, J.; Qiu, Z.; Li, C.; Ren, H.; Li, Y.; He, X.; et al. miR-27b synergizes with anticancer drugs via p53 activation and CYP1B1 suppression. *Cell Res.* **2015**, *25*, 477–495. [CrossRef] [PubMed]
10. Ju, Q.; Zhao, L.; Gao, J.; Zhou, L.; Xu, Y.; Sun, Y.; Zhao, X. Mutant p53 increases exosome-mediated transfer of miR-21-3p and miR-769-3p to promote pulmonary metastasis. *Chin. J. Cancer Res.* **2019**, *31*, 533–546. [CrossRef]
11. Bednarczyk, M.; Fatyga, E.; Dzięgielewska-Gęsiak, S.; Waniczek, D.; Grabarek, B.; Zmarzły, N.J.; Anikowska, G.; Muc-Wierzgoń, M. The Expression Paterns of BECN1, LAMP2, and PINK1 Genes in Colorectal Cancer Are Potentially Regu-lated by Micrornas and CpG Islands: An in Silico Study. *J. Clin. Med.* **2020**, *9*, 4020. [CrossRef]
12. Saito, Y.; Liang, G.; Egger, G.; Friedman, J.M.; Chuang, J.C.; Coetzee, G.A.; Jones, P.A. Specific activation of microRNA-127 with downregulation of the proto-oncogene BCL6 by chromatin-modifying drugs in human cancer cells. *Cancer Cell* **2006**, *9*, 435–443. [CrossRef]
13. Liang, Y.-J.; Wang, Q.-Y.; Zhou, C.-X.; Yin, Q.-Q.; He, M.; Yu, X.-T.; Cao, D.-X.; Chen, G.-Q.; He, J.-R.; Zhao, Q. MiR-124 targets Slug to regulate epithelial–mesenchymal transition and metastasis of breast cancer. *Carcinogenesis* **2012**, *34*, 713–722. [CrossRef]
14. Tsai, K.-W.; Wu, C.-W.; Hu, L.-Y.; Li, S.-C.; Liao, Y.-L.; Lai, C.-H.; Kao, H.-W.; Fang, W.-L.; Huang, K.-H.; Chan, W.-C.; et al. Epigenetic regulation of miR-34b and miR-129 expression in gastric cancer. *Int. J. Cancer* **2011**, *129*, 2600–2610. [CrossRef]
15. Wegert, J.; Ishaque, N.; Vardapour, R.; Georg, C.; Gu, Z.; Bieg, M.; Ziegler, B.; Bausenwein, S.; Nourkami, N.; Ludwig, N.; et al. Mutations in the SIX1/2 pathway and the DROSHA/DGCR8 miRNA microprocessor complex underlie high-Risk blastemal typeWilms tumors. *Cancer Cell* **2015**, *27*, 298–311. [CrossRef]
16. Hata, A.; Kashima, R. Dysregulation of microRNA biogenesis machinery in cancer. *Crit. Rev. Biochem. Mol. Biol.* **2016**, *51*, 121–134. [CrossRef]
17. De Kock, L.; Rivera, B.; Revil, T.; Thorner, P.; Goudie, C.; Soglio, D.B.-D.; Choong, C.S.; Priest, J.R.; Van Diest, P.J.; Tanboon, J.; et al. Sequencing of DICER1 in sarcomas identifies biallelic somatic DICER1 mutations in an adult-onset embryonal rhabdomyosarcoma. *Br. J. Cancer* **2017**, *116*, 1621–1626. [CrossRef]
18. Murray, M.J.; Bailey, S.M.; Raby, K.L.; Saini, H.K.; De Kock, L.A.A.; Burke, G.; Foulkes, W.D.; Enright, A.J.; Coleman, N.; Tischkowitz, M. Serum levels of mature microRNAs in DICER1-mutated pleuropulmonary blastoma. *Oncog.* **2014**, *3*, e87. [CrossRef]
19. Waldman, S.A.; Terzic, A. A study of microRNAs in silico and in vivo: Diagnostic and therapeutic applications in cancer. *FEBS J.* **2009**, *276*, 2157–2164. [CrossRef]
20. Lee, Y.S.; Dutta, A. MicroRNAs in Cancer. *Annu. Rev. Pathol. Mech. Dis.* **2009**, *4*, 199–227. [CrossRef]
21. Siegel, R.L.; Miller, K.D.; Goding Sauer, A.; Fedewa, S.A.; Butterly, L.F.; Anderson, J.C.; Cercek, A.; Smith, R.A.; Jemal, A. Colorectal cancer statistics. *CA Cancer J. Clin.* **2020**, *70*, 145–164. [CrossRef]
22. Dave, V.P.; Ngo, T.A.; Pernestig, A.-K.; Tilevik, D.; Kant, K.; Nguyen, T.; Wolff, A.; Bang, D.D. MicroRNA amplification and detection technologies: Opportunities and challenges for point of care diagnostics. *Lab. Investig.* **2018**, *99*, 452–469. [CrossRef] [PubMed]

23. Cheng, L.; Geng, L.; Dai, B.; Zheng, T.; Fuv, J.; Qiao, L.; Cai, W.; Wang, Y.; Yang, J. Repression of let-7a cluster prevents ad-hesion of colorectal cancer cells by enforcing a mesenchymal phenotype in presence of liver inflammation. *Cell Death. Dis.* **2018**, *9*, 489. [CrossRef] [PubMed]
24. Chen, J.-F.; Mandel, E.M.; Thomson, J.M.; Wu, Q.E.; Callis, T.; Hammond, S.M.; Conlon, F.L.; Wang, D.-Z. The Role of MicroRNA-1 and MicroRNA-133 in Skeletal Muscle Proliferation and Differentiation. *Nat. Genet.* **2005**, *38*, 228–233. [CrossRef]
25. Chen, W.-S.; Leung, C.-M.; Pan, H.-W.; Hu, L.-Y.; Li, S.-C.; Ho, M.-R.; Tsai, K.-W. Silencing of miR-1-1 and miR-133a-2 cluster expression by DNA hypermethylation in colorectal cancer. *Oncol. Rep.* **2012**, *28*, 1069–1076. [CrossRef] [PubMed]
26. Lei, X.; Li, L.; Duan, X. Long non-coding RNA ABHD11-AS1 promotes colorectal cancer development through regulation of miR-133a/SOX4 axis. *Biosci. Rep.* **2018**, *38*, 20181386. [CrossRef] [PubMed]
27. Yu, X.; Wang, D.; Wang, X.; Sun, S.; Zhang, Y.; Wang, S.; Miao, R.; Xu, X.; Qu, X. CXCL12/CXCR4 promotes inflamma-tion-driven colorectal cancer progression through activation of RhoA signaling by sponging miR-133a-3p. *J. Exp. Clin. Cancer Res.* **2019**, *38*, 32. [CrossRef]
28. Li, Z.; Liu, J.; Chen, H.; Zhang, Y.; Shi, H.; Huang, L.; Tao, J.; Shen, R.; Wang, T. Ferritin Light Chain (FTL) competes with long noncoding RNA Linc00467 for miR-133b binding site to regulate chemoresistance and metastasis of colorectal cancer. *Carcinogenesis* **2019**, *41*, 467–477. [CrossRef]
29. Zhu, D.; Sun, Y.; Zhang, D.; Dong, M.; Jiang, G.; Zhang, X.; Zhou, J. miR-1 inhibits the progression of colon cancer by regulating the expression of vascular endothelial growth factor. *Oncol. Rep.* **2018**, *40*, 589–598. [CrossRef]
30. Furukawa, S.; Kawasaki, Y.; Miyamoto, M.; Hiyoshi, M.; Kitayama, J.; Akiyama, T. The miR-1-NOTCH3-Asef Pathway Is Important for Colorectal Tumor Cell Migration. *PLoS ONE* **2013**, *8*, e80609. [CrossRef]
31. Xu, W.; Zhang, Z.; Zou, K.; Cheng, Y.; Yang, M.; Chen, H.; Wang, H.; Zhao, J.; Chen, P.; He, L.; et al. MiR-1 suppresses tumor cell proliferation in colorectal cancer by inhibition of Smad3-mediated tumor glycolysis. *Cell Death Dis.* **2017**, *8*, e2761. [CrossRef] [PubMed]
32. Zhou, G.-Q.; Han, F.; Shi, Z.-L.; Yu, L.; Li, X.-F.; Yu, C.; Shen, C.-L.; Wan, D.-W.; Zhu, X.-G.; Li, R.; et al. miR-133a-3p Targets SUMO-Specific Protease 1 to Inhibit Cell Proliferation and Cell Cycle Progress in Colorectal Cancer. *Oncol. Res. Featur. Preclin. Clin. Cancer Ther.* **2018**, *26*, 795–800. [CrossRef]
33. Meng, X.; Fu, R. miR-206 regulates 5-FU resistance by targeting Bcl-2 in colon cancer cells. *OncoTargets Ther.* **2018**, *ume 11*, 1757–1765. [CrossRef]
34. Sun, L.N.; Zhi, Z.; Chen, L.Y.; Zhou, Q.; Li, X.M.; Gan, W.J.; Chen, S.; Yang, M.; Liu, Y.; Shen, T.; et al. SIRT1 suppresses col-orectal cancer metastasis by transcriptional repression of miR-15b-5p. *Cancer. Lett.* **2017**, *409*, 104–115. [CrossRef]
35. Wang, L.; Zhang, X.; Sheng, L.; Qiu, C.; Luo, R. LINC00473 promotes the Taxol resistance via miR-15a in colorectal cancer. *Biosci. Rep.* **2018**, *38*, 20180790. [CrossRef]
36. Liu, Y.; Zhou, J.; Wang, S.; Song, Y.; Zhou, J.; Ren, F. Long non-coding RNA SNHG12 promotes proliferation and invasion of colorectal cancer cells by acting as a molecular sponge of microRNA-16. *Exp. Ther. Med.* **2019**, *18*, 1212–1220. [CrossRef]
37. Dai, L.; Wang, W.; Zhang, S.; Jiang, Q.; Wang, R.; Dai, L.; Cheng, L.; Yang, Y.; Wei, Y.Q.; Deng, H.X. Vector-based miR-15a/16-1 plasmid inhibits colon cancer growth in vivo. *Cell Biol. Int.* **2012**, *36*, 765–770. [CrossRef]
38. Gopalan, V.; Ebrahimi, F.; Islam, F.; Vider, J.; Qallandar, O.B.; Pillai, S.; Lu, C.-T.; Lam, A.K.-Y. Tumour suppressor properties of miR-15a and its regulatory effects on BCL2 and SOX2 proteins in colorectal carcinomas. *Exp. Cell Res.* **2018**, *370*, 245–253. [CrossRef] [PubMed]
39. Shi, L.; Jackstadt, R.; Siemens, H.; Li, H.; Kirchner, T.; Hermeking, H. p53-induced miR-15a/16-1 and AP4 form a dou-ble-negative feedback loop to regulate epithelial-mesenchymal transition and metastasis in colorectal cancer. *Cancer Res.* **2014**, *74*, 532–542. [CrossRef]
40. Fesler, A.; Liu, H.; Ju, J. Modified miR-15a has therapeutic potential for improving treatment of advanced stage colorectal cancer through inhibition of BCL2, BMI1, YAP1 and DCLK. *Oncotarget* **2017**, *9*, 2367–2383. [CrossRef]
41. You, C.; Liang, H.; Sun, W.; Li, J.; Liu, Y.; Fan, Q.; Zhang, H.; Yue, X.; Li, J.; Chen, X.; et al. Deregulation of the miR-16-KRAS axis promotes colorectal cancer. *Sci. Rep.* **2016**, *6*, 37459. [CrossRef]
42. Yang, I.-P.; Tsai, H.-L.; Huang, C.-W.; Lu, C.-Y.; Miao, Z.-F.; Chang, S.-F.; Juo, S.-H.H.; Wang, J.-Y. High blood sugar levels significantly impact the prognosis of colorectal cancer patients through down-regulation of microRNA-16 by targeting Myb and VEGFR. *Oncotarget* **2016**, *7*, 18837–18850. [CrossRef] [PubMed]
43. Zhang, Z.; Feng, L.; Liu, P.; Duan, W. ANRIL promotes chemoresistance via disturbing expression of ABCC1 by regulating the expression of Let-7a in colorectal cancer. *Biosci. Rep.* **2018**, *38*, 20180620. [CrossRef] [PubMed]
44. Chen, H.; Xu, Z. Hypermethylation-Associated Silencing of miR-125a and miR-125b: A Potential Marker in Colorectal Cancer. *Dis. Markers* **2015**, *2015*, 1–7. [CrossRef]
45. Chen, D.; Sun, Q.; Zhang, L.; Zhou, X.; Cheng, X.; Zhou, D.; Ye, F.; Lin, J.; Wang, W. The lncRNA HOXA11-AS functions as a competing endogenous RNA to regulate PADI2 expression by sponging miR-125a-5p in liver metastasis of colorectal cancer. *Oncotarget* **2017**, *8*, 70642–70652. [CrossRef] [PubMed]
46. Li, W.; Yuan, F.; Zhang, X.; Chen, W.; Tang, X.; Lu, L. Elevated MIR100HG promotes colorectal cancer metastasis and is as-sociated with poor prognosis. *Oncol. Lett.* **2019**, *18*, 6483–6490.

47. Sureban, S.M.; May, R.; Ramalingam, S.; Subramaniam, D.; Natarajan, G.; Wyche, J.H.; Anant, S.; Houchen, C.W. S1942 Selective Blockade of DCAMKL-1 Results in Tumor Growth Arrest by a Let-7a MicroRNA Dependent Mechanism. *Gastroenterology* **2009**, *136*, A–298. [CrossRef]
48. Wang, F.; Zhang, P.; Ma, Y.; Yang, J.; Moyer, M.P.; Shi, C.; Peng, J.; Qin, H. NIRF is frequently upregulated in colorectal cancer and its oncogenicity can be suppressed by let-7a microRNA. *Cancer Lett.* **2012**, *314*, 223–231. [CrossRef]
49. Li, B.; Chen, P.; Chang, Y.; Qi, J.; Fu, H.; Guo, H. Let-7a inhibits tumor cell growth and metastasis by directly targeting RTKN in human colon cancer. *Biochem. Biophys. Res. Commun.* **2016**, *478*, 739–745. [CrossRef]
50. Han, H.B.; Gu, J.; Zuo, H.J.; Chen, Z.G.; Zhao, W.; Li, M.; Ji, D.B.; Lu, Y.Y.; Zhang, Z.Q. Let-7c functions as a metastasis sup-pressor by targeting MMP11 and PBX3 in colorectal cancer. *J. Pathol.* **2012**, *226*, 544–555. [CrossRef]
51. Kohlan, A.K.; Saidijam, M.; Amini, R.; Samadi, P.; Najafi, R. Induction of let-7e gene expression attenuates oncogenic phenotype in HCT-116 colorectal cancer cells through targeting of DCLK1 regulation. *Life Sci.* **2019**, *228*, 221–227. [CrossRef] [PubMed]
52. Li, Z.; Pan, W.; Shen, Y.; Chen, Z.; Zhang, L.; Zhang, Y.; Luo, Q.; Ying, X. IGF1/IGF1R and microRNA let-7e down-regulate each other and modulate proliferation and migration of colorectal cancer cells. *Cell Cycle* **2018**, *17*, 1212–1219. [CrossRef]
53. Zhu, P.; Liu, J.; Lu, M.; Wu, G.; Lin, X.; Cai, L.; Zhang, X. Influence and mechanism of miR-99a suppressing development of colorectal cancer (CRC) with diabetes mellitus (DM). *OncoTargets Ther.* **2019**, *ume 12*, 10311–10321. [CrossRef]
54. Tong, Z.; Liu, N.; Lin, L.; Guo, X.; Yang, D.; Zhang, Q. miR-125a-5p inhibits cell proliferation and induces apoptosis in colon cancer via targeting BCL2, BCL2L12 and MCL. *Biomed. Pharmacother.* **2015**, *75*, 129–136. [CrossRef]
55. Yang, X.; Qiu, J.; Kang, H.; Wang, Y.; Qian, J. miR-125a-5p suppresses colorectal cancer progression by targeting VEGFA. *Cancer Manag. Res.* **2018**, *ume 10*, 5839–5853. [CrossRef]
56. Li, D.; Xu, X.; Miao, J.; Cai, J. MicroRNA-125a inhibits tumorigenesis by targeting Smurf1 in colorectal carcinoma. *FEBS Open Bio* **2019**, *9*, 1305–1314. [CrossRef]
57. Barta, J.A.; Powell, C.A.; Wisnivesky, J.P. Global Epidemiology of Lung Cancer. *Ann. Glob. Health* **2019**, *85*. [CrossRef]
58. Hwang, H.-W.; Mendell, J.T. MicroRNAs in cell proliferation, cell death, and tumorigenesis. *Br. J. Cancer* **2006**, *94*, 776–780. [CrossRef]
59. Navarro, F.; Lieberman, J. miR-34 and p53: New Insights into a Complex Functional Relationship. *PLoS ONE* **2015**, *10*, e0132767. [CrossRef] [PubMed]
60. Banzhaf-Strathmann, J.; Edbauer, D. Good guy or bad guy: The opposing roles of microRNA 125b in cancer. *Cell Commun. Signal.* **2014**, *12*, 30. [CrossRef] [PubMed]
61. Bublik, D.R.; Bursa´c, S.; Sheffer, M.; Oršoli´c, I.; Shalit, T.; Tarcic, O.; Kotler, E.; Mouhadeb, O.; Hoffman, Y.; Fuchs, G.; et al. Regulatory module involving FGF13, miR-504, and p53 regulates ribosomal biogenesis and supports cancer cell survival. *Proc. Natl. Acad. Sci. USA* **2017**, *114*, E496–E505. [CrossRef]
62. Chaffer, C.L.; Weinberg, R.A. A Perspective on Cancer Cell Metastasis. *Science* **2011**, *331*, 1559–1564. [CrossRef] [PubMed]
63. Wu, S.-G.; Chang, T.-H.; Liu, Y.-N.; Shih, J.-Y. MicroRNA in Lung Cancer Metastasis. *Cancers* **2019**, *11*, 265. [CrossRef] [PubMed]
64. Claperon, A.; Mergey, M.; Ho-Bouldoires, T.H.N.; Vignjevic, D.; Wendum, D.; Chretien, Y.; Merabtene, F.; Frazao, A.; Paradis, V.; Housset, C.; et al. EGF/EGFR axis contributes to the progression of cholangiocarcinoma through the induction of an epi-thelial-mesenchymal transition. *J. Hepatol.* **2014**, *61*, 325–332. [CrossRef] [PubMed]
65. Thiery, J.P.; Acloque, H.; Huang, R.Y.; Nieto, M.A. Epithelial-Mesenchymal Transitions in Development and Disease. *Cell* **2009**, *139*, 871–890. [CrossRef] [PubMed]
66. Ma, L.; Teruya-Feldstein, J.; Weinberg, R.A. Tumour invasion and metastasis initiated by microRNA-10b in breast cancer. *Nat. Cell Biol.* **2007**, *449*, 682–688. [CrossRef]
67. Li, Y.; Li, Y.; Liu, J.; Fan, Y.; Li, X.; Dong, M.; Liu, H.; Chen, J. Expression levels of microRNA-145 and microRNA-10b are associated with metastasis in non-small cell lung cancer. *Cancer Biol. Ther.* **2016**, *17*, 272–279. [CrossRef]
68. Lou, W.; Liu, J.; Gao, Y.; Zhong, G.; Chen, D.; Shen, J.; Bao, C.; Xu, L.; Pan, J.; Cheng, J.; et al. MicroRNAs in cancer metastasis and angiogenesis. *Oncotarget* **2017**, *8*, 115787–115802. [CrossRef]
69. Mao, G.; Liu, Y.; Fang, X.; Liu, Y.; Fang, L.; Lin, L.; Liu, X.; Wang, N. Tumor-derived microRNA-494 promotes angiogenesis in non-small cell lung cancer. *Angiogenesis* **2015**, *18*, 373–382. [CrossRef]
70. Hu, J.; Cheng, Y.; Li, Y.; Jin, Z.; Pan, Y.; Liu, G.; Fu, S.; Zhang, Y.; Feng, K.; Feng, Y. microRNA-128 plays a critical role in human non-small cell lung cancer tumourigenesis, angiogenesis and lymphangiogenesis by directly targeting vascular endothelial growth factor-C. *Eur. J. Cancer* **2014**, *50*, 2336–2350. [CrossRef]
71. Xue, D.; Yang, Y.; Liu, Y.; Wang, P.; Dai, Y.; Liu, Q.; Chen, L.; Shen, J.; Ju, H.; Li, Y.; et al. MicroRNA-206 attenuates the growth and angiogenesis in non-small cell lung cancer cells by blocking the 14-3-3ζ/STAT3/HIF-1α/VEGF signaling. *Oncotarget* **2016**, *7*, 79805–79813. [CrossRef] [PubMed]
72. Zhou, Y.; Li, S.; Li, J.; Wang, D.; Li, Q. Effect of microRNA-135a on Cell Proliferation, Migration, Invasion, Apoptosis and Tumor Angiogenesis Through the IGF-1/PI3K/Akt Signaling Pathway in Non-Small Cell Lung Cancer. *Cell Physiol. Biochem.* **2017**, *42*, 1431–1446. [CrossRef] [PubMed]
73. Liu, R.; Zhang, Y.-S.; Zhang, S.; Cheng, Z.-M.; Yu, J.-L.; Zhou, S.; Song, J. MiR-126-3p suppresses the growth, migration and invasion of NSCLC via targeting CCR. *Eur. Rev. Med. Pharmacol. Sci.* **2019**, *23*, 679–689. [PubMed]
74. Xing, Y.; Meng, Q.; Chen, X.; Zhao, Y.; Liu, W.; Hu, J.; Xue, F.; Wang, X.; Cai, L. TRIM44 promotes proliferation and metastasis in non-small cell lung cancer via mTOR signaling pathway. *Oncotarget* **2016**, *7*, 30479–30491. [CrossRef] [PubMed]

75. Xiao, H. MiR-7-5p suppresses tumor metastasis of non-small cell lung cancer by targeting NOVA. *Cell. Mol. Biol. Lett.* **2019**, *24*, 1–13. [CrossRef]
76. Liao, M.; Peng, L. MiR-206 may suppress non-small lung cancer metastasis by targeting CORO1C. *Cell. Mol. Biol. Lett.* **2020**, *25*, 1–13. [CrossRef]
77. Du, W.; Tang, H.; Lei, Z.; Zhu, J.; Zeng, Y.; Liu, Z.; Huang, J.-A. miR-335-5p inhibits TGF-β1-induced epithelial–mesenchymal transition in non-small cell lung cancer via ROCK. *Respir. Res.* **2019**, *20*, 1–11. [CrossRef] [PubMed]
78. Jiang, F.; Yu, Q.; Chu, Y.; Zhu, X.; Lu, W.; Liu, Q.; Wang, Q. MicroRNA-98-5p inhibits proliferation and metastasis in non-small cell lung cancer by targeting TGFBR. *Int. J. Oncol.* **2018**, *54*, 128–138. [CrossRef]
79. Chen, Q.; Chen, S.; Zhao, J.; Zhou, Y.; Xu, L. MicroRNA-126: A new and promising player in lung cancer (Review). *Oncol. Lett.* **2020**, *21*, 1. [CrossRef]
80. Chen, Z.; Wang, D.; Gu, C.; Liu, X.; Pei, W.; Li, J.; Cao, Y.; Jiao, Y.; Tong, J.; Nie, J. Down-regulation of let-7 microRNA increased K-ras expression in lung damage induced by radon. *Environ. Toxicol. Pharmacol.* **2015**, *40*, 541–548. [CrossRef]
81. Song, L.; Li, D.; Gu, Y.; Wen, Z.-M.; Jie, J.; Zhao, D.; Peng, L.-P. MicroRNA-126 Targeting PIK3R2 Inhibits NSCLC A549 Cell Proliferation, Migration, and Invasion by Regulation of PTEN/PI3K/AKT Pathway. *Clin. Lung Cancer* **2016**, *17*, e65–e75. [CrossRef] [PubMed]
82. Globocan, Estimated Cancer Incidence, Mortality et Prevalence Worldwide. Available online: http://globocan.iarc.fr/Default.aspx (accessed on 23 December 2020).
83. Kasiappan, R.; Rajarajan, D. Role of MicroRNA Regulation in Obesity-Associated Breast Cancer: Nutritional Perspectives. *Adv. Nutr.* **2017**, *8*, 868–888. [CrossRef] [PubMed]
84. Cantini, L.; Bertoli, G.; Cava, C.; Dubois, T.; Zinovyev, A.; Caselle, M.; Castiglioni, I.; Barillot, E.; Martignetti, L. Identification of microRNA clusters cooperatively acting on epithelial to mesenchymal transition in triple negative breast cancer. *Nucleic Acids Res.* **2019**, *47*, 2205–2215. [CrossRef] [PubMed]
85. Li, P.; Sheng, C.; Huang, L.; Zhang, H.; Huang, L.; Cheng, Z.; Zhu, Q. MiR-183/-96/-182 cluster is up-regulated in most breast cancers and increases cell proliferation and migration. *Breast Cancer Res.* **2014**, *16*, 473. [CrossRef]
86. Dambal, S.; Shah, M.; Mihelich, B.L.; Nonn, L. The microRNA-183 cluster: The family that plays together stays together. *Nucleic Acids Res.* **2015**, *43*, 7173–7188. [CrossRef]
87. Ma, Y.; Liang, A.-J.; Fan, Y.-P.; Huang, Y.-R.; Zhao, X.-M.; Sun, Y.; Chen, X.-F. Dysregulation and functional roles of miR-183-96-182 cluster in cancer cell proliferation, invasion and metastasis. *Oncotarget* **2016**, *7*, 42805–42825. [CrossRef]
88. Cao, D.; Di, M.; Liang, J.; Shi, S.; Tan, Q.; Wang, Z. MicroRNA-183 in Cancer Progression. *J. Cancer* **2020**, *11*, 1315–1324. [CrossRef]
89. Cheng, Y.; Xiang, G.; Meng, Y.; Dong, R. MiRNA-183-5p promotes cell proliferation and inhibits apoptosis in human breast cancer by targeting the PDCD. *Reprod. Biol.* **2016**, *16*, 225–233. [CrossRef] [PubMed]
90. Hong, Y.; Liang, H.; Rehman, U.-U.; Wang, Y.; Zhang, W.; Zhou, Y.; Chen, S.; Yu, M.; Cui, S.; Liu, M.; et al. miR-96 promotes cell proliferation, migration and invasion by targeting PTPN9 in breast cancer. *Sci. Rep.* **2016**, *6*, 37421. [CrossRef] [PubMed]
91. Guttilla, I.K.; White, B.A. Coordinate Regulation of FOXO1 by miR-27a, miR-96, and miR-182 in Breast Cancer Cells. *J. Biol. Chem.* **2009**, *284*, 23204–23216. [CrossRef]
92. Song, J.; Ouyang, Y.; Che, J.; Li, X.; Zhao, Y.; Yang, K.; Zhao, X.; Chen, Y.; Fan, C.; Yuan, W. Potential value of miR-221/222 as diagnostic, prognostic, and therapeutic biomarkers for diseases. *Front. Immunol.* **2017**, *8*, 1–9. [CrossRef]
93. Di Leva, G.; Gasparini, P.; Piovan, C.; Ngankeu, A.; Garofalo, M.; Taccioli, C.; Iorio, M.V.; Li, M.; Volinia, S.; Alder, H.; et al. MicroRNA Cluster 221-222 and Estrogen Receptor α Interactions in Breast Cancer. *J. Natl. Cancer Inst.* **2010**, *102*, 706–721. [CrossRef]
94. Li, B.; Lu, Y.; Wang, H.; Han, X.; Mao, J.; Li, J.; Yu, L.; Wang, B.; Fan, S.; Yu, X.; et al. MiR-221/222 enhance the tumorigenicity of human breast cancer stem cells via modulation of PTEN/Akt pathway. *Biomed. Pharmacother.* **2016**, *79*, 93–101. [CrossRef]
95. Alimirah, F.; Peng, X.; Gupta, A.; Yuan, L.; Welsh, J.E.; Cleary, M.; Mehta, R.G. Crosstalk between the vitamin D receptor (VDR) and miR-214 in regulating SuFu, a hedgehog pathway inhibitor in breast cancer cells. *Exp. Cell Res.* **2016**, *349*, 15–22. [CrossRef]
96. Derfoul, A.; Juan, A.H.; Difilippantonio, M.J.; Palanisamy, N.; Ried, T.; Sartorelli, V. Decreased microRNA-214 levels in breast cancer cells coincides with increased cell proliferation, invasion and accumulation of the Polycomb Ezh2 methyltransferase. *Carcinogenesis* **2011**, *32*, 1607–1614. [CrossRef]
97. Yi, S.-J.; Li, L.-L.; Tu, W.-B. MiR-214 negatively regulates proliferation and WNT/β-catenin signaling in breast cancer. *Eur. Rev. Med. Pharmacol. Sci.* **2016**, *20*, 5148–5154.
98. Liu, B.; Tian, Y.; Like, Z.; Zhao, Z.; Jiang, X.; Zhai, C.; Han, X.; Zhang, L. Tumor-suppressing roles of miR-214 and miR-218 in breast cancer. *Oncol. Rep.* **2016**, *35*, 3178–3184. [CrossRef] [PubMed]
99. Wang, Y.; Liang, H.; Zhou, G.; Hu, X.; Liu, Z.; Jin, F.; Yu, M.; Sang, J.; Zhou, Y.; Fu, Z.; et al. HIC1 and miR-23~27~24 clusters form a double-negative feedback loop in breast cancer. *Cell Death Differ.* **2017**, *24*, 421–432. [CrossRef] [PubMed]
100. Hannafon, B.N.; Cai, A.; Calloway, C.L.; Xu, Y.-F.; Zhang, R.; Fung, K.-M.; Ding, W.-Q. miR-23b and miR-27b are oncogenic microRNAs in breast cancer: Evidence from a CRISPR/Cas9 deletion study. *BMC Cancer* **2019**, *19*, 1–12. [CrossRef] [PubMed]
101. Ell, B.; Qiu, Q.; Wei, Y.; Mercatali, L.; Ibrahim, T.; Amadori, D.; Kang, Y. The MicroRNA-23b/27b/24 Cluster Promotes Breast Cancer Lung Metastasis by Targeting Metastasis-suppressive Gene Prosaposin. *J. Biol. Chem.* **2014**, *289*, 21888–21895. [CrossRef] [PubMed]

102. Li, M.; Han, Y.; Zhou, H.; Li, X.; Lin, C.; Zhang, E.; Chi, X.; Hu, J.; Xu, H. Transmembrane protein 170B is a novel breast tu-morigenesis suppressor gene that inhibits the Wnt/β-catenin pathway. *Cell Death Dis.* **2018**, *9*, 91. [CrossRef]
103. Pellegrino, L.; Stebbing, J.; Braga, V.M.; Frampton, A.E.; Jacob, J.; Buluwela, L.; Jiao, L.R.; Periyasamy, M.; Madsen, C.D.; Caley, M.P.; et al. miR-23b regulates cytoskeletal remodeling, motility and metastasis by directly targeting multiple transcripts. *Nucleic Acids Res.* **2013**, *41*, 5400–5412. [CrossRef]
104. Shiels, M.S.; O'Brien, T.R. Recent Decline in Hepatocellular Carcinoma Rates in the United States. *Gastroenterology* **2020**, *158*, 1503–1505. [CrossRef] [PubMed]
105. Chandan, K.; Sarwat, M. Role of microRNA and Long Non-Coding RNA in Hepatocellular Carcinoma. *Curr. Pharm. Des.* **2020**, *26*, 415–428. [CrossRef]
106. Ludwig, N.; Leidinger, P.; Becker, K.; Backes, C.; Fehlmann, T.; Pallasch, C.; Rheinheimer, S.; Meder, B.; Stähler, C.; Meese, E.; et al. Distribution of miRNA expression across human tissues. *Nucleic Acids Res.* **2016**, *44*, 3865–3877. [CrossRef] [PubMed]
107. Meng, F.; Henson, R.; Wehbe-Janek, H.; Ghoshal, K.; Jacob, S.T.; Patel, T. MicroRNA-21 regulates expression of the PTEN tumor suppressor gene in human hepatocellular cancer. *Gastroenterology* **2007**, *133*, 647–658. [CrossRef]
108. Liu, C.; Yu, J.; Yu, S.; Lavker, R.M.; Cai, L.; Liu, W.; Yang, K.; He, X.; Chen, S. MicroRNA-21 acts as an oncomir through multiple targets in human hepatocellular carcinoma. *J. Hepatol.* **2010**, *53*, 98–107. [CrossRef]
109. Zhou, J.; Yu, L.; Gao, X.; Hu, J.; Wang, J.; Dai, Z.; Wang, J.-F.; Zhang, Z.; Lu, S.; Huang, X.; et al. Plasma MicroRNA Panel to Diagnose Hepatitis B Virus–Related Hepatocellular Carcinoma. *J. Clin. Oncol.* **2011**, *29*, 4781–4788. [CrossRef]
110. Pineau, P.; Volinia, S.; McJunkin, K.; Marchio, A.; Battiston, C.; Terris, B.; Mazzaferro, V.; Lowe, S.W.; Croce, C.M.; Dejean, A. miR-221 overexpression contributes to liver tumorigenesis. *Proc. Natl. Acad. Sci. USA* **2009**, *107*, 264–269. [CrossRef] [PubMed]
111. Fornari, F.; Gramantieri, L.; Ferracin, M.; Veronese, A.; Sabbioni, S.; Calin, G.A.; Grazi, G.L.; Giovannini, C.; Croce, C.M.; Bolondi, L.; et al. MiR-221 controls CDKN1C/p57 and CDKN1B/p27 expression in human hepatocellular carcinoma. *Oncogene* **2008**, *27*, 5651–5661. [CrossRef]
112. Jiang, J.; Gusev, Y.; Aderca, I.; Mettler, T.A.; Nagorney, D.M.; Brackett, D.J.; Roberts, L.R.; Schmittgen, T.D. Association of MicroRNA Expression in Hepatocellular Carcinomas with Hepatitis Infection, Cirrhosis, and Patient Survival. *Clin. Cancer Res.* **2008**, *14*, 419–427. [CrossRef]
113. Zhao, Y.-J.; Ju, Q.; Li, G.-C. Tumor markers for hepatocellular carcinoma. *Mol. Clin. Oncol.* **2013**, *1*, 593–598. [CrossRef]
114. Roderburg, C.; Urban, G.-W.; Bettermann, K.; Vucur, M.; Zimmermann, H.W.; Schmidt, S.; Janssen, J.; Koppe, C.; Knolle, P.; Castoldi, M.; et al. Micro-RNA profiling reveals a role for miR-29 in human and murine liver fibrosis. *Hepatology* **2010**, *53*, 209–218. [CrossRef]
115. Xiong, Y.; Fang, J.H.; Yun, J.P.; Yang, J.; Zhang, Y.; Jia, W.H.; Zhuang, S.M. Effects of microRNA-29 on apoptosis, tumor-igenicity, and prognosis of hepatocellular carcinoma. *Hepatology* **2010**, *51*, 836–845.
116. Xu, Y.; An, Y.; Wang, Y.; Zhang, C.; Zhang, H.; Huang, C.; Jiang, H.; Wang, X.; Li, X. miR-101 inhibits autophagy and enhances cisplatin-induced apoptosis in hepatocellular carcinoma cells. *Oncol. Rep.* **2013**, *29*, 2019–2024. [CrossRef] [PubMed]
117. Gong, J.; Zhang, J.-P.; Li, B.; Zeng, C.; You, K.; Chen, M.-X.; Yuan, Y.; Zhuang, S.-M. MicroRNA-125b promotes apoptosis by regulating the expression of Mcl-1, Bcl-w and IL-6R. *Oncogene* **2013**, *32*, 3071–3079. [CrossRef] [PubMed]
118. Girard, M.; Jacquemin, E.; Munnich, A.; Lyonnet, S.; Henrion-Caude, A. miR-122, a paradigm for the role of microRNAs in the liver. *J. Hepatol.* **2008**, *48*, 648–656. [CrossRef] [PubMed]
119. Xu, H.; He, J.-H.; Xiao, Z.-D.; Zhang, Q.-Q.; Chen, Y.-Q.; Zhou, H.; Qu, L.-H. Liver-enriched transcription factors regulate MicroRNA-122 that targets CUTL1 during liver development. *Hepatology* **2010**, *52*, 1431–1442. [CrossRef] [PubMed]
120. Szabo, G.; Bala, S. MicroRNAs in liver disease. *Nat. Rev. Gastroenterol. Hepatol.* **2013**, *10*, 542–552. [CrossRef] [PubMed]
121. NCI. Surveillance, Epidemiology, and End Results Program. Cancer Stat Facts: Ovarian Cancer. Available online: https://seer.cancer.gov/statfacts/html/ovary.html (accessed on 14 October 2019).
122. Bray, F.; Ferlay, J.; Soerjomataram, I.; Siegel, R.L.; Torre, L.A.; Jemal, A. Global cancer statistics 2018: GLOBOCAN estimates of incidence and mortality worldwide for 36 cancers in 185 countries. *CA Cancer J. Clin.* **2018**, *68*, 394–424. [CrossRef] [PubMed]
123. Ehdaivand, S.; WHO. Classification of Ovarian Neoplasms. Available online: https://www.pathologyoutlines.com/topic/ovarytumorwhoclassif.html (accessed on 14 October 2019).
124. Berman, T.A.; Schiller, J.T. Human papillomavirus in cervical cancer and oropharyngeal cancer: One cause, two diseases. *Cancer* **2017**, *123*, 2219–2229. [CrossRef] [PubMed]
125. Pardini, B.; De Maria, D.; Francavilla, A.; Di Gaetano, C.; Ronco, G.; Naccarati, A. MicroRNAs as markers of progression in cervical cancer: A systematic review. *BMC Cancer* **2018**, *18*, 1–17. [CrossRef] [PubMed]
126. Flores, C.P.; García-Vázquez, R.; Rincón, D.G.; Ruiz-García, E.; De La Vega, H.A.; Marchat, L.A.; Vera, Y.M.S.; López-Camarillo, C. MicroRNAs driving invasion and metastasis in ovarian cancer: Opportunities for translational medicine (Review). *Int. J. Oncol.* **2017**, *50*, 1461–1476. [CrossRef] [PubMed]
127. Fang, Y.; Xu, C.; Fu, Y. MicroRNA-17-5p induces drug resistance and invasion of ovarian carcinoma cells by targeting PTEN signaling. *J. Biol. Res.* **2015**, *22*, 12. [CrossRef] [PubMed]
128. Yang, H.; Kong, W.; He, L.; Zhao, J.-J.; O'Donnell, J.D.; Wang, J.; Wenham, R.M.; Coppola, D.; Kruk, P.A.; Nicosia, S.V.; et al. MicroRNA Expression Profiling in Human Ovarian Cancer: miR-214 Induces Cell Survival and Cisplatin Resistance by Targeting PTEN. *Cancer Res.* **2008**, *68*, 425–433. [CrossRef] [PubMed]

129. Shishodia, G.; Shukla, S.; Srivastava, Y.; Masaldan, S.; Mehta, S.; Bhambhani, S.; Sharma, S.; Mehrotra, R.; Das, B.C.; Bharti, A.C. Alterations in microRNAs miR-21 and let-7a correlate with aberrant STAT3 signaling and downstream effects during cervical carcinogenesis. *Mol. Cancer* **2015**, *14*, 1–13. [CrossRef]
130. Bumrungthai, S.; Ekalaksananan, T.; Evans, M.F.; Chopjitt, P.; Tangsiriwatthana, T.; Patarapadungkit, N.; Kleebkaow, P.; Luanratanakorn, S.; Kongyingyoes, B.; Worawichawong, S.; et al. Up-Regulation of miR-21 Is Associated with Cervicitis and Human Papillomavirus Infection in Cervical Tissues. *PLoS ONE* **2015**, *10*, e0127109. [CrossRef]
131. Xu, L.; Xu, Q.; Li, X.; Zhang, X. MicroRNA-21 regulates the proliferation and apoptosis of cervical cancer cells via tumor ne-crosis factor-alpha. *Mol. Med. Rep.* **2017**, *16*, 4659–4663. [CrossRef]
132. Zhang, Y.; Wang, Z.-C.; Zhang, Z.-S.; Chen, F. MicroRNA-155 regulates cervical cancer via inducing Th17/Treg imbalance. *Eur. Rev. Med. Pharmacol. Sci.* **2018**, *22*, 3719–3726.
133. Vang, S.; Wu, H.-T.; Fischer, A.; Miller, D.H.; MacLaughlan, S.; Douglass, E.; Steinhoff, M.; Collins, C.; Smith, P.J.S.; Brard, L.; et al. Identification of Ovarian Cancer Metastatic miRNAs. *PLoS ONE* **2013**, *8*, e58226. [CrossRef]
134. Jin, M.; Yang, Z.; Ye, W.; Xu, H.; Hua, X. MicroRNA-150 Predicts a Favorable Prognosis in Patients with Epithelial Ovarian Cancer, and Inhibits Cell Invasion and Metastasis by Suppressing Transcriptional Repressor ZEB1. *PLoS ONE* **2014**, *9*, e103965. [CrossRef]
135. Lü, X.; Li, J.; Liang, S.; Jin, H.; Xu, C.; Ma, D. Tiam1, negatively regulated by miR-22, miR-183 and miR-31, is involved in migration, invasion and viability of ovarian cancer cells. *Oncol. Rep.* **2012**, *27*, 1835–1842. [CrossRef]
136. Loganathan, S.; Kandala, P.K.; Gupta, P.; Srivastava, S.K. Inhibition of EGFR-AKT Axis Results in the Suppression of Ovarian Tumors In Vitro and in Preclinical Mouse Model. *PLoS ONE* **2012**, *7*, e43577. [CrossRef] [PubMed]
137. Chen, Z.; Han, Y.; Song, C.; Wei, H.; Chen, Y.; Huang, K.; Li, S.; Ma, D.; Wang, S.; Wang, J. Systematic review and me-ta-analysis of the prognostic significance of microRNAs in cervical cancer. *Oncotarget* **2018**, *9*, 17141–17148. [CrossRef] [PubMed]
138. Fan, J.Y.; Fan, Y.J.; Wang, X.L.; Xie, H.; Gao, H.J.; Zhang, Y.; Liu, M.; Tang, H. miR-429 is involved in regulation of NF-kB activity by targeting IKKbeta and suppresses oncogenic activity in cervical cancer cells. *FEBS Lett.* **2017**, *591*, 118–128. [CrossRef] [PubMed]
139. Huang, F.; Lin, C.; Shi, Y.H.; Kuerban, G. MicroRNA-101 inhibits cell proliferation, invasion, and promotes apoptosis by reg-ulating cyclooxygenase-2 in Hela cervical carcinoma cells. *Asian Pac. J. Cancer Prev.* **2013**, *14*, 5915–5920. [CrossRef] [PubMed]
140. Chandrasekaran, K.S.; Sathyanarayanan, A.; Karunagaran, D. Downregulation of HMGB1 by miR-34a is sufficient to suppress proliferation, migration and invasion of human cervical and colorectal cancer cells. *Tumor Biol.* **2016**, *37*, 13155–13166. [CrossRef] [PubMed]
141. Chen, J.; Li, G. MiR-1284 enhances sensitivity of cervical cancer cells to cisplatin via downregulating HMGB1. *Biomed. Pharmacother.* **2018**, *107*, 997–1003. [CrossRef]
142. Jiang, D.; Wang, H.; Li, Z.; Li, Z.; Chen, X.; Cai, H. MiR-142 inhibits the development of cervical cancer by targeting HMGB1. *Oncotarget* **2016**, *8*, 4001–4007. [CrossRef]
143. Sun, L.; Wang, D.; Li, H.; She, H. Significance of high YKL-40 expression regulated by miR-24 in cervical cancer progression and prognosis. *Int. J. Clin. Exp. Pathol.* **2016**, *9*, 5128–5137.
144. Liu, D.; Liu, C.; Wang, X.; Ingvarsson, S.; Chen, H. MicroRNA-451 suppresses tumor cell growth by down-regulating IL6R gene expression. *Cancer Epidemiol.* **2014**, *38*, 85–92. [CrossRef]
145. Fan, Z.; Cui, H.; Xu, X.; Lin, Z.; Zhang, X.; Kang, L.; Han, B.; Meng, J.; Yan, Z.; Yan, X.; et al. MiR-125a suppresses tumor growth, invasion and metastasis in cervical cancer by targeting STAT3. *Oncotarget* **2015**, *6*, 25266–25280. [CrossRef] [PubMed]
146. Chen, J.; Hu, C.; Pan, P. Extracellular Vesicle MicroRNA Transfer in Lung Diseases. *Front. Physiol.* **2017**, *8*, 1028. [CrossRef] [PubMed]
147. Behera, J.; Tyagi, N. Exosomes: Mediators of bone diseases, protection, and therapeutics potential. *Oncoscience* **2018**, *5*, 181–195. [CrossRef] [PubMed]
148. Mao, L.; Li, X.; Gong, S.; Yuan, H.; Jiang, Y.; Huang, W.; Sun, X.; Dang, X. Serum exosomes contain ECRG4 mRNA that suppresses tumor growth via inhibition of genes involved in inflammation, cell proliferation, and angiogenesis. *Cancer Gene Ther.* **2018**, *25*, 248–259. [CrossRef] [PubMed]
149. Wu, H.; Zhou, J.; Mei, S.; Wu, D.; Mu, Z.; Chen, B.; Xie, Y.; Ye, Y.; Liu, J. Circulating exosomal microRNA-96 promotes cell proliferation, migration and drug resistance by targeting LMO7. *J. Cell. Mol. Med.* **2016**, *21*, 1228–1236. [CrossRef]
150. Hsu, Y.-L.; Hung, J.-Y.; Chang, W.-A.; Lin, Y.-S.; Pan, Y.-C.; Tsai, P.-H.; Wu, C.-Y.; Kuo, P.-L. Hypoxic lung cancer-secreted exosomal miR-23a increased angiogenesis and vascular permeability by targeting prolyl hydroxylase and tight junction protein ZO-1. *Oncogene* **2017**, *36*, 4929–4942. [CrossRef]
151. Wei, F.; Ma, C.; Zhou, T.; Dong, X.; Luo, Q.; Geng, L.; Ding, L.; Zhang, Y.; Zhang, L.; Li, N.; et al. Exosomes derived from gemcitabine-resistant cells transfer malignant phenotypic traits via delivery of miRNA-222-3p. *Mol. Cancer* **2017**, *16*, 1–14. [CrossRef]
152. Jing, C.; Cao, H.; Qin, X.; Yu, S.; Wu, J.; Wang, Z.; Ma, R.; Feng, J. Exosome-mediated gefitinib resistance in lung cancer HCC827 cells via delivery of miR-21. *Oncol. Lett.* **2018**, *15*, 9811–9817. [CrossRef]
153. Aushev, V.N.; Zborovskaya, I.B.; Laktionov, K.K.; Girard, N.; Cros, M.-P.; Herceg, Z.; Krutovskikh, V. Comparisons of microRNA Patterns in Plasma before and after Tumor Removal Reveal New Biomarkers of Lung Squamous Cell Carcinoma. *PLoS ONE* **2013**, *8*, e78649. [CrossRef]

154. Jin, X.; Chen, Y.; Chen, H.; Fei, S.; Chen, D.; Cai, X.; Liu, L.; Lin, B.; Su, H.; Zhao, L.; et al. Evaluation of Tumor-Derived Exosomal miRNA as Potential Diagnostic Biomarkers for Early-Stage Non–Small Cell Lung Cancer Using Next-Generation Sequencing. *Clin. Cancer Res.* **2017**, *23*, 5311–5319. [CrossRef] [PubMed]
155. Syn, N.L.; Wang, L.; Chow, E.K.-H.; Lim, C.T.; Goh, B.-C. Exosomes in Cancer Nanomedicine and Immunotherapy: Prospects and Challenges. *Trends Biotechnol.* **2017**, *35*, 665–676. [CrossRef] [PubMed]
156. Monfared, H.; Jahangard, Y.; Nikkhah, M.; Mirnajafi-Zadeh, S.J.; Mowla, S.J. Potential Therapeutic Effects of Exosomes Packed With a miR-21-Sponge Construct in a Rat Model of Glioblastoma. *Front. Oncol.* **2019**, *9*, 782. [CrossRef] [PubMed]

biomedicines

Review

The Multifaceted Role and Utility of MicroRNAs in Indolent B-Cell Non-Hodgkin Lymphomas

Pinelopi I. Artemaki [1], Petros A. Letsos [1], Ioanna C. Zoupa [1], Katerina Katsaraki [1], Paraskevi Karousi [1], Sotirios G. Papageorgiou [2], Vasiliki Pappa [2], Andreas Scorilas [1] and Christos K. Kontos [1,*]

[1] Department of Biochemistry and Molecular Biology, Faculty of Biology, National and Kapodistrian University of Athens, 15701 Athens, Greece; partemaki@biol.uoa.gr (P.I.A.); petr.sletsos@gmail.com (P.A.L.); joanz3999@gmail.com (I.C.Z.); kkatsaraki@biol.uoa.gr (K.K.); pkarousi@biol.uoa.gr (P.K.); ascorilas@biol.uoa.gr (A.S.)

[2] Second Department of Internal Medicine and Research Unit, University General Hospital "Attikon", 12462 Athens, Greece; sotirispapageorgiou@hotmail.com (S.G.P.); vaspappa@med.uoa.gr (V.P.)

* Correspondence: chkontos@biol.uoa.gr; Tel.: +30-210-727-4616

Abstract: Normal B-cell development is a tightly regulated complex procedure, the deregulation of which can lead to lymphomagenesis. One common group of blood cancers is the B-cell non-Hodgkin lymphomas (NHLs), which can be categorized according to the proliferation and spread rate of cancer cells into indolent and aggressive ones. The most frequent indolent B-cell NHLs are follicular lymphoma and marginal zone lymphoma. MicroRNAs (miRNAs) are small non-coding RNAs that can greatly influence protein expression. Based on the multiple interactions among miRNAs and their targets, complex networks of gene expression regulation emerge, which normally are essential for proper B-cell development. Multiple miRNAs have been associated with B-cell lymphomas, as the deregulation of these complex networks can lead to such pathological states. The aim of the present review is to summarize the existing information regarding the multifaceted role of miRNAs in indolent B-cell NHLs, affecting the main B-cell subpopulations. We attempt to provide insight into their biological function, the complex miRNA-mRNA interactions, and their biomarker utility in these malignancies. Lastly, we address the limitations that hinder the investigation of the role of miRNAs in these lymphomas and discuss ways that these problems could be overcome in the future.

Keywords: miRNAs; prognosis; follicular lymphoma; marginal zone lymphoma; Waldenström's macroglobulinemia; hairy cell leukemia; primary cutaneous follicle center lymphoma; normal B-cell development; therapeutic target; diagnosis

Citation: Artemaki, P.I.; Letsos, P.A.; Zoupa, I.C.; Katsaraki, K.; Karousi, P.; Papageorgiou, S.G.; Pappa, V.; Scorilas, A.; Kontos, C.K. The Multifaceted Role and Utility of MicroRNAs in Indolent B-Cell Non-Hodgkin Lymphomas. *Biomedicines* **2021**, *9*, 333. https://doi.org/10.3390/biomedicines9040333

Academic Editor: Francesca Lovat

Received: 8 February 2021
Accepted: 18 March 2021
Published: 25 March 2021

1. Introduction

B-cell non-Hodgkin lymphomas (NHLs) are one of the most common malignancies. They consist of different types of lymphomas, which are characterized by great heterogeneity. Their common feature is the absence of Reed-Sternberg cells, which in contrast are present in Hodgkin Lymphomas. B-cell NHLs are more common than the Hodgkin lymphomas and are classified according to the proliferating and spread rate of cancer cells into indolent or slowly growing and aggressive or quickly growing lymphomas. Recent advances have assisted in the elucidation of the etiology and the molecular background of these lymphomas; however, several questions remain unanswered and thus hinder the in-depth understanding of the pathogenesis of these lymphomas and the adoption of a personalized treatment approach. Two of the most common indolent B-cell NHLs are marginal zone lymphoma (MZL) and follicular lymphoma (FL), which sometimes could transform into more aggressive types, such as diffuse large B-cell lymphoma (DLBCL) [1,2]. Besides these two lymphoma types, Waldenström's macroglobulinemia (WM), hairy cell leukemia (HCL), and primary cutaneous follicle center lymphoma (PCFCL), which occur more rarely, are considered as indolent B-cell NHLs as well (Figure 1).

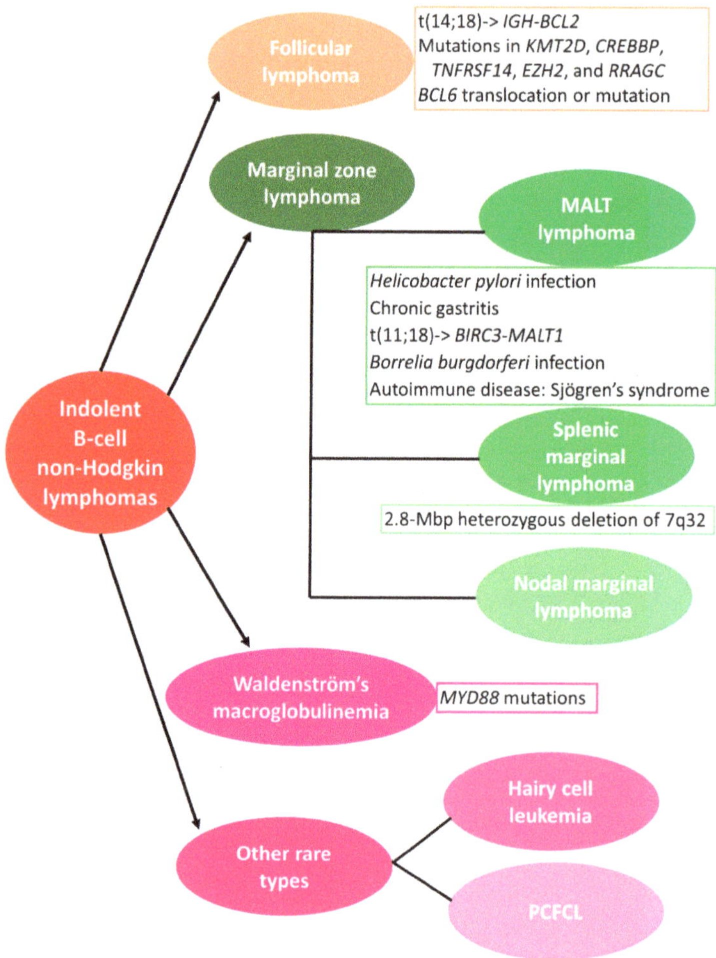

Figure 1. The main types of indolent B-cell non-Hodgkin lymphomas and some initiation factors of these malignancies. Abbreviations: MALT, mucosa-associated lymphoid tissue; PCFCL, primary cutaneous follicle center lymphoma.

FL is the most common indolent B-cell NHL and shows great heterogeneity. Its diagnosis is based on the detection of malignant centrocytes and centroblasts that resemble germinal center B cells within lymphoid follicles [3,4]. MZL is less investigated in comparison with FL, while various treatment options are available. It derives from malignant marginal zone B cells and persistent immune system stimulation triggered by infections or autoimmune diseases, which constitutes a major driving factor of lymphomagenesis. MZL is divided into three main subtypes: Extranodal MZL or mucosa-associated lymphoid tissue (MALT) lymphoma, splenic MZL, and nodal MZL, depending on the site where the malignancy originates [1,2,5].

MicroRNAs (miRNAs) are small non-coding RNAs of approximately 22 nucleotides long that can greatly influence protein expression. RNA polymerase II–mediated transcription gives rise to pri-miRNA, which is subsequently cleaved by the nuclear RNase III DROSHA. The resulting pre-miRNA is exported from the nucleus to the cytoplasm via exportin 5 (EXP5), and then the cytoplasmic RNase III DICER1 cleaves the pre-miRNA near the terminal loop. Ultimately, one of the two strands prevails and interacts with proteins of the Argonaute (AGO) family to form the RNA-induced silencing complex (RISC).

Targeting of a specific mRNA molecule through base-pairing between the miRNA and the 3′ untranslated region (3′ UTR) results in translational repression or mRNA degradation and, therefore, diminished protein levels [6].

Considering that a miRNA can target multiple mRNAs, that a single mRNA can be targeted by several miRNAs, and that miRNA transcription can be regulated, complex gene expression networks emerge. As will be thoroughly discussed below, some of these networks have been proved to be essential for proper B-cell development and therefore their deregulation can lead to B-cell lymphomas. Multiple miRNAs have been associated with such pathological states, such as those of the miR-17/92 cluster, miR-155-5p, and miR-150-5p, while their targets involve the transcription factors FOXP1, MYC, and MYB, and the affected signaling pathways include the BCR, NFkB, and PI3K/AKT [7,8]. Besides their regulatory potential in B-cell lymphomas, miRNAs have been proposed as potential biomarkers, due to their relatively high stability in biological samples, including bodily fluids and fixed tissues, and their high specificity and sensitivity. Therefore, they could be used for personalized prognosis, prediction of therapeutic response, and as an additional tool for differential diagnosis [9].

In this review, we attempt to summarize the existing information regarding the multifaceted role of miRNAs in the indolent B-cell NHLs. On the one hand, most of the current studies investigate the expression profiles of miRNAs, an endeavor that could lead to novel biomarkers discovery. These biomarkers could be utilized to predict the transformation to a more aggressive entity of the disease, to assist in the correct differential diagnosis and to choose the optional treatment, as well as to monitor the therapeutic response. On the other hand, we attempt to provide insight into the biological function and the complex miRNA-mRNA interactions, and to further elucidate the molecular mechanisms underlying disease progression. Lastly, we address the limitations that hinder the investigation of the role of miRNAs in these lymphomas and discuss ways that these problems could be overcome in the future.

2. miRNAs in Normal B-Cell Development

Normal B-cell development is a tightly regulated complex procedure. Briefly, B cells derive from hematopoietic stem cells (HSCs) in the bone marrow, where the first steps of differentiation occur. V(D)J recombination of immunoglobulin (Ig) heavy (IgH) and light (IgL) chain genes facilitate the differentiation from pro-B cell to pre-B cell and can lead to the formation of an immature B cell that expresses a functional B-cell receptor (BCR) with unique specificity. In case these gene rearrangements are unproductive or the BCR binds strongly to presented self-antigens, the B cell cannot complete the central tolerance checkpoint, and therefore B cells are eliminated. After that, migration of B cells to the spleen takes place. In the spleen, naïve B cells can be activated by foreign-antigen recognition. Provided that they do not become autoreactive through somatic hypermutation, they differentiate into follicular or marginal zone B cells. This differentiation strongly depends on BCR signaling. Next, marginal zone B cells reside in the marginal zone, while follicular B cells enter germinal centers, forming three distinct zones: the dark, light, and mantle zone. Finally, germinal center B cells differentiate into memory or plasma cells [10].

2.1. miRNAs in Primary Lymphoid Tissue B-Cell Development

The role of miRNAs in normal B-cell development has been variously described. Most interactions between miRNAs and mRNAs involved in normal B-cell development have been investigated in mouse models; however, the vast majority of them have been predicted and/or validated in human cells as well, as miRNAs are highly conserved among species. The great effect of miRNA function in the development of B cells is prominent, as their absence completely abolishes this process; specifically, in Dicer-deficient mice, the developmental procedure was arrested, while mice lacking Dgcr8 showed elevated early B-cell apoptosis [11,12]. Interestingly, most miRNAs show a stage-specific expression pattern, indicating their stage-specific function [13]; relative examples of miRNAs showing stage-specific expression are those of miR-150-5p, miR-181a-5p, miR-126-3p, and miR-132-3p [14–16].

The developmental procedure is mainly dictated by transcription factors. Some of them show a stage-related expression as well, while others are essential in every developmental stage. TCF3 suggests a transcription factor vital for the whole developmental process, while EBF1 and PAX5 are essential for specific steps of it, as they are involved in particular processes, including BCR formation [17–20]. However, EBF1 deficiency does not lead to the eradication of the development of B cells, as the process is rescued by miR-126-3p, which was shown to stimulate the expression of *RAG1* and *RAG2* recombinases that mediate VDJ recombination [15,21]. This explains the necessity for high miR-126-3p levels in the early steps of the process. RAG1 and RAG2 expression is also regulated by the transcription factor FOXP1 [22]; in this sense, murine *Foxp1* and *Tcf3* suggest miR-191-5p targets, and so does *Egr1*, another transcription factor vital for the maturation of B cells, also targeted by miR-146a-5p [23,24]. miR-191-5p has been characterized as a rheostat for the process, as both its higher and lower levels disrupt B-cell development, due to the subsequent changes in transcription factor levels [24].

The above data delineate a miRNA-transcription factor network, showing a great impact on the developmental procedure. Disruption of this network has been variously witnessed to abolish B-cell development at pro- to pre-B cell differentiation stage. A typical such paradigm is that of miR-132-3p, a miRNA normally expressed in late developmental stages. Under physiological circumstances, miR-132-3p expression is BCR-dependent, thus it is abundant after the pro-B stage, when a functional BCR has been developed [25]. When overexpressed in the early stages, the process stopped at pro- to pre-B-cell transition due to *Sox4* transcription factor deficiency, which regulates Rag1 expression, as it was shown in xenografts. In this context, another miRNA showing stage-specific expression, namely miR-150-5p, blocks the developmental process at the same point when expressed prematurely, through *MYB* transcription factor inhibition [14,26]. This transcription factor participates in the proliferation and differentiation of hematopoietic progenitor cells.

Moreover, miR-24-3p, a member of the miR-23a cluster that promotes HSC differentiation towards common myeloid progenitors rather than lymphoid progenitors [27], functions as a *MYC* inhibitor, leading to inhibition of pro- to pre- B-cell transition [28]. MYC transcription factor is considered as a key molecule for B-cell development, as it regulates the expression of the miR-17/92 cluster; members of this cluster target *BCL2L11*, which encodes a pro-apoptotic protein, as well as *PTEN*, a key molecule for the PI3K pathway with inhibitory role. Therefore, high levels of miRNAs of the miR-17/92 cluster have been shown to block pro- to pre-B-cell differentiation and also advocate immature B-cell survival [29–32]. Immature B-cell survival is also advocated by miR-148a-3p. This further leads to self-reactive antibody production and subsequently B-cell elimination [32]. These data delineate the significant role of miRNAs in the primary steps of B-cell development.

2.2. miRNAs in Secondary Lymphoid Tissue B-Cell Development

Besides playing an important role in the development in primary lymphoid tissues, miRNAs have been reported to affect B-cell maturation in secondary lymphoid tissues as well. Specifically, a lower marginal zone B-cell number is observed upon miR-146a-5p expression, due to its binding to *NUMB* [33]. NUMB protects TP53 from degrading and advocates the Notch signaling pathway to enhance marginal zone B-cell formation [33–35]. On the contrary, miR-142-5p is crucial for marginal zone B-cell development, as it targets Tnfrs13c, (also known as Baff-R), which is required for B-cell maturation. Mice lacking miR-142-5p showed vigorous proliferation of B cells, due to high Tnfrs13c levels [36].

Concerning follicular B-cell maturation, plasma cell formation is the most frequently reported to be affected. More specifically, miRNAs affect the class-switch recombination, which includes further Ig gene rearrangements, leading to plasma cell formation. A relevant example is miR-181b-5p, another member of the miR-181 family; this miRNA targets *AID*, which is crucial for class-switch recombination, leading to inhibition of plasma cell formation [37,38]. In the same context, miR-125b-5p inhibits *PRDM1* and *IRF4* transcription factor expression, both of which stimulate class-switch recombination [39]. Therefore, repression of this miRNA is required during normal B-cell development; otherwise, B-cell malignancies may occur [40,41]. miR-30b-5p, miR-30d-5p, and miR-9-5p also attenuate *PRDM1* expression [42]. On the other hand, miR-148a-3p attenuates the expression of Bach2 and Mitf transcription factors, and consequently induces the expression of their downstream targets, Prdm1 and Irf4, leading to the terminal differentiation of B cells [43]. This fact highlights once again the necessity for stage-specific expression, as the expression of miR-148a-3p in early B-cell development leads to the arrest of this process [32]. In addition, miR-155-5p targets *SPI1* mRNA and hence reduces PAX5 expression levels, as *PAX5* expression is induced by the transcription factor SPI1 in both human and murine plasma cells [44]. In this way, the transition of germinal center B cells to plasma cells is advocated, as PAX5 downregulation is a necessity for that. Those miRNAs highly affecting B-cell development are presented in Figure 2.

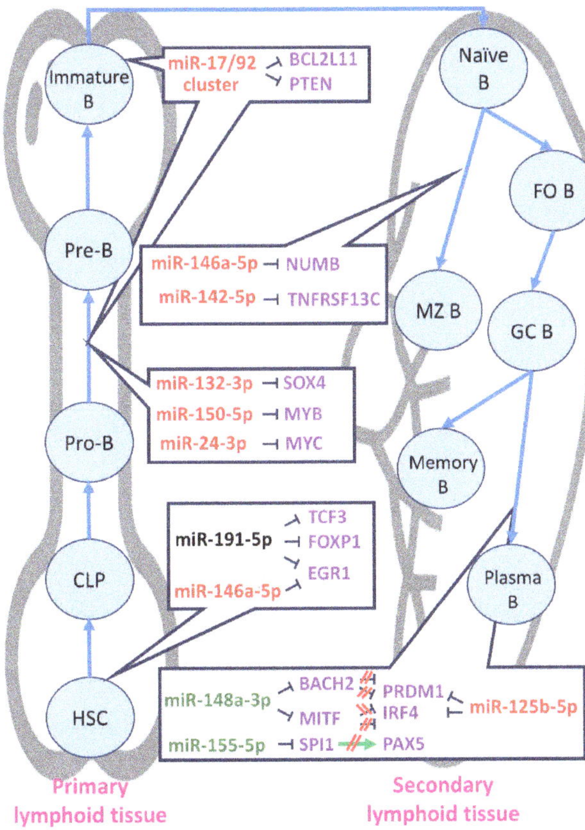

Figure 2. Brief illustration of B-cell development. miRNAs with a positive impact on the procedure are shown in green font, while those with a negative impact are shown in red. Black color indicates a miRNA with a controversial impact on B-cell development. miRNA targets are shown in purple font. Light blue arrows indicate the transition to the next developmental stage of B cells; dark blue "reverse tau" symbols (⊥) indicate attenuation of expression, whereas green arrows indicate induction of expression. Abbreviations: CLP, common lymphoid progenitor; FO B, follicular B cell; GC B, germinal center B cell; HSC, hematopoietic stem cell; MZ B, marginal zone B cell.

3. miRNAs in Follicular Lymphoma

Follicular lymphoma (FL) is one of the most common types of NHLs deriving from B cells, as aforementioned. It usually is an indolent lymphoma; however, there is the possibility to transform into an aggressive type, namely diffuse large B-cell lymphoma (DLBCL) [45,46].

FL is a broad and extremely complex clinical entity. Many genes and cellular pathways participate in the emergence and transformation of FL. In the majority of affected tissues, a t(14;18) chromosomal translocation occurs, placing *BCL2* locus next to the immunoglobulin heavy-chain enhancer and resulting in the constitutive expression of this anti-apoptotic protein [47]. However, FL development requires the acquisition of additional aberrations that enable proliferation, immune evasion, and support from microenvironmental factors. This is usually achieved by acquired aberrations in genes that control normal germinal center B-cell development.

Precisely, in the early stages of development, FL cells acquire aberrations that enable them to (a) persist in germinal centers; (b) increase BCR signaling; (c) confer a "sustainable" level of genomic instability; and (d) inhibit apoptosis. These characteristics are achieved through mutations occurring in a set of genes (*KMT2D, CREBBP, TNFRSF14, EZH2, RRAGC*). However, these FL cells usually resemble centrocytes and, similar to their normal counterparts, have a relatively low level of proliferation. The acquisition of aberrations that enable rapid proliferation, including MYC, FOXO1, BCL6, and the BCR, TLR, and TP53 pathways, alters the tumor nature, frequently leading to histological transformation. Particularly, mutations and/or translocations in the *BCL6* genomic locus are quite important in B-cell lymphomas, since BCL6 is a transcription repressor targeting many genes, including *PRDM1, TP53, CDKN1A* and *BCL2*, thus controlling the germinal B-cell formation, cell cycle, and differentiation [48,49].

Thus far, none of the current scoring systems and therapeutic approaches have been able to mitigate the risk of early progression or histologic transformation to DLBCL. Therefore, the discovery of novel biomarkers is of significant importance.

3.1. miRNAs as Potential Regulators and Biomarkers in FL

Several studies support that miRNA expression profiles can serve as signatures to differentiate between different FL subtypes, which express distinctive genes and molecular markers. Different FL subtypes can divergently progress to an aggressive type. One FL subtype which has not been well-studied is t(14;18)–negative FLs, and hence the molecular events triggering FL development in cases without a t(14;18) chromosomal translocation and without high expression of *BCL2* remain largely unknown. An interesting study analyzing t(14;18)–negative FL patients and t(14;18)–positive FL patients showed that miRNA expression between these subtypes was different. Additionally, this distinct miRNA expression was reflected in the expression of their mRNA targets. One of the miRNAs with the most robust expression changes in its potential targets was miR-16-5p. More specifically, miR-16-5p was significantly decreased in t(14;18)–negative FL patients. The decreased expression levels of this miRNA were also observed in chronic lymphocytic leukemia (CLL) patients compared to non-cancerous individuals [50]. miR-16-5p has been associated with repression of the expression of *BCL2* and hence induction of apoptosis. Although in t(14;18)–negative FL patients miR-16-5p expression is also associated with apoptosis, miR-16-5p exerts its role via an alternative regulatory network. More specifically, the decreased expression levels of this miRNA lead to increased expression of its target genes *CHEK1*, which encodes an apoptosis inhibitor and DNA repair monitor, and *CDK6*, which encodes a cyclin-dependent kinase and promoter of the cell cycle (Figure 3) [48]. These findings suggest a potential mechanism which could contribute to the pro-proliferative phenotype of t(14;18)–negative FLs.

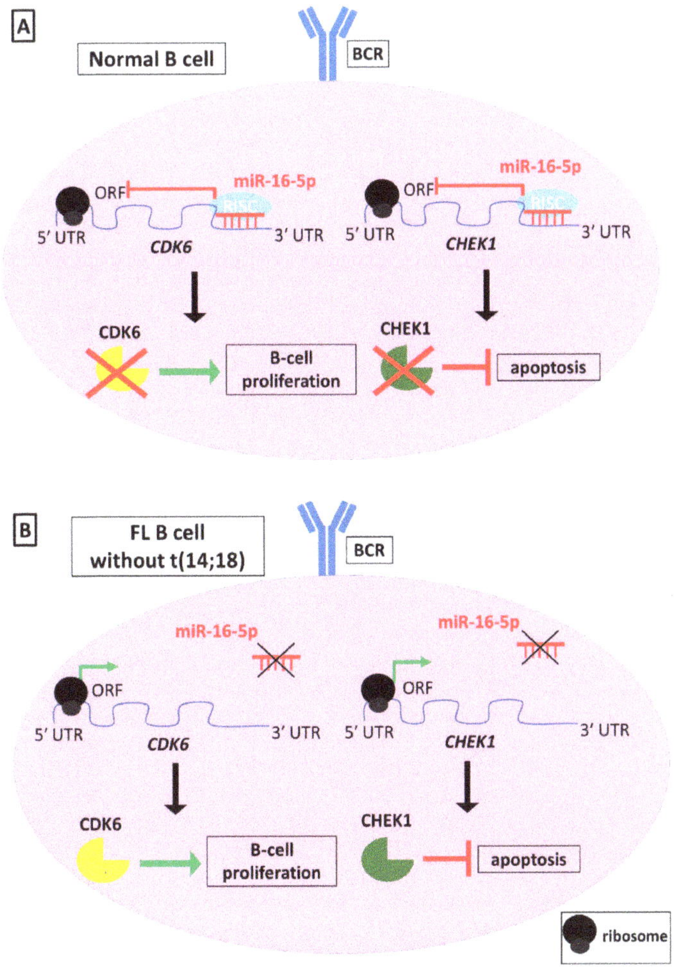

Figure 3. One of the potential effects of miR-16-5p in normal B cells (**A**) and follicular lymphoma (FL) B cells without t(14;18) (**B**). In the physiological state, its expression levels are high; miR-16-5p suppresses the expression levels of its target genes, *CHEK1* and *CDK6*, leading to apoptosis and cell cycle arrest. On the contrary, in a FL B cell, the expression levels of miR-16-5p are low, leading to increased expression of its targets. This results in B-cell apoptosis inhibition and increased proliferation rate, two hallmarks of a malignant cell. Black arrows indicate the transition to the next step; red "reverse tau" symbols (⊥) indicate inhibition, whereas green arrows indicate promotion of a cellular process. Abbreviations: BCR, B-cell receptor; RISC, RNA-induced silencing complex.

Despite the extensive investigation concerning the *BCL2* translocation in FL, *BCL6* translocation is not well-studied. Interestingly, Gebauer et al. attempted to find unique miRNA signatures between typical FL with translocation in *BCL2*, but not in *BCL6* (BCL2+/BCL6− FL) and FL with translocation in *BCL6*, but with or without translocation in *BCL2* (BCL2+/BCL6+ or BCL2−/BCL6+ FL). More specifically, in BCL2+/BCL6+ and BCL2−/BCL6+ FL patients, 21 miRNAs were significantly upregulated, and 12 miRNAs were significantly downregulated compared to BCL6− ones. Even though the functional role of these miRNAs was not further investigated and future validation of these results in a larger patient cohort is required, these results underline the differential molecular background of these subtypes and pave the way for the potential integration of miRNA signatures in FL molecular classification [51].

Additionally, the current system of differential diagnosis between distinct types of B-cell NHLs is not effective, and novel biomarkers are required. miRNAs, due to the plethora of advantages by which are characterized, have emerged as promising candidates for this aim. Especially, the discrimination between DLBCL and FL lymphoma is quite significant, since DLBCL shows high phenotypic diversity and de novo DLBCL is not easily distinguished from transformed FL [52]. Two independent analyses examined the expression pattern of miRNAs in these two different types of B-cell lymphoma; however, they resulted in distinct molecular signatures. More specifically, in the first research study, miR-200c-3p and miR-638 along with members of the miR-17/92 cluster were among the most highly expressed miRNAs in DLBCL compared to FL [53]. The second study led to different results, with limited coverage with the miRNAs of the first research study. However, this study also designated the distinct expression levels of miRNAs of the miR-17/92 cluster [54].

Particularly, the miR-17/92 cluster (genomic locus: 13q31.3) and its two paralogs, namely the miR-106b-25 and miR-106a-363 clusters, have been reported to be involved in several hematological malignancies, and more specifically in the most aggressive ones. This could be partly attributed to enhanced transcription of the miR-17/92 cluster host gene (*MIR17HG*) by MYC oncoprotein and, thus, the upregulation of 6 oncogenic miRNAs (miR-17-5p, miR-18a-5p, miR-19a-3p, miR-19b-3p, miR-20a-5p, and miR-92a-3p). miR-17-5p inactivates CDKN1A, leading to deregulation of the cell cycle and increased cell proliferation. miR-18a-5p and miR-19a-3p repress CD32 (Fc fragment of IgG receptor IIb, FCGR2B) and CD22, respectively, resulting in the upregulation of the BCR signaling pathway, and hence in elevated B-cell activation and division [55]. Moreover, miR-19b-3p inhibits PTEN phosphatase [56], a regulator of cell cycle and growth, which subsequently suppresses the oncogenic PI3K/AKT signaling pathway [57,58]. Considering all these findings, further investigation of this miRNA cluster role in FL is considered fruitful.

An additional miRNA that is implicated in the high-grade transformation of FL is miR-150-5p. This miRNA plays, also, a key role in normal B-cell development via targeting *MYB* [14]. In FL, its expression is repressed by MYC, leading to its decreased expression levels in FL cells and, consequently, high expression levels of another one of its targets, the transcription factor *FOXP1*. The elevated expression of the latter has been linked to lower survival rates and high-grade malignant transformation. This could be attributed to its role, since FOXP1 regulates the expression of many genes involved in cell survival and cell cycle activation, and promotes BCR signaling, while it is also critical for normal B-cell development [57,59].

Furthermore, in a patient cohort study it has been observed that as the disease progresses from FL to aggressive DLBCL, miR-31-5p expression levels decrease. This miRNA has attracted researchers' interest due to its multifaceted role. Depending on its specific targets in distinct cell types, miR-31-5p can exert either an oncogenic or an onco-suppressive role in several malignant states. The low expression of this miRNA has also been observed in a cohort study with DLBCL patients [60]. This low expression can be achieved either by loss of the gene locus of this miRNA or by hypermethylation of its promoter, while both mechanisms have been detected in different malignancies. The *MIR31* gene is located on chromosome band 9p21.3, $\sim$500 kb from the locus of the well-known tumor suppressors CDKN2A and CDKN2B. Due to their proximity, it is reasonable to suppose that *MIR31* would be lost together with *CDKN2A* [61]. The deletion of the latter has been associated with poor prognosis of DLBCL patients. In a recent study regarding FL transformation, it was observed that *E2F2* and *PIK3C2A* could be direct targets of miR-31-5p. E2F2 is a transcription factor of the E2F family, which permits the entry of cells to the S-phase of the cell cycle, thus promoting the cell cycle, and PIK3C2A, which is a catalytic subunit of the PI3K family, is involved in cell migration, survival, and proliferation. High levels of E2F2 and elevated activity of the PI3K/AKT signaling pathway have been observed in DLBCL, while the latter has been associated with poor outcome of DLBCL patients, as well. Therefore, low miR-31-5p expression levels could result in a B-cell high-grade tumor, via the increased levels of the aforementioned proteins [62]. Additionally, the same study uncovered the increase of the expression levels of miR-17-5p during FL transformation, a finding which is consistent with the oncogenic role of this miRNA in several other malignancies.

Besides miRNA signatures with the potential to discriminate between different B-cell NHLs, an interesting study revealed a miRNA signature capable of distinguishing FL cells from normal germinal center B cells in follicular hyperplasia. The most highly expressed miRNAs in FL included miR-20a-5p, miR-20b-5p, and miR-194-5p. The first two miRNAs have been proved to target *CDKN1A*, which partly accounts for cell cycle arrest, while miR-194-5p controls the expression of SOCS2, a suppressor of the JAK/STAT signaling pathway, which participates in cell proliferation and survival [63].

miRNAs exerting a regulatory role in indolent B-cell NHLs, along with their targets and effect in malignant B cells, are summarized in Table 1, while Table 2 highlights those miRNAs showing a potential clinical utility as candidate biomarkers in FL.

Table 1. miRNAs with a regulatory effect in indolent B-cell non-Hodgkin lymphomas (NHLs).

Disease	miRNAs	Expression in Lymphomas	Targets	Effect	References
Follicular lymphoma (FL)	miR-150-5p	Decreased	*FOXP1*	Inhibition of B-cell survival	[59]
	miR-31-5p		*E2F2, PIK3C2A*	Inhibition of cell cycle, survival, and migration	[62]
	miR-202-3p		*DICER1*	Regulation of biogenesis of miRNAs	[64]
			SKP2	Regulation of cell cycle transition	
	miR-618		*HDAC3*	Inhibition of cell cycle	[65]
			CUL4A	Inhibition of DNA damage response	
	miR-155-5p		*INPP5D*	Promotion of anti-tumor immune responses	[66]
	miR-16-5p		*CHEK1*	Promotion of B-cell apoptosis	[48]
			CDK6	Inhibition of B-cell proliferation	
	miR-20a-5p, miR-20b-5p	Increased	*CDKN1A*	Promotion of cell cycle	[63]
	miR-194-5p		*SOCS2*	Promotion of B-cell proliferation and survival	
	miR-93-5p		*MICA, MICB*	Inhibition of B-cell cytotoxicity	[67]
Gastric MALT lymphoma	miR-34a-5p	Decreased	*FOXP1*	Inhibition of B-cell survival	[22,68–71]
	miR-383-5p		*ZEB2*	Inhibition of epithelial-to-mesenchymal transition (EMT)	[72]
	miR-203a-3p		*ABL1*	Inhibition of B-cell proliferation	[73,74]
	miR-155-5p, miR-142-5p	Increased	*TP53INP1*	Inhibition of apoptosis	[74–76]
Splenic MZL	miR-26b-5p	Decreased	*NEK6*	Inhibition of mitosis-cell division	[77,78]
Waldenström's macroglobulinemia (WM)	miR-9-3p	Decreased	*HDAC4, HDAC5*	Regulation of histone acetylation; Induction of WM cell cytotoxicity; promotion of WM cell autophagy and apoptosis	[79]
	miR-23b-3p		*SP1*	Suppression of NFkB signaling; Inhibition of cell proliferation and survival	[80]
	miR-155-5p	Increased	-	Promotion of MAPK/ERK, PI3K/AKT, and NFkB signaling; promotion of cell proliferation, adhesion, and migration	[81,82]
			FOXO3, BCL2L11	Inhibition of apoptosis	[83]
	miR-206-3p		*KAT6A*	Regulation of histone acetylation	[79]

Abbreviations: MALT, mucosa-associated lymphoid tissue; MZL, marginal zone lymphoma.

3.2. Genetic Polymorphisms of miRNA Genes in FL

Genetic variation in miRNA regulatory pathways in distinct malignancies has raised researchers' interest, as well. These polymorphisms can be developed in miRNA-binding site target genes, in miRNA biogenesis pathway genes, and in different regions in miRNA genes. Therefore, they can change the function of the respective miRNA and serve as potential indicators for diagnosis and prognosis in clinical practice. Although these miRNA variants or single nucleotide polymorphisms (SNPs) have been extensively examined in a wide range of malignancies, the knowledge of miRNA SNPs in FL and generally in B-cell NHLs remains poor [84].

miR-202-3p and miR-618 have been implicated in FL, while the SNPs in their precursor sequences are associated with FL, via the impact on the levels of the target gene expression. More specifically, the presence of these miRNA SNPs has been linked with elevated risk for FL. A possible explanation of this finding is that the presence of a SNP in mir-202 and mir-618 could lead to a decrease in the expression levels of miR-202-3p and miR-618, respectively, which seem to exert onco-suppressive roles. miR-202-3p has been shown to target *DICER1*, which is essential for the proper biogenesis and function of all miRNAs, while its high expression has been associated with B-cell lymphoma development and survival. Additionally, miR-202-3p targets *SKP2*, which encodes a regulator of G1 to S-phase transition of the cell cycle and inhibitor of CDKN1B [64]. Concerning miR-618, it targets genes encoding histone deacetylase, HDAC3, which represses the function of the onco-suppressor, TP53, and CUL4A, a protein that is involved in the degradation of DNA damage-response proteins, TP53 and TP73 [65].

3.3. miRNAs and the Immune System

The immune system activation plays a critical role in inhibiting the progression of nascent tumors by recognizing specific antigens on the surface of malignant cells, while it has been investigated in the context of FL as well. For instance, the natural killer (NK) cell receptor KLRK1 was found to bind both MICA and MICB, thus leading to suppression of B-cell lymphomas by inducing cell cytotoxicity. This finding is consistent with the high expression levels of the aforementioned KLRK1 ligands in low-grade FL and their low expression levels in high-grade FL. The increased expression levels of miR-93-5p in high-grade FL are associated with the lower expression levels of MICA and MICB, since the respective mRNAs constitute direct targets of miR-93-5p [67].

Immunotherapy and specifically monoclonal antibodies (mAb) have been introduced in FL treatment arsenal. Particularly, Obinutuzumab, a humanized anti-CD20 mAb, is approved as a treatment for FL. This mAb increases the affinity between the CD20 receptor of malignant B cells and the CD16 receptor of NK cells, leading to increased INF-γ levels. Interestingly, miR-155-5p is implicated in this procedure, as was shown in a recent study. More specifically, it was observed that obinutuzumab-induced CD16 stimulation led to overexpression of miR-155-5p, which targeted INPP5D inositol phosphatase, a regulator of the PI3K pathway. This resulted in the activation of the downstream target of PI3K, MTOR, and the stimulation of IFNγ production from NK cells, triggering anti-tumor immune responses [85]. The NK cells are pivotal components of innate immunosurveillance against malignancies and represent a particularly attractive tool in the context of anti-tumor immunotherapy. NK cells rapidly recognize and destroy many tumor cell types and also play an immunoregulatory role in the instruction of adaptive anti-tumor responses [86]. Therefore, deciphering the cascade following the treatment with mAb is critical, since it is expected to contribute utmost to the optimization of treatment strategies against malignancies.

Table 2. miRNAs as candidate biomarkers in follicular lymphoma (FL).

miRNAs	Sample Origin	Expression	Potential Biomarker	References
miRNAs of miR-17/92 and miR-106a-363 clusters, miR-200c-3p, miR-638, miR-518a-3p	FFPE tissues	Upregulated in DLBCL vs. FL	Differential diagnosis	[53]
miR-17-5p				[54]
miR-217-5p, miR-634, miR-26b-5p		Upregulated in FL vs. DLBCL		[53]
miR-330-3p, miR-106a-5p, miR-210-3p		Upregulated in FL vs. DLBCL		[54]
miR-612, miR-188-5p, miR-302c-3p, miR-433-3p, miR-584-5p		Upregulated in BCL2+/BCL6+ and BCL2−/BCL6+ FL vs. BCL2+/BCL6− FL		[51]
miR-200a-3p, miR-135a-5p, miR-375-3p, miR-138-5p, miR-517 isomiRs		Downregulated in BCL2+/BCL6+ and BCL2−/BCL6+ FL vs. BCL2+/BCL6− FL		
miR-16-5p, miR-138-5p, miR-26a-5p, miR-29c-3p		Downregulated in t(14;18)–negative FL vs. t(14;18)–positive FL		[48]
miR-193a-5p, miR-345-5p, miR-574-3p, miR-1287-5p, miR-1471	Enriched FL cells	Upregulated in FL vs. with follicular hyperplasia		[63]
miR-570-3p, miR-205-5p, miR-222-3p, miR-30a-5p, miR-301b-3p, miR-141-3p		Downregulated in FL vs. with follicular hyperplasia		
miR-20b-5p, miR-26a-5p, miR-92b-3p, miR-487b-3p	Cancer cell lines	Upregulated in FL cell lines vs. DLBCL cell lines		[87]
miR-330-3p, miR-106a-5p, miR-210-3p, miR-301 isomiRs, miR-338-5p	FFPE tissues	Upregulated in FL vs. non- neoplastic lymph nodes	Diagnosis	[54]
miR-149-5p, miR-139-5p		Downregulated in FL vs. non- neoplastic lymph nodes		
miR-16-5p, miR-17-5p, miR-26a-5p, miR-29a-3p, let-7d-5p, let-7g-5p, let-7i-5p	Cancer cell lines	Downregulated in FL cell lines vs. CD19+ negatively- selected B cells		[87]
miR-144-3p, miR-431-5p	FL biopsies and blood samples	Upregulated in relapsed FL patients	Prognosis, prediction of disease progression	[88]
miR-376b-3p		Downregulated in non-relapsed FL patients		

Abbreviations: DLBCL, diffuse large B-cell lymphoma; FFPE, formalin-fixed, paraffin-embedded.

4. miRNAs in Marginal Zone Lymphoma

4.1. Extranodal Marginal Zone Lymphoma or Mucosa-Associated Lymphoid Tissue (MALT) Lymphoma

Extranodal MZL, also known as MALT lymphoma, is the most common type of indolent MZL and it starts at places where malignant marginal zone B cells initially infiltrate MALTs, other than the lymph nodes (hence the name extranodal). Stomach is the most common organ where this malignancy can arise, accounting for almost half the incidences. This type of lymphoma is known as MALT gastric lymphoma (MALT GL). Less frequently, MALT lymphoma can start at organs other than the stomach (non-gastric), such as the skin. The exact underlying mechanisms of this disease are not yet known, but clinical and epidemiological data have profoundly associated the high risk of MALT lymphoma development with certain chronic infections and autoimmune diseases. The *Helicobacter pylori* infection and Sjögren's syndrome constitute the most important leading factors, respectively [89–92].

4.2. Gastric MALT Lymphoma and H. pylori

A major leading factor of MALT GL is chronic inflammation triggered by persistent *H. pylori* infection. As a proof, *H. pylori* eradication can fully treat MALT GL in a large number of cases [90,93]. The majority of patients with MALT GL resistant to *H. pylori* eradication therapy, have in common a t(11;18)(q21;q21) chromosomal translocation [94]. This translocation leads to the production of a fusion gene, consisting of the apoptosis inhibitor *BIRC3* and the caspase-like protease *MALT1*. The encoded fusion protein enhances NFkB signaling and hence leads to inhibition of apoptosis [95–97].

Saito et al. contributed greatly to the elucidation of the mechanisms involved in MALT GL progression. Firstly, they indicated that miR-142-5p and miR-155-5p are overexpressed in MALT GL compared to non-tumor gastric mucosa, while their levels were significantly higher in patients unresponsive to *H. pylori* eradication treatment, compared to the responsive ones, implying their potential prognostic and predictive utility as biomarkers. Considering the aforementioned critical role of these two miRNAs in normal B-cell development, their further investigation is quite significant. Interestingly, two of the patients in this study were resistant to *H. pylori* eradication therapy but lacked the *BIRC3-MALT1* fusion gene. Nevertheless, they were also characterized by increased expression of miR-142-5p and miR-155-5p, suggesting that these molecules could be used as additional biomarkers in MALT GL [75]. Moreover, it was shown that they target the pro-apoptotic gene *TP53INP1* (Figure 4). This interaction could lead to inhibition of apoptosis and acceleration of MALT GL cell proliferation, designating miR-142-5p and miR-155-5p as potential therapeutic targets [75]. Additionally, a former research study had revealed that *TP53INP1* transcription is activated by TP73, and therefore cell cycle arrest is facilitated [98]. Considering this additional regulation level of *TP53INP1*, it would be interesting to be further investigated in the context of miR-142-5p and miR-155-5p expression.

Another relevant study concluded that miR-383-5p is downregulated in MALT GL patients infected with *H. pylori*, compared to normal non-tumor gastric mucosae tissues, and determined *ZEB2* as its target [72]. High levels of ZEB2 have previously been reported to promote epithelial-mesenchymal transition (EMT) of gastric cancer cells via regulation of expression of CDH1 (E-cadherin) and other EMT markers, such as VIM (vimentin) and matrix metallopeptidases (MMP2 and MMP9) [99]. Thus, the effect of reduced expression of miR-383-5p and the high expression of *ZEB2* could assist in the understanding of the role of *H. pylori* infection in MALT GL development.

4.3. *From Chronic Gastritis to MALT GL*

Chronic gastritis can also progress to MALT GL development; however, the molecular basis of this transformation remains unknown. Craig et al. identified miR-203a-3p to be significantly underexpressed in MALT GL compared to normal lymphoid tissue. Furthermore, *MIR203A* promoter was hypermethylated, and its target protein, ABL1, was overexpressed in MALT GL in comparison with gastritis, indicating ABL inhibitors as a novel therapeutic approach in MALT GL [73]. ABL1 is a tyrosine kinase that activates various signaling pathways, including BCR, leading to cell proliferation. The BCR signaling activity is elevated in several hematological malignancies, while its targeting constitutes a therapeutic approach in several cancers. In accordance with the aforementioned observations regarding miR-155-5p overexpression in *H. pylori*–positive MALT GL patients [75] and the miR-203a-3p underexpression in MALT GL, findings from an independent study showed a decrease in miR-203a-3p expression levels and a concomitant increase in miR-155-5p expression levels in MALT GL patients, compared to chronic gastritis patients [74]. The same study designated the high expression levels of miR-142-3p in MALT GL compared to chronic gastritis (Table 3). These findings are quite important since the morphological diagnosis of MALT GL is still hampered by overlapping histological features with advanced chronic gastritis. Considering that either the 3′ or the 5′ of a miRNA stem-loop is expressed under certain circumstances, the overexpression of miR-142-5p and miR-142-3p in MALT GL patients would be interesting to examine further.

The role of miRNAs in malignant transformation from chronic gastritis to MALT GL has been investigated in other studies, as well. Another miRNA with a critical role in normal B-cell development, via targeting the transcription factor *MYB* [14], and with deregulated expression levels in MALT GL, is miR-150-5p. More precisely, it was found to be significantly overexpressed in MALT GL in comparison with chronic gastritis [76,100]. On the contrary, miR-150-5p expression levels were low during FL transformation [59].

Considering these findings, it could be fruitful to investigate the functional role of miR-150-5p in MALT GL. Additionally, findings deriving from gastric cancer research revealed that miR-150-5p inhibits apoptosis in gastric cancer cells by targeting the pro-apoptotic gene *EGR2* [101]. This molecule has been investigated in hematological malignancies and its deregulated expression or potential mutations have been associated with tumorigenesis [102,103]. Therefore, this could be a potential mechanism of action via which miR-150-5p could exert its role in MALT GL.

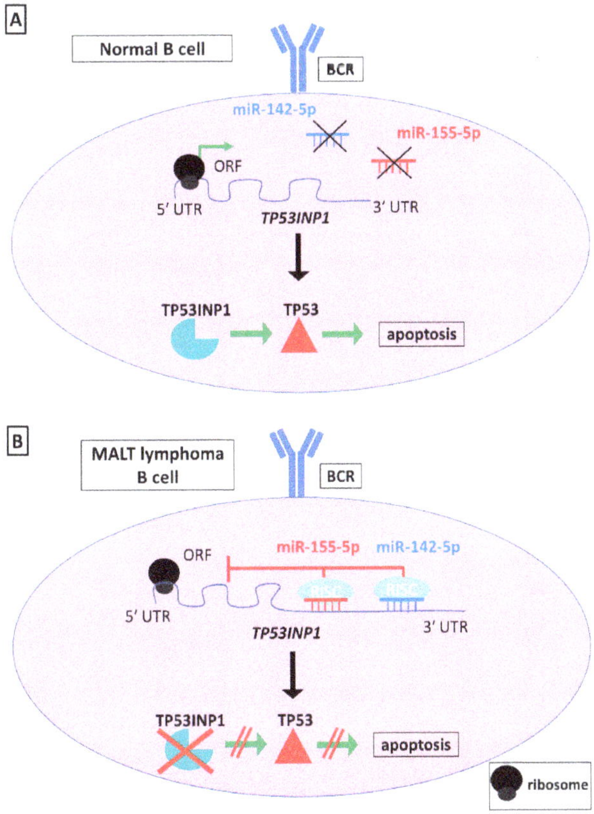

Figure 4. One of the potential effects of miR-155-5p and miR-142-5p in a normal B cell (**A**) and a mucosa-associated lymphoid tissue (MALT) lymphoma B cell (**B**). In physiological state, their expression levels are low and their target gene, *TP53INP1*, is expressed, leading to activation of TP53 and, subsequently, to apoptosis. On the contrary, in a MALT lymphoma B cell, the expression levels of miR-155-5p and miR-142-5p are increased, leading to decreased expression of their target. This results in apoptosis inhibition, one of the hallmarks of a malignant cell. Black arrows indicate transition to the next step; red "reverse tau" symbols (⊥) indicate inhibition, whereas green arrows indicate promotion of a cellular process. Abbreviations: BCR, B-cell receptor; RISC, RNA-induced silencing complex.

Additionally, Blosse et al. showed that miR-150-5p, miR-155-5p, miR-196a-5p, and miR-138-5p were upregulated and miR-7-5p and miR-153-3p were downregulated in MALT GL patients compared to gastritis control patients [76]. miR-150-5p and miR-155-5p have been associated with MALT GL in other studies, as aforementioned. The convergence of several research studies in the deregulated expression of these two miRNAs highlights their critical role in MALT GL. Regarding the deregulated expression levels of the rest miRNAs and their significant role in other malignancies, the investigation of their function in MALT GL is critical, as well. Herein, we propose potential mechanisms of action of these miRNAs, based on current literature. Interestingly, miR-196a-5p is highly expressed in gastric cancer cells and targets the cell cycle inhibitor *CDKN1B* (*p27^{kip1}*), leading to increased cell proliferation [104], while miR-153-3p acts on *AKT3* in lung cancer reducing cell proliferation rate [105]. Moreover, *EGFR* and *IGF1R* have been proposed as targets of miR-7-5p in gastric cancer cells, suggesting a way that miR-7-5p can suppress the invasion and metastasis of these malignant cells [106,107]. Finally, conflicting studies have been made about the role of miR-138-5p in gastric cancer cells, as some support its role as onco-suppressive or as oncogenic miRNA [108,109], so further investigation is required to properly decipher its role.

4.4. From MALT GL to Gastric DLBCL

An alarming situation arises when MALT GL is transformed to gastric DLBCL through mechanisms which are not well understood. A microarray analysis between these two pathological states revealed 27 underexpressed miRNAs in gastric DLBCL compared to MALT GL [68]. These miRNAs were transcriptionally repressed by MYC, as previously shown in a B-cell lymphoma mouse model [69], while miR-34a-5p possessed the most tumor-suppressive properties [68]. In the same study, MYC was found to be greatly overexpressed in gastric DLBCL in comparison with MALT GL, and the MIR34A promoter was also found to be hypermethylated only in gastric DLBCL. Moreover, the validated target of miR-34a-5p in the aforementioned study was *FOXP1* [68]. During normal B-cell development, constitutive expression of miR-34a-5p can result in a block in B cell development at the pro-B cell to pre-B cell transition, leading to a reduction in mature B cells. This block appeared to be mediated primarily by inhibited expression of the *FOXP1* [110]. Several previous studies highlighted that FOXP1 is necessary for normal B-cell differentiation [22] and has been reported to predict the transformation of MALT GL to gastric DLBCL [70]. Indeed, data from another clinical study confirm that miR-34a-5p could be utilized as a prognostic biomarker to investigate MALT GL to gastric DLBCL transformation [71]. These results suggest a novel way that FOXP1 can lead to MALT GL progression, besides the t(3;14)(p14;q32) chromosomal translocation that results in a *IGH-FOXP1* fusion gene and, therefore, in elevated levels of FOXP1 in MALT lymphomas [111,112]. Moreover, FOXP1 was found to be elevated in high-grade lymphomas resulting from transformation of FL [59]. All these findings highlight the pivotal role of *FOXP1* in the development of B-cell malignancies and hence the role of miR-34a-5p as one of its potential regulators.

In addition, Gu et al. demonstrated that miR-16-5p had higher expression levels in MALT GL patients than those with gastric DLBCL and could be used as another biomarker predicting MALT GL transformation [113]. miR-16-5p is a key tumor-suppressive miRNA and has repeatedly been associated with CLL and, as aforementioned, with FL. Its more well-known target is the anti-apoptotic gene *BCL2*, via the suppression of which can inhibit apoptosis in CLL [114]. Even though its potential functional role in MALT GL transformation has not been unraveled, the existing data regarding its function in other hematological malignancies highlight its further investigation in this cancer, as well.

4.5. Non-Gastric MALT Lymphoma

Studies investigating the role of miRNAs in non-gastric MALT lymphomas are far less frequent. Cutaneous marginal zone B-cell lymphoma is another extranodal MZL, which has been associated with *Borrelia burgdorferi* infection [115,116]. Reduced expression of miR-150-5p and miR-155-5p in primary cutaneous MZL in comparison with primary cutaneous centrofollicular lymphoma has been linked to disease deterioration and lower survival rates only in primary cutaneous MZL [117]. Therefore, these two miRNAs could be used to predict the outcome of this lymphoma type. Comparing the aforementioned results which propose that miR-150-5p and miR-155-5p are elevated in gastric MALT lymphoma and could contribute to this lymphoma progression [75,76,100] and the results regarding the expression levels of these two miRNAs in primary cutaneous MZL, a conflict in their role emerges. Probably this could be attributed to the involvement of these miRNAs in several networks only certain of which could prevail depending on the organ and the microenvironment (stomach or skin). The fact that the expression levels of miR-150-5p were decreased during FL transformation supports the aforementioned hypothesis [59]. However, further investigation is needed.

Ocular adnexal lymphoma (OAL) is another less common type of extranodal MZL affecting tissues surrounding the eye, though the driving mechanisms of this disease are still under investigation. The only study to date to have examined the miRNA expression profiles in OAL has revealed that let-7g-5p, miR-16-5p, members of the miR-29 family, miR-199a-5p, and miR-222-3p were overexpressed in OAL in comparison with ocular DLBCL, the aggressive transformed malignancy which can arise in some patients [118]. Strikingly, transcription of many of these miRNAs is suppressed by MYC, which usually drives B-cell proliferation [69]. Thus, transcriptional repression of miRNA host genes, mediated by MYC, most likely contributes to the transformation of OAL to ocular DLBCL. The majority of these miRNAs play a critical role in hematological malignancies and solid tumors, necessitating their further investigation in B-cell NHLs, as well. Below, some potential modes of action of these miRNAs are proposed. One of the most critical miRNAs for further investigation is miR-16-5p, as it has been repeatedly characterized as a pivotal tumor suppressor in B-cell malignancies [113]. Furthermore, the members of the let-7 family can act as regulators of stem-cell differentiation and have also been implicated in tumor suppression in several ways. Interestingly, some members of this family suppress the acquisition and utilization of key nutrients, which are essential for B-cell activation. Additionally, members of the miR-29 family have been characterized as tumor suppressors in other malignancies, including mantle cell lymphoma, Burkitt lymphoma, and FL. This miRNA family is implicated in the regulation of several key pathways in carcinogenesis. Some of its main target genes are *CDK6*, *DNMT3B*, *TCL1A*, and *MCL1*, which are involved in cell cycle control, DNA methylation, and apoptosis inhibition, respectively [119]. Regarding miR-199a-5p, its high expression has been associated with a better outcome in DLBCL patients, while one of its potential roles is the suppression of the NFkB signaling pathway, a critical pathway for the development of this malignancy [120,121]. Finally, miR-222-3p is another miRNA with a contradictory role since it has been characterized both as oncomiR and a tumor suppressor in lymphomas, highlighting the complex regulatory roles and networks of miRNAs.

Besides chronic bacterial infections, Sjögren's syndrome (SS) is a chronic autoimmune disease affecting predominantly exocrine glands, in which a considerable percentage of patients are at high risk of developing B-cell NHL, with MALT lymphoma being the most frequent subtype [91,122]. miR-200b-5p was found to be significantly underexpressed in minor salivary glands of SS patients with MALT lymphoma compared to SS patients without lymphoma [123]. Interestingly, in another study, it has been revealed that low expression levels of this miRNA in minor salivary glands could predict SS patients who are at high risk of B-cell NHL development, even before the appearance of clinical symptoms of the disease. However, these results necessitate validation in a larger cohort of patients [124]. The discovery of a potential implication of this miRNA in MALT lymphoma is quite important since until recently it was believed that it was degraded and the miR-200b-3p prevailed. However, recent data support the synergistic action of both miRNAs in the inhibition of EMT [125]. Therefore, the functional investigation of miR-200b-5p in MALT lymphoma and B-cell development in general could be interesting.

4.6. Splenic Marginal Zone Lymphoma

Splenic marginal zone lymphoma (SMZL) is an indolent B-cell NHL, but the possibility to be transformed eventually to more aggressive type lymphomas is quite high. It originates in the spleen and lacks a clear etiology, which necessitates the discovery of the implicated molecular mechanisms and of indicators for disease development and progression. Due to the multifaceted role of miRNAs in normal B-cell development, miRNAs can prove to be beneficial in the aforementioned endeavor [126]. The most common chromosomal abnormality, present in approximately 40% of SMZL cases, is a characteristic 2.8-Mbp 7q32 heterozygous deletion [127,128]. This genomic locus comprises the host genes of several miRNAs that are underexpressed in SMZL, including *MIR593*, *MIR129-1*, *MIR182*, *MIR96*, *MIR183*, *MIR335*, *MIR29A*, and *MIR29B1* [128]. These miRNAs are involved in various regulatory networks affecting cell differentiation, proliferation, and apoptosis, and hence their specific role in SMZL requires further investigation. However, indicative mechanisms of action have been proposed for the most significantly underexpressed miRNAs, mainly in other types of malignancies [126]. In detail, miRNAs of the miR-29a/b1 cluster are believed to act as tumor suppressors by inhibiting the expression of the *TCL1A* oncogene, one of the most overexpressed genes in SMZL [129]. This interaction has already been observed in CLL [130]. Additionally, miR-129-5p has been shown to target the notorious anti-apoptotic *BCL2* mRNA in colorectal cancer, both in vitro and in vivo, leading to apoptosis and enhancing the cytotoxic effect of 5-fluorouracil [131]. On the contrary, miR-182-5p and miR-183-5p have been characterized as oncogenic miRNAs. More specifically, their overexpression in mesothelioma leads to enhanced cell proliferation and invasion. This function is established mainly by preventing the expression of *FOXO1* transcription factor, which in turn facilitates the expression of *CDKN1B*, a key inhibitor of CDKs [132]. Therefore, their downregulation in the present research study raises questions regarding their function.

Another study showed that miR-21-5p overexpression in SMZL is linked to an aggressive transformation type of the disease [133]. Even though the role of miR-21-5p has not been investigated in SMZL, there are several studies that characterize it as oncomiR in NHLs, while there are studies, which examine this miRNA as a therapeutic target. More precisely, it inhibits the expression of PTEN and FOXO3, which are molecules with a critical role in normal B-cell development, in human B-cell NHL cell lines. This inhibition activates the PI3K/AKT pathway and renders a human DLBCL cell line resistant to chemotherapy [134].

Studies demonstrate that hepatitis C virus (HCV) infection is a risk factor for SMZL development, but the underlying mechanisms leading to this condition are poorly understood. A solid argument supporting this notion is that HCV-positive SMZL patients who received antiviral treatment achieved complete or partial remission [135–137]. An extensive miRNA profiling in SMZL revealed five miRNAs to be overexpressed and seven miRNAs having decreased expression in SMZL compared to the non-tumor splenic marginal zone. Following the stratification of SMZL patients to HCV-positive and -negative ones, miR-26b-5p was proved to be significantly underexpressed in HCV-positive SMZL patients compared to negative ones, and *NEK6* is a predicted target of miR-26b-5p [77]. NEK kinases, including NEK6, facilitate many mitotic events and, subsequently, cell division, while they are critical for STAT3 phosphorylation and hence JAK/STAT signaling pathway activation [78]. Although the overexpression of *NEK6* has not been associated with SMZL, it has been associated with the development of other malignancies [138]. Thus, the decreased expression of miR-26b-5p and the subsequently increased expression of *NEK6* in HCV-positive patients suggest a molecular mechanism of action through which HCV infection could lead to SMZL.

4.7. Nodal Marginal Zone Lymphoma

Nodal marginal zone lymphoma (NMZL) is a rare MZL subtype initiating in the lymph nodes which is challenging to differentiate diagnostically, due to the lack of specific indicators for it. Considering the decisive role of miRNAs in B-cell developmental stages, potential distinct expression patterns of these regulatory molecules could serve as useful biomarkers [139,140]. Intriguingly, Arribas et al. conducted a miRNA expression analysis accompanied by an analysis of their targets, in order to conclude in a miRNA signature able to distinguish NMZL from FL [139]. There are some cases of NMZL that are challenging to be distinguished and a combination of clinical, histological, immunohistochemical, and molecular data is required, so these findings could be an additional tool for classifying patients standing in the diagnostic grey zone [141]. Although further investigation is required, they showed that miR-223-3p and let-7f-5p were the most highly expressed miRNAs in NMZL compared to FL. Interestingly, miR-223-3p has been proved to regulate naïve to germinal center B-cell transition and germinal center to memory B-cell transition, via the repression of the key protein for hematopoietic development, LMO2 [42]. FL and germinal center cells are distinguished by an increased expression of LMO2, and a diminished expression of miR-223-3p. In the aforementioned study, it was shown that the expression of LMO2 was low and the expression of miR-223-3p was high in NMZL patients, implicating a potential role of these molecules in NMZL development. Although it is known that let-7f-5p is a member of the let-7 miRNA family, which has been shown to target various oncogenes and is usually underexpressed in many malignancies [142], a functional explanation of the differential expression of this miRNA among these malignancies is not provided. Another study, conducted by Gebauer et al., identified several miRNAs differentially expressed between transformed NMZL, which is characterized by the presence of larger cells under histopathological examination, and DLBCL. These distinct miRNA signatures support the notion that transformed NMZL is biologically a distinct disease entity, while the presence of large cells in some cases of NMZL does not correspond to an aggressive type transformation into DLBCL [143,144].

Table 3. miRNAs as candidate biomarkers in marginal zone lymphoma (MZL).

Disease	miRNAs	Sample Origin	Expression	Potential Biomarker	References
Gastric MALT lymphoma	miR-142-3p, miR-155-5p	FFPE tissues	Upregulated in gastric MALT lymphoma vs. chronic gastritis	Differential diagnosis	[74]
	miR-203a-3p		Downregulated in gastric MALT lymphoma vs. chronic gastritis		
Ocular adnexal lymphoma (OAL)	let-7g-5p, miR-16-5p, miR-29 family, miR-199a-5p, miR-222-3p	FFPE tissues	Upregulated in OAL vs. ocular DLBCL	Differential diagnosis	[118]
Sjögren's syndrome (SS) associated with MALT lymphoma	miR-200b-5p	Minor salivary glands and PBMCs	Downregulated in SS-associated MALT lymphoma vs. SS	Prognosis, prediction of patients' relapse	[123,124]
Splenic MZL (SMZL)	miR-21-5p	Fresh frozen and FFPE tissues	Upregulated in aggressive SMZL vs. indolent SMZL	Prognosis, prediction of patients' relapse	[133]
Nodal MZL (NMZL)	miR-223-3p, let-7f-5p	FFPE tissues	Upregulated in NMZL vs. FL	Differential diagnosis	[139]

Abbreviations: DLBCL, diffuse large B-cell lymphoma; FFPE, formalin-fixed, paraffin-embedded; MALT, mucosa-associated lymphoid tissue; PBMCs, peripheral blood mononuclear cells.

5. miRNAs in Rare Types of Indolent B-Cell NHLs

5.1. Waldenström's Macroglobulinemia or Lymphoplasmacytic Lymphoma

Waldenström's macroglobulinemia (WM) or lymphoplasmacytic lymphoma is an indolent B-cell NHL, characterized by an arrest of B cells after somatic hypermutation and prior to isotype class switching. These cancer cells produce large amounts of immunoglobulin M (IgM). The accumulation of these malignant cells can result in an indirect reduction of red and white blood cells in the bone marrow, leading potentially to hyperviscosity due to the IgM aggregation, anemia, and to attenuated functionality of the immune system.

A common genetic alteration in patients with WM, which is present in more than 90% of patients, is a mutation in the *MYD88* gene; this mutation is also abundant in other B-cell malignancies with a different frequency. In WM, this mutation leads to an activation of the NFkB signaling pathway, resulting in the growth and survival of the WM cancer cells. Considering the highly regulatory potential of miRNAs, their role in this pathway could be critical for WM. More specifically, miR-23b-3p was found downregulated in patients with

WM, and transfection experiments with miR-23b-3p mimics resulted in a decrease in WM cell proliferation and survival. In a functional experiment, miR-23b-3p was found to target *SP1* 3′ UTR, which is an overly activated transcriptional factor positively affecting the NFkB signaling pathway in WM and multiple myeloma. Moreover, the *MIR23B* promoter was found to be under transcriptional control by MYC. The downregulation of MYC resulted in increased levels of miR-23b-3p, proposing an important MYC/miR-23b-3p/SP1 regulatory axis with a significant role in the proliferation and survival of WM cells [80]. Furthermore, another genetic alteration has been found in cells of WM patients, associating with miRNA functions. Specifically, a study showed a deletion in the 13q14 chromosomal region in 10% of WM patients [145]. This region includes the genomic location from which miR-15a-5p and miR-16-5p are generated. These two miRNAs which possess a critical regulatory role in other malignancies such as CLL, downregulate BCL2 anti-apoptotic protein leading to apoptosis of cancer cells. Moreover, these two miRNAs have been characterized as negative regulators of the NFkB signaling pathway. As a result, the downregulation of these two miRNAs in WM patients could lead to increased proliferation and survival of the malignant cells. All the aforementioned information shows a downregulation of these miRNAs in WM cells, leading to activation of NFkB signaling and enhanced cancer cell properties.

Another important miRNA in WM is miR-155-5p, which has also a significant impact on other B-cell malignancies and normal B-cell development. In WM, this miRNA was found upregulated in comparison with cells from healthy individuals. Functional studies, in which miR-155-5p was knocked down, found a critical involvement of this miRNA in the PI3K/AKT and the NFkB signaling pathways in WM. Moreover, miR-155-5p was found to dispose a positive regulation in the proliferation, adhesion, and migration of WM cells [81]. Furthermore, in WM cells with augmented miR-155-5p expression, a significant downregulation of FOXO3 transcription factor and BCL2L11 pro-apoptotic BCL2 family member was observed, resulting in abrogation of the activation of apoptosis [83]. Additionally, another study enforced the knowledge about the impact of this miRNA in WM. More specifically, in experiments where WM cells with knocked-down miR-155-5p were treated with everolimus, an MTOR inhibitor, inhibition of cytotoxicity was observed, in comparison with WM cells with normal levels of miR-155-5p [82]. Furthermore, in another experiment, the levels of miR-155-5p were found downregulated in a dose-dependent manner as everolimus concentration augmented. All the aforementioned information highlights the significance of miR-155-5p in WM, not only with regard to the way it affects important pathways and cell properties, but its therapeutic potency, as well.

miRNAs have also been reported as potential biomarkers for WM (Table 4) [81,83], with a part of them also possessing a regulatory role in this malignancy, as summarized in Table 1. After the observation that miR-206-3p expression levels are elevated and miR-9-3p expression levels are lower in WM cells compared to cells from normal individuals, a study showed a change in the levels of histone deacetylases (HDACs) and histone acetyltransferases (HATs) in WM cells, following the downregulation of miR-206-3p and the upregulation of the miR-9-3p [79]. Specifically, miR-206-3p was found to downregulate the histone acetyltransferase KAT6A and miR-9-3p was found to downregulate the histone deacetylases HDAC4 and HDAC5. This epigenetic regulation which is driven by miRNAs is of high importance, as deregulation of HDACs and HATs is a common phenomenon in numerous malignancies. However, further research is essential to shed more light on the role of miRNAs in this malignancy, as current knowledge remains limited.

5.2. Other Rare Types of B-Cell NHLs

Hairy cell leukemia (HCL) is a rare type of leukemia with an incidence of 0.3/100,000 people. The malignant cells are a type of B lymphocyte, but they're different from those seen in CLL. They possess projections coming off them that give them a "hairy" appearance, hence the name of the disease. Specific miRNAs have been identified as biomarkers in this malignancy, as well. In 2011, Moussay et al. found a downregulation in the levels of circulating miR-363-3p and miR-708-5p in plasma samples of patients with HCL in comparison with patients with CLL [146]. This observation may be of high importance for the discrimination between these two malignancies. In another study, six miRNAs were found upregulated in patients with HCL in comparison with normal samples or patients with other malignant B cells [147]. Target prediction of these six molecules revealed the regulatory impact on MAPK pathways, mainly via targeting of the molecules which contribute to the activation of the JNK signaling pathway, which has an apoptotic effect in HCL. This negative regulation of the pro-apoptotic JNK signaling pathway may possess an anti-apoptotic effect in HCL cells, leading to prolonged survival of the cancer cells.

miRNAs also appear as promising molecules in PCFCL, another type of indolent B-cell NHLs, with limited information regarding its pathogenesis. As previously described in this review, Monsalvez et al. uncovered the differential expression of miR-150-5p in comparison with primary cutaneous marginal zone B-cell lymphoma. This miRNA, which targets the transcription factor MYB, a factor that participates in the proliferation and differentiation of hematopoietic progenitor cells, has been found downregulated in PCFCL [117]. Moreover, the fact that this miRNA is differentially expressed in other types of indolent B-cell NHLs points out its high significance in these malignancies. Additionally, another study comparing PCFCL and primary cutaneous DLBCL-leg type, highlighted four other miRNAs (miR-9-5p, miR-31-5p, miR-129-2-3p, and miR-214-3p) which could be used for the distinction between these two malignancies [148]. These miRNAs have been acknowledged both as tumor-suppressors and as oncogenes in distinct malignancies. Moreover, miR-9-5p, miR-31-5p, and miR-214-3p were found to regulate the activity of signaling pathways such as the NFkB and PI3K/AKT.

Specific miRNAs with potential biomarker utility in the aforementioned rare types of indolent B-cell NHLs are summarized in Table 4. The identification of specific miRNA signatures with biomarker utility that can be used in order to distinguish specific rare types of indolent B-cell NHLs from other types of leukemia, is of high importance for the timely and optimal management of patients. Moreover, discovering miRNAs with biomarker utility may reveal other promising molecules.

Although all the aforementioned information is promising, further research is essential to elucidate the involvement of miRNAs in rare types of indolent B-cell NHLs. Elucidating the regulatory effect of miRNAs with different levels in rare indolent B-cell NHLs may reveal novel candidates that participate in pathogenic events that lead to these distinct malignancies. Characteristically, as previously mentioned, in all three types of rare indolent B-cell NHLs which are presented in this review, there are paradigms of miRNAs, having a potential biomarker utility, which can also have an oncogenic or an onco-suppressive role in each disease. Moreover, predicted targets of miR-363-3p, miR-494-3p, miR-184-3p, and miR-542-3p, which are increased in WM patients include tumor suppressors, cell-cycle inhibitors, cytokine signaling suppressors, and tyrosine phosphatases [81]. miR-9-3p, which acts as a onco-suppressor and is decreased in WM patients, targets protein kinases, oncogenes, and transcription factors enhancing apoptosis and inhibiting B-cell differentiation and proliferation [149]. Additionally, some members of the let-7 and miR-9 families with decreased levels in WM patients, in comparison with normal individuals, downregulate *PRDM1*, a significant regulator of B-cell development. Other miRNAs with increased levels in this malignancy such as miR-125b-5p and miR-181a-5p also target *PRDM1* and other factors contributing to B-cell development, including *IRF4* [149]. Let-7a-5p with lower levels in WM, compared to normal individuals, acts as an onco-suppressor by regulating different oncogenes such as *MYC* [142]. Conversely, miR-21-5p with increased levels in WM acts as an oncogene as it regulates numerous tumor suppressors, including *PTEN* and *PDCD4* [150]. In PCFCL, miR-9-5p, miR-129-2-3p, and miR-155-5p with upregulated levels in comparison to primary cutaneous DLBCL-leg type and cutaneous MZL are involved in normal B-cell development by targeting *PRDM1*, *SOX4*, and *SPI1*, respectively [148]. SOX4 is involved in the transition step from pro- to pre-B cell, whereas SPI1 and PRDM1 are transcription factors acting at later stages of B-cell development and regulating plasma cell differentiation. Therefore, all the aforementioned information highlights miRNAs with deregulated levels in these malignancies, appearing to have also a significant regulatory effect in disease onset and progression.

Knowing the miRNA targets and the respective effect of their activity is important as it sheds light on a part of the regulatory network. However, it is of high importance to u the regulatory network of as many miRNAs or other small non-coding RNAs with specific mRNAs as possible, in order to identify unique and significant pathogenic effects. In this approach, an extensive "screening" would reveal specific interactions between miRNAs and mRNAs that could be assessed in order to diagnose, predict, and cure the disease. Nevertheless, it is the balanced outcome of all distinct regulatory steps that determine the final outcome at a cellular level.

Table 4. miRNAs as candidate biomarkers in rare indolent B-cell NHLs.

Disease	miRNAs	Sample Origin	Expression	Potential Biomarker	References
WM	miR-363-3p, miR-206-3p, miR-494-3p, miR-155-5p, miR-184-3p, miR-542-3p	Bone marrow $CD19^+$ B cells	Upregulated in WM vs. normal $CD19^+$ B cells	Diagnosis	[81]
	miR-9-3p		Downregulated in WM vs. normal $CD19^+$ B cells		
	miR-193b-3p, miR-126-3p, miR-181a-5p, miR-125b-5p, miR-451a	Bone marrow or peripheral blood $CD19^+$ and $CD138^+$ B cells	Upregulated in WM vs. CLL	Differential diagnosis	[149]
	miR-92a-3p, miR-223-3p, miR-92b-3p, miR-363-3p		Upregulated in WM vs. MM		
	miR-9-3p, miR-193b-3p, miR-182-5p, miR-152-3p		Downregulated in WM vs. MM		
	miR-21-5p, miR-142-3p		Upregulated in WM vs. normal B-lineage cells	Diagnosis	
	miR-182-5p, miR-152-3p, miR-373-5p, miR-575-3p		Downregulated in WM vs. normal B-lineage cells		
	Combination of miR-320a-3p and miR-320b-3p	Serum	Downregulated in WM vs. normal blood donors; downregulated in WM vs. MGUS and MM	Diagnosis, differential diagnosis	[151]
	miR-151-5p, let-7a-5p		Downregulated in WM vs. normal blood donors; downregulated in WM vs. MGUS		
	miR-21-5p, miR-192-5p, miR-320b-3p	Exosomes	Increases with disease progression	Prediction of disease progression	[152]
	let-7d-5p		Decreases with disease progression		
HCL	miR-363-3p, miR-708-5p	Peripheral blood B cells	Downregulated in HCL vs. CLL	Differential diagnosis	[146]
	miR-221-3p, miR-222-3p, miR-22-3p, miR-24-3p, miR-27a-3p, let-7b-5p	Peripheral blood $CD19^+$ B cells	Upregulated in HCL vs. other B-cell malignancies; upregulated in HCL vs. normal B-lineage cells	Diagnosis, differential diagnosis	[147]
PCFCL	miR-150-5p, miR-155-5p	FFPE tissues	Upregulated in PCFCL vs. cutaneous MZL	Differential diagnosis	[117]
	miR-129-2-3p, miR-214-3p, miR-31-5p, miR-9-5p		Upregulated in PCFCL vs. primary cutaneous DLBCL-leg type		[148]

Abbreviations: CLL, chronic lymphocytic leukemia; DLBCL, diffuse large B-cell lymphoma; FFPE, formalin-fixed, paraffin-embedded; HCL, hairy cell leukemia; MGUS, monoclonal gammopathy of undetermined significance; MM, multiple myeloma; MZL, marginal zone lymphoma; PCFCL, primary cutaneous follicle center lymphoma; WM, Waldenström's macroglobulinemia.

6. Interplay between Cytokines and miRNAs in B-Cell Malignancies

Cytokines compose a broad category of small proteins, including chemokines, interferons, interleukins (IL), lymphokines, and tumor necrosis factors (TNFs), which are important in cell signaling. They are produced and secreted by a variety of cells including stromal cells, fibroblasts, and endothelial cells. In the immune system, they are produced by leukocytes and exert their function on other leukocytes or tissues that express cytokine receptors. Several cytokines act on B cells and play key roles in the development, survival, differentiation, and/or proliferation of B cells. Additionally, certain chemokines are implicated in B-cell function, namely in antibody production, while the chemokine signaling regulates adhesion and migration, and hence, it is vital for B-cell survival and development [153].

Considering the key role of cytokines in normal physiology, their deregulation can assist in the development of B-cell malignancies. An interesting example is provided by the CXCR4/CXCL12 axis in FL. More specifically, this axis is especially important, since it regulates normal B-cell recirculation between GC zones, the bone marrow, and peripheral blood [154]. CXCR4 is a G-protein coupled chemokine receptor, to which the chemokine CXCL12 binds. A frequent characteristic of FL cells is the high expression of CXCR4. These elevated CXCR4 levels could be attributed to and/or explain the increased activity of proteins such as HIF1A, VEGFA, and signaling pathways including the PI3K/AKT, NFkB, and NOTCH. For instance, FL is often characterized by high levels of the transcription factor HIF1A, whose target genes include CXCR4 and the angiogenesis regulator, VEGF. Moreover, activation of the CXCR4-CXCL12 axis, in turn, promotes signaling through the PI3K/AKT and MAPK pathways and affects surface levels of CD20 and BCR signaling, leading to a proliferative and antiapoptotic phenotype of FL cells. MYC is one of the targets of PI3K/AKT and MAPK pathways that is activated via the aforementioned axis, which in turn has multiple and key effects in miRNA expression in FL, as previously discussed [47].

miRNAs have been shown to respond to dynamic micro-environmental cues and to regulate multiple functions of B-cell populations, including their survival, development, and activation. Thus, miRNA functions contribute not only to immune homeostasis, but also to the control of immune tolerance. Among the most important proteins whose expression is targeted by miRNAs, are the cytokines. Cytokines act as both key upstream signals and major functional outputs, and therefore, can affect miRNA levels, as well [155]. Every cell procedure takes place in the context of a regulatory network rather than a regulatory axis and these networks alter under pathological states. For instance, miR-21-5p is a well-known oncomiR, which is highly expressed in several B-cell NHLs and has been associated with resistance to apoptosis. During plasma cell differentiation, miR-21-5p expression is downregulated by PRDM1, a key molecule in the terminal differentiation of B cells and a tumor suppressor in several lymphoid neoplasms. The expression of *PRDM1* is upregulated by the transcription factor STAT3, which has previously been activated by IL21. Although STAT3 also enhances *MIR21* expression, it is not able to counteract the repression of the latter by PRDM1 [156]. However, in multiple myeloma, a different phenomenon is observed. More specifically, IL6 activates STAT3, which in turn promotes *MIR21* expression. The high levels of miR-21-5p contribute to the high proliferative rate and anti-apoptotic phenotype of malignant cells [157,158].

Another interesting example of cytokine-induced miRNAs has been observed in DLBCL. More specifically, it was shown that miR-155-5p levels were increased by TNFA, even though the molecular background behind this interaction is not known. In turn, TNFA-induced miR-155-5p inhibits the expression of *INPP5D*, a suppressor of the PI3K/AKT signaling pathway. The elevated activity of PI3K/AKT pathway can lead to increased cell proliferation and growth in malignant cells [159]. However, the interplay between cytokines and miRNAs has not been thoroughly investigated in indolent B-cell NHLs. Considering the key role of both cytokines and miRNAs in the development and progression of these malignancies, the investigation of their interactions is critical. Moreover, it would shed light in the molecular base of these diseases assisting in the development of targeted and more efficient therapeutic approaches.

7. Limitations

As previously analyzed, miRNAs are characterized by a great regulatory potential in indolent B-cell NHLs. However, our knowledge regarding their function in these malignancies remains limited and derives from individual research studies, since the majority of studies focus on the expression profiling of miRNAs. Therefore, a massive functional analysis is critical and will assist in the unraveling of the role of these tiny regulators in indolent B-cell NHLs. Moreover, it would be helpful if the miRNAs with a validated role in normal B-cell development were investigated in the context of indolent B-cell NHLs. Based on the current literature, we suggest potential regulatory interactions via which miRNAs with deregulated expression patterns in indolent B-cell NHLs can exert their role in these malignancies (Table 5).

Table 5. Potential interactions between miRNAs that are deregulated in indolent B-cell NHLs and their validated targets in other malignancies.

Disease	miRNAs	Potential Target in Lymphomas	Potential Effect in B Cells	References
Gastric MALT lymphoma	miR-150-5p	*MYB*	Regulation of B-cell development	[14,76,100,101]
		EGR2	Inhibition of apoptosis	
	miR-196a-5p	*CDKN1B*	Promotion of cell cycle	[76,104]
	miR-153-3p	*AKT3*	Inhibition of cell proliferation	[76,105]
	miR-7-5p	*EGFR* and *IGF1R*	Inhibition of metastasis	[76,106,107]
	miR-16-5p	*BCL2*	Promotion of apoptosis	[113,114]
Splenic MZL	miR-29a/b1 cluster	*TCL1A*	Inhibition of cell proliferation	[126,128–131]
	miR-129-5p	*BCL2*	Promotion of apoptosis	
	miR-21-5p	*PTEN, FOXO3*	Promotion of cell proliferation, inhibition of apoptosis	[133,134]
Nodal MZL	miR-223-3p	*LMO2*	Inhibition of B-cell differentiation	[42,139]
OAL	miR-29 family	*TCL1A*	Inhibition of cell proliferation	[118,119]
		CDK6		
		DNMT3B	Inhibition of DNA methylation	
		MCL1	Promotion of apoptosis	
	miR-199a-5p	*IKBKB*	Promotion of apoptosis	[118,120,121]
WM	miR-9-3p	*PDRM1*	Inhibition of B-cell differentiation	[149]
	miR-125b-5p and miR-181a-5p	*PRDM1, IRF4*		
	let-7a-5p	*MYC*	Inhibition of cell proliferation	[142]
	miR-21-5p	*PTEN, PDCD4*	Promotion of cell proliferation, regulation of apoptosis	[150]
PCFCL	miR-9-5p	*PRDM1*	Inhibition of B-cell differentiation	[148]
	miR-129-2-3p	*SOX4*		
	miR-155-5p	*SPI1*		

Abbreviations: MALT, mucosa-associated lymphoid tissue; MZL, marginal zone lymphoma; OAL, Ocular adnexal lymphoma; PCFCL, primary cutaneous follicle center lymphoma; WM, Waldenström's macroglobulinemia.

So far, the mouse and other animal models provide important insights into human B-cell development and disease. However, several studies report intrinsic differences in gene expression and gene regulation between the human system and mouse model and more prominently in the immune system [160]. An interesting example of such differences is observed in gene expression, early after T-cell activation, under the effect of IL2. More precisely, differential IL2 transcription kinetics can inhibit splicing in mouse models, but not in humans [161]. Regarding B-cell development, it has been clearly stated that B-cell populations exist in different abundances between human and murine organisms, while they can have additional differences, including localization. Precisely, the identification of differences in the non-memory B-cell pools is important for understanding the differences in mechanisms that contribute to B-cell homeostasis in the two species and in translating information obtained from mouse models to studies of human disease [162]. However, the existing comparative studies of mouse and human B-cell development have focused on B-cell precursor populations and activated B cells [163]. Considering these differences between these organisms, the distinct expression pattern of miRNAs and regulatory networks can exist, as well. Therefore, a critical consideration is required when extrapolating mouse data to the human system in basic and translational research.

One of the major obstacles in the research of miRNAs is the fact that they act not in a regulatory axis, but as part of a complex regulatory network. More precisely, one miRNA is able to bind to multiple target genes, which subsequently affect several pathways, while simultaneously one target gene can be targeted by multiple miRNAs. Such examples are miR-150-5p and miR-155-5p, which have been shown to target multiple genes that regulate lymphomagenesis, creating an interaction network, while they have been correlated with several hematological malignancies, as well. Particularly, miR-155-5p has been characterized both as oncogenic and tumor-suppressive miRNA, depending on the cellular context, the intermolecular interactions, and the type of malignancy. Besides miR-155-5p, other miRNAs have also been characterized as double-edged swords complicating the miRNA functional research and pointing out the complexity of cell homeostasis. Due to this complex miRNA regulatory network, there are contradictory findings regarding the function of miRNAs in normal and pathological states; therefore, it is difficult for the researchers to reach a conclusion. The phenomenon of crosstalk between different signaling pathways further complicates the regulation of cellular processes in lymphocytes and consequently in lymphomas. B-cell lymphomas are often characterized by elevated molecular and phenotypical heterogeneity, even among the malignant cells comprising the tumor. This high heterogeneity could also be reflected in differential expression patterns of miRNAs even among cells of the same tumor and provides another potential explanation regarding the contradictory results concerning the role of miRNAs in malignant conditions [164,165].

Moreover, as aforementioned, miRNAs are ideal biomarker candidates; however, research in this field is still in its infancy, especially due to the lack of an efficient and cost-effective method for the accurate detection of miRNAs. One of the reasons why this has not yet been achieved is that features such as detection limits, range of concentrations in bodily fluids, and modulation depending on various parameters (age, gender, health/disease) have not clearly been established, yet. Additionally, the findings generally lack reproducibility. There are several discordances reported between different teams that have analyzed the same malignancy types. In order to resolve this issue, standardized protocols must be developed both for the initial stages of the process, like sample collection, transport, and storage, as well as data analysis for the diversity of technological methods used. Particularly for sample collection, it is critical that the sample size is large enough so that the result can be characterized as statistically valid [9,166]. Finally, in the majority of studies, it was not explicitly stated whether miRNAs were 5p or 3p, and their sequence was not provided. This creates ambiguity in future research, necessitating the implementation of the current nomenclature system in all future studies.

8. Future Perspectives

Although the expression profiles of miRNAs have been greatly investigated, further research is necessary to unravel the complex functional networks. This endeavor shall aid in the utilization of miRNAs as therapeutic targets (Figure 5). miRNAs can either promote tumor cell proliferation and hence act as oncogenic miRNAs, or suppress uncontrolled cell division, acting as tumor suppressors. According to these distinct properties, two main therapeutic strategies involving miRNAs have been developed. The first one introduces single-stranded antisense oligonucleotides, known as antimiRs or antagomiRs, that target an oncogenic miRNA, into the cell. The aforementioned interaction prevents the miRNA from binding to its target mRNA, resulting in unaffected protein expression levels. The second approach provides an artificial double-stranded RNA molecule, known as miRNA mimic, that imitates the naturally occurring pre-miRNA. This strategy attempts to restore the reduced innate expression levels of a tumor-suppressive miRNA. A major obstacle hindering translation into the clinic is the possible degradation of these agents by RNases. Therefore, chemically modified RNA nucleotides and molecules are being tested to increase stability and efficacy. Locked nucleic acid (LNA) nucleotides are most widely used, followed by the addition of 2′-O-methyl groups or phosphorothioate-like groups. Safe and efficient delivery inside the desirable cells, without triggering an immune response and by minimizing potential endosomal escape, is of equal importance, and a variety of liposomes and nanoparticles are being tested for this purpose. Besides stand-alone therapies, it would be interesting to investigate whether the combination of chemotherapy, radiotherapy, or immunotherapy with a miRNA-based therapy could be more beneficial for patients and/or overcome resistance to currently established therapeutic regimens [7,167–169].

A quite hot research topic is the reciprocal regulation between miRNAs and the epigenetic machinery. More specifically, miRNAs as a component of the epigenetic machinery are implicated in epigenetic regulation. At the same time, RNA and histone modifications and DNA methylation regulate miRNA expression, while epigenetic-related enzymes can be the target of miRNAs. All these findings have enlightened the researchers regarding the miRNA-epigenetic feedback loop. Several studies have associated the dysregulation of this miRNA-epigenetic feedback loop with the initiation and development of various diseases, including B-cell NHLs, and have demonstrated its potential for application in clinical diagnosis and prognosis. Particularly, the determination of the methylation profile of miRNA genes and the quantification of the expression of enzymes involved in epigenetic mechanisms affecting miRNA expression could constitute a powerful approach for diagnosis and prognosis, while drugs targeting epigenetic regulators have become a promising therapeutic strategy for several malignancies, including leukemia. Although this research field is still in its infancy, and further study is required for establishing miRNAs as pivotal modulators of epigenetic effects in clinical practice, it seems to be a promising research field with great potential [170].

Lastly, several challenges need to be overcome so the miRNAs are widely used as efficient biomarkers, as it has already been addressed in the limitations section. Furthermore, it would be very useful if some extracellular and/or circulating miRNAs found in bodily fluids were associated strongly with pathological states. The introduction of circulating miRNAs in clinical research as non-invasive biomarkers would be quite beneficial since non-invasive procedures are relatively convenient, fast, and not painful for the patients. Extracellular miRNAs can be stabilized via protein interactions, particularly AGO2, or via inclusion within extracellular vesicles, such as exosomes and microvesicles, and apoptotic bodies, that are thoroughly investigated for non-invasive biomarkers discovery. However, further research is required to uncover the exact secretion and stabilization mechanisms involved in each case, as well as the establishment of standardized detection and quantification protocols [171]. This is of major importance, because miRNAs show great promise in personalized medicine and could probably assist in patient stratification, selection of optimal treatment, and monitoring of therapeutic response for each individual [171].

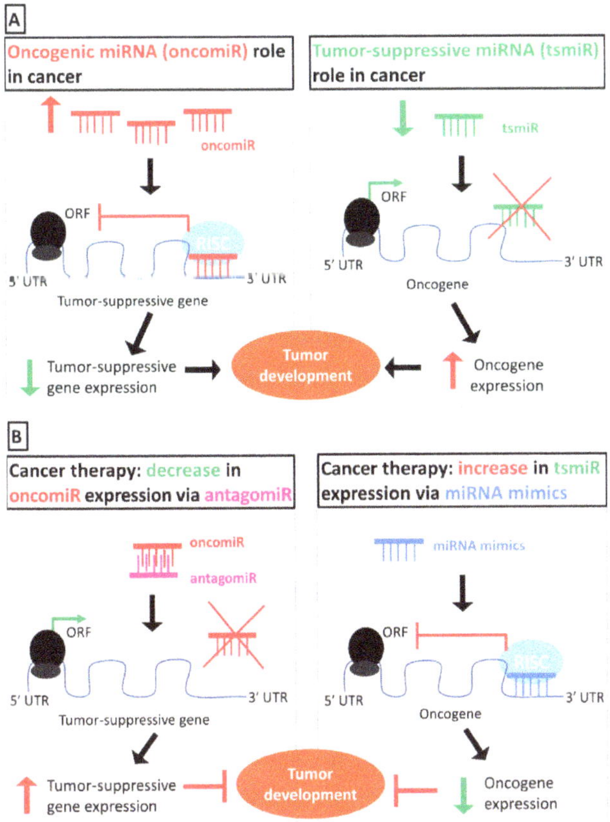

Figure 5. The role of oncogenic and tumor-suppressive miRNAs in cancer, and therapeutic strategies against cancer, based on miRNA targeting. (**A**) The oncogenic miRNAs are highly expressed in cancer; they bind to the 3′ UTR of their target genes (tumor suppressors), recruit RISC complex and suppress the expression of their target genes, leading to decreased levels of the respective proteins. In cancer, the tumor-suppressive miRNAs are expressed at low levels. Therefore, they are not able to suppress the expression levels of their target genes (oncogenes), leading to high levels of their proteins. (**B**) For the downregulation of oncogenic miRNAs and the subsequent attenuation of their harmful impact on cellular function, antagomiRs are used. They bind complementarily to the oncogenic miRNA and hence inhibit the binding of the latter to its target. For the upregulation of the tumor-suppressive miRNAs and the subsequent promotion of their beneficial impact on cell function, miRNA mimics are used. They have the same sequence as the specific tumor-suppressive miRNAs and, therefore, are able to bind to the targets of the latter and exert their function. Black arrows indicate the transition to the next step; red lines (⊥) indicate an inhibitory effect; upstream red arrows indicate an increase in the expression levels, while downstream green arrows indicate a decrease in the expression levels; horizontal green arrows indicate promotion of gene expression.

9. Conclusions

Indolent B-cell NHLs compose a highly heterogeneous group of lymphomas with a high occurrence rate worldwide. Therefore, the elucidation of its molecular background and pathogenesis, in general, is considered quite critical. The recent advances have assisted towards this direction, but this research field is still in its infancy. miRNAs have been repeatedly investigated in the context of their usage as biomarkers or therapeutic targets of several malignancies, including indolent B-cell NHLs, due to their regulatory potential.

These tiny regulators play a vital role in B-cell development and normal B-cell function, in general, and their deregulation could lead to fatal consequences for the cell. Among the most intriguing miRNAs are miR-150-5p, miR-155-5p, and those of the miR-17/92 cluster, as they are not only necessary for normal B-cell development, but are also implicated in the pathogenesis of the majority of the aforementioned malignancies. Several studies have attempted to analyze the expression levels of miRNAs in indolent B-cell NHLs, aiming to establish a miRNA signature, distinct for each malignant state. Even though progress has been achieved in this field, additional research is necessary in order to lead to more solid conclusions. Additionally, the regulatory networks via miRNAs function have to be unraveled. Despite all the aforementioned difficulties and limitations in miRNA research, their great regulatory potential is quite promising regarding the deciphering of cancer development and progression as well as the potential exploitation of miRNAs in therapy.

Author Contributions: All authors have read and agreed to the published version of the manuscript.

Funding: No external funding.

Conflicts of Interest: The authors declare no conflict of interest.

References

1. Swerdlow, S.H.; Campo, E.; Pileri, S.A.; Harris, N.L.; Stein, H.; Siebert, R.; Advani, R.; Ghielmini, M.; Salles, G.A.; Zelenetz, A.D.; et al. The 2016 revision of the World Health Organization classification of lymphoid neoplasms. *Blood* **2016**, *127*, 2375–2390. [CrossRef]
2. Armitage, J.O.; Gascoyne, R.D.; Lunning, M.A.; Cavalli, F. Non-Hodgkin lymphoma. *Lancet* **2017**, *390*, 298–310. [CrossRef]
3. Freedman, A.; Jacobsen, E. Follicular lymphoma: 2020 update on diagnosis and management. *Am. J. Hematol.* **2020**, *95*, 316–327. [CrossRef]
4. Carbone, A.; Roulland, S.; Gloghini, A.; Younes, A.; von Keudell, G.; Lopez-Guillermo, A.; Fitzgibbon, J. Follicular lymphoma. *Nat. Rev. Dis. Primers* **2019**, *5*, 83. [CrossRef]
5. Sindel, A.; Al-Juhaishi, T.; Yazbeck, V. Marginal Zone Lymphoma: State-of-the-Art Treatment. *Curr. Treat. Options Oncol.* **2019**, *20*, 90. [CrossRef]
6. Ha, M.; Kim, V.N. Regulation of microRNA biogenesis. *Nat. Rev. Mol. Cell Biol.* **2014**, *15*, 509–524. [CrossRef]
7. Musilova, K.; Mraz, M. MicroRNAs in B-cell lymphomas: How a complex biology gets more complex. *Leukemia* **2015**, *29*, 1004–1017. [CrossRef] [PubMed]
8. Katsaraki, K.; Karousi, P.; Artemaki, P.I.; Scorilas, A.; Pappa, V.; Kontos, C.K.; Papageorgiou, S.G. MicroRNAs: Tiny Regulators of Gene Expression with Pivotal Roles in Normal B-Cell Development and B-Cell Chronic Lymphocytic Leukemia. *Cancers* **2021**, *13*, 593. [CrossRef]
9. Condrat, C.E.; Thompson, D.C.; Barbu, M.G.; Bugnar, O.L.; Boboc, A.; Cretoiu, D.; Suciu, N.; Cretoiu, S.M.; Voinea, S.C. miRNAs as Biomarkers in Disease: Latest Findings Regarding Their Role in Diagnosis and Prognosis. *Cells* **2020**, *9*, 276. [CrossRef] [PubMed]
10. Melchers, F. Checkpoints that control B cell development. *J. Clin. Investig.* **2015**, *125*, 2203–2210. [CrossRef]
11. Koralov, S.B.; Muljo, S.A.; Galler, G.R.; Krek, A.; Chakraborty, T.; Kanellopoulou, C.; Jensen, K.; Cobb, B.S.; Merkenschlager, M.; Rajewsky, N.; et al. Dicer ablation affects antibody diversity and cell survival in the B lymphocyte lineage. *Cell* **2008**, *132*, 860–874. [CrossRef]
12. Brandl, A.; Daum, P.; Brenner, S.; Schulz, S.R.; Yap, D.Y.; Bosl, M.R.; Wittmann, J.; Schuh, W.; Jack, H.M. The microprocessor component, DGCR8, is essential for early B-cell development in mice. *Eur. J. Immunol.* **2016**, *46*, 2710–2718. [CrossRef] [PubMed]
13. Spierings, D.C.; McGoldrick, D.; Hamilton-Easton, A.M.; Neale, G.; Murchison, E.P.; Hannon, G.J.; Green, D.R.; Withoff, S. Ordered progression of stage-specific miRNA profiles in the mouse B2 B-cell lineage. *Blood* **2011**, *117*, 5340–5349. [CrossRef]
14. Xiao, C.; Calado, D.P.; Galler, G.; Thai, T.H.; Patterson, H.C.; Wang, J.; Rajewsky, N.; Bender, T.P. MiR-150 controls B cell differentiation by targeting the transcription factor c-Myb. *Cell* **2007**, *131*, 146–159. [CrossRef]
15. Okuyama, K.; Ikawa, T.; Gentner, B.; Hozumi, K.; Harnprasopwat, R.; Lu, J.; Yamashita, R.; Ha, D.; Toyoshima, T.; Chanda, B.; et al. MicroRNA-126-mediated control of cell fate in B-cell myeloid progenitors as a potential alternative to transcriptional factors. *Proc. Natl. Acad. Sci. USA* **2013**, *110*, 13410–13415. [CrossRef]
16. Chen, C.Z.; Li, L.; Lodish, H.F.; Bartel, D.P. MicroRNAs modulate hematopoietic lineage differentiation. *Science* **2004**, *303*, 83–86. [CrossRef]
17. Hagman, J.; Lukin, K. Transcription factors drive B cell development. *Curr. Opin. Immunol.* **2006**, *18*, 127–134. [CrossRef]
18. Medvedovic, J.; Ebert, A.; Tagoh, H.; Busslinger, M. Pax5: A master regulator of B cell development and leukemogenesis. *Adv. Immunol.* **2011**, *111*, 179–206. [CrossRef]
19. Kwon, K.; Hutter, C.; Sun, Q.; Bilic, I.; Cobaleda, C.; Malin, S.; Busslinger, M. Instructive role of the transcription factor E2A in early B lymphopoiesis and germinal center B cell development. *Immunity* **2008**, *28*, 751–762. [CrossRef] [PubMed]

20. Hagman, J.; Ramirez, J.; Lukin, K. B lymphocyte lineage specification, commitment and epigenetic control of transcription by early B cell factor 1. *Curr. Top. Microbiol. Immunol.* **2012**, *356*, 17–38. [CrossRef] [PubMed]
21. Nechanitzky, R.; Akbas, D.; Scherer, S.; Gyory, I.; Hoyler, T.; Ramamoorthy, S.; Diefenbach, A.; Grosschedl, R. Transcription factor EBF1 is essential for the maintenance of B cell identity and prevention of alternative fates in committed cells. *Nat. Immunol.* **2013**, *14*, 867–875. [CrossRef]
22. Hu, H.; Wang, B.; Borde, M.; Nardone, J.; Maika, S.; Allred, L.; Tucker, P.W.; Rao, A. Foxp1 is an essential transcriptional regulator of B cell development. *Nat. Immunol.* **2006**, *7*, 819–826. [CrossRef]
23. Contreras, J.R.; Palanichamy, J.K.; Tran, T.M.; Fernando, T.R.; Rodriguez-Malave, N.I.; Goswami, N.; Arboleda, V.A.; Casero, D.; Rao, D.S. MicroRNA-146a modulates B-cell oncogenesis by regulating Egr1. *Oncotarget* **2015**, *6*, 11023–11037. [CrossRef]
24. Blume, J.; Zietara, N.; Witzlau, K.; Liu, Y.; Sanchez, O.O.; Puchalka, J.; Winter, S.J.; Kunze-Schumacher, H.; Saran, N.; Duber, S.; et al. miR-191 modulates B-cell development and targets transcription factors E2A, Foxp1, and Egr1. *Eur. J. Immunol.* **2019**, *49*, 121–132. [CrossRef]
25. Mehta, A.; Mann, M.; Zhao, J.L.; Marinov, G.K.; Majumdar, D.; Garcia-Flores, Y.; Du, X.; Erikci, E.; Chowdhury, K.; Baltimore, D. The microRNA-212/132 cluster regulates B cell development by targeting Sox4. *J. Exp. Med.* **2015**, *212*, 1679–1692. [CrossRef]
26. Zhou, B.; Wang, S.; Mayr, C.; Bartel, D.P.; Lodish, H.F. miR-150, a microRNA expressed in mature B and T cells, blocks early B cell development when expressed prematurely. *Proc. Natl. Acad. Sci. USA* **2007**, *104*, 7080–7085. [CrossRef]
27. Kurkewich, J.L.; Bikorimana, E.; Nguyen, T.; Klopfenstein, N.; Zhang, H.; Hallas, W.M.; Stayback, G.; McDowell, M.A.; Dahl, R. The mirn23a microRNA cluster antagonizes B cell development. *J. Leukoc. Biol.* **2016**, *100*, 665–677. [CrossRef]
28. Kong, K.Y.; Owens, K.S.; Rogers, J.H.; Mullenix, J.; Velu, C.S.; Grimes, H.L.; Dahl, R. MIR-23A microRNA cluster inhibits B-cell development. *Exp. Hematol.* **2010**, *38*, 629–640.e1. [CrossRef]
29. Lai, M.; Gonzalez-Martin, A.; Cooper, A.B.; Oda, H.; Jin, H.Y.; Shepherd, J.; He, L.; Zhu, J.; Nemazee, D.; Xiao, C. Regulation of B-cell development and tolerance by different members of the miR-17 approximately 92 family microRNAs. *Nat. Commun.* **2016**, *7*, 12207. [CrossRef]
30. Benhamou, D.; Labi, V.; Getahun, A.; Benchetrit, E.; Dowery, R.; Rajewsky, K.; Cambier, J.C.; Melamed, D. The c-Myc/miR17-92/PTEN Axis Tunes PI3K Activity to Control Expression of Recombination Activating Genes in Early B Cell Development. *Front. Immunol.* **2018**, *9*, 2715. [CrossRef]
31. Xiao, C.; Srinivasan, L.; Calado, D.P.; Patterson, H.C.; Zhang, B.; Wang, J.; Henderson, J.M.; Kutok, J.L.; Rajewsky, K. Lymphoproliferative disease and autoimmunity in mice with increased miR-17-92 expression in lymphocytes. *Nat. Immunol.* **2008**, *9*, 405–414. [CrossRef]
32. Benhamou, D.; Labi, V.; Novak, R.; Dai, I.; Shafir-Alon, S.; Weiss, A.; Gaujoux, R.; Arnold, R.; Shen-Orr, S.S.; Rajewsky, K.; et al. A c-Myc/miR17-92/Pten Axis Controls PI3K-Mediated Positive and Negative Selection in B Cell Development and Reconstitutes CD19 Deficiency. *Cell Rep.* **2016**, *16*, 419–431. [CrossRef]
33. King, J.K.; Ung, N.M.; Paing, M.H.; Contreras, J.R.; Alberti, M.O.; Fernando, T.R.; Zhang, K.; Pellegrini, M.; Rao, D.S. Regulation of Marginal Zone B-Cell Differentiation by MicroRNA-146a. *Front. Immunol.* **2016**, *7*, 670. [CrossRef] [PubMed]
34. He, Y.; Pear, W.S. Notch signalling in B cells. *Semin Cell Dev. Biol.* **2003**, *14*, 135–142. [CrossRef]
35. Luo, Z.; Mu, L.; Zheng, Y.; Shen, W.; Li, J.; Xu, L.; Zhong, B.; Liu, Y.; Zhou, Y. NUMB enhances Notch signaling by repressing ubiquitination of NOTCH1 intracellular domain. *J. Mol. Cell Biol.* **2020**, *12*, 345–358. [CrossRef]
36. Kramer, N.J.; Wang, W.L.; Reyes, E.Y.; Kumar, B.; Chen, C.C.; Ramakrishna, C.; Cantin, E.M.; Vonderfecht, S.L.; Taganov, K.D.; Chau, N.; et al. Altered lymphopoiesis and immunodeficiency in miR-142 null mice. *Blood* **2015**, *125*, 3720–3730. [CrossRef]
37. De Yebenes, V.G.; Belver, L.; Pisano, D.G.; Gonzalez, S.; Villasante, A.; Croce, C.; He, L.; Ramiro, A.R. miR-181b negatively regulates activation-induced cytidine deaminase in B cells. *J. Exp. Med.* **2008**, *205*, 2199–2206. [CrossRef] [PubMed]
38. Zhu, W.; Shan, X.; Wang, T.; Shu, Y.; Liu, P. miR-181b modulates multidrug resistance by targeting BCL2 in human cancer cell lines. *Int. J. Cancer* **2010**, *127*, 2520–2529. [CrossRef] [PubMed]
39. Gururajan, M.; Haga, C.L.; Das, S.; Leu, C.M.; Hodson, D.; Josson, S.; Turner, M.; Cooper, M.D. MicroRNA 125b inhibition of B cell differentiation in germinal centers. *Int. Immunol.* **2010**, *22*, 583–592. [CrossRef] [PubMed]
40. Li, G.; So, A.Y.; Sookram, R.; Wong, S.; Wang, J.K.; Ouyang, Y.; He, P.; Su, Y.; Casellas, R.; Baltimore, D. Epigenetic silencing of miR-125b is required for normal B-cell development. *Blood* **2018**, *131*, 1920–1930. [CrossRef]
41. Enomoto, Y.; Kitaura, J.; Hatakeyama, K.; Watanuki, J.; Akasaka, T.; Kato, N.; Shimanuki, M.; Nishimura, K.; Takahashi, M.; Taniwaki, M.; et al. Emu/miR-125b transgenic mice develop lethal B-cell malignancies. *Leukemia* **2011**, *25*, 1849–1856. [CrossRef]
42. Zhang, J.; Jima, D.D.; Jacobs, C.; Fischer, R.; Gottwein, E.; Huang, G.; Lugar, P.L.; Lagoo, A.S.; Rizzieri, D.A.; Friedman, D.R.; et al. Patterns of microRNA expression characterize stages of human B-cell differentiation. *Blood* **2009**, *113*, 4586–4594. [CrossRef] [PubMed]
43. Porstner, M.; Winkelmann, R.; Daum, P.; Schmid, J.; Pracht, K.; Corte-Real, J.; Schreiber, S.; Haftmann, C.; Brandl, A.; Mashreghi, M.F.; et al. miR-148a promotes plasma cell differentiation and targets the germinal center transcription factors Mitf and Bach2. *Eur. J. Immunol.* **2015**, *45*, 1206–1215. [CrossRef] [PubMed]
44. Lu, D.; Nakagawa, R.; Lazzaro, S.; Staudacher, P.; Abreu-Goodger, C.; Henley, T.; Boiani, S.; Leyland, R.; Galloway, A.; Andrews, S.; et al. The miR-155-PU.1 axis acts on Pax5 to enable efficient terminal B cell differentiation. *J. Exp. Med.* **2014**, *211*, 2183–2198. [CrossRef] [PubMed]

45. Bouska, A.; Zhang, W.; Gong, Q.; Iqbal, J.; Scuto, A.; Vose, J.; Ludvigsen, M.; Fu, K.; Weisenburger, D.D.; Greiner, T.C.; et al. Combined copy number and mutation analysis identifies oncogenic pathways associated with transformation of follicular lymphoma. *Leukemia* **2017**, *31*, 83–91. [CrossRef]
46. Fassina, A.; Marino, F.; Siri, M.; Zambello, R.; Ventura, L.; Fassan, M.; Simonato, F.; Cappellesso, R. The miR-17-92 microRNA cluster: A novel diagnostic tool in large B-cell malignancies. *Lab. Invest.* **2012**, *92*, 1574–1582. [CrossRef]
47. Devan, J.; Janikova, A.; Mraz, M. New concepts in follicular lymphoma biology: From BCL2 to epigenetic regulators and non-coding RNAs. *Semin. Oncol.* **2018**, *45*, 291–302. [CrossRef]
48. Leich, E.; Zamo, A.; Horn, H.; Haralambieva, E.; Puppe, B.; Gascoyne, R.D.; Chan, W.C.; Braziel, R.M.; Rimsza, L.M.; Weisenburger, D.D.; et al. MicroRNA profiles of t(14;18)-negative follicular lymphoma support a late germinal center B-cell phenotype. *Blood* **2011**, *118*, 5550–5558. [CrossRef]
49. Wagner, S.D.; Ahearne, M.; Ko Ferrigno, P. The role of BCL6 in lymphomas and routes to therapy. *Br. J. Haematol.* **2011**, *152*, 3–12. [CrossRef]
50. Braga, T.V.; Evangelista, F.C.G.; Gomes, L.C.; Araujo, S.; Carvalho, M.D.G.; Sabino, A.P. Evaluation of MiR-15a and MiR-16-1 as prognostic biomarkers in chronic lymphocytic leukemia. *Biomed. Pharmacother.* **2017**, *92*, 864–869. [CrossRef] [PubMed]
51. Gebauer, N.; Gollub, W.; Stassek, B.; Bernard, V.; Rades, D.; Feller, A.C.; Thorns, C. MicroRNA signatures in subtypes of follicular lymphoma. *Anticancer Res.* **2014**, *34*, 2105–2111.
52. Lu, J.; Getz, G.; Miska, E.A.; Alvarez-Saavedra, E.; Lamb, J.; Peck, D.; Sweet-Cordero, A.; Ebert, B.L.; Mak, R.H.; Ferrando, A.A.; et al. MicroRNA expression profiles classify human cancers. *Nature* **2005**, *435*, 834–838. [CrossRef]
53. Lawrie, C.H.; Chi, J.; Taylor, S.; Tramonti, D.; Ballabio, E.; Palazzo, S.; Saunders, N.J.; Pezzella, F.; Boultwood, J.; Wainscoat, J.S.; et al. Expression of microRNAs in diffuse large B cell lymphoma is associated with immunophenotype, survival and transformation from follicular lymphoma. *J. Cell. Mol. Med.* **2009**, *13*, 1248–1260. [CrossRef]
54. Roehle, A.; Hoefig, K.P.; Repsilber, D.; Thorns, C.; Ziepert, M.; Wesche, K.O.; Thiere, M.; Loeffler, M.; Klapper, W.; Pfreundschuh, M.; et al. MicroRNA signatures characterize diffuse large B-cell lymphomas and follicular lymphomas. *Br. J. Haematol.* **2008**, *142*, 732–744. [CrossRef]
55. Psathas, J.N.; Doonan, P.J.; Raman, P.; Freedman, B.D.; Minn, A.J.; Thomas-Tikhonenko, A. The Myc-miR-17-92 axis amplifies B-cell receptor signaling via inhibition of ITIM proteins: A novel lymphomagenic feed-forward loop. *Blood* **2013**, *122*, 4220–4229. [CrossRef]
56. Olive, V.; Bennett, M.J.; Walker, J.C.; Ma, C.; Jiang, I.; Cordon-Cardo, C.; Li, Q.J.; Lowe, S.W.; Hannon, G.J.; He, L. miR-19 is a key oncogenic component of mir-17-92. *Genes Dev.* **2009**, *23*, 2839–2849. [CrossRef]
57. Filip, D.; Mraz, M. The role of MYC in the transformation and aggressiveness of 'indolent' B-cell malignancies. *Leuk. Lymphoma* **2020**, *61*, 510–524. [CrossRef]
58. Yan, S.; Jia, C.; Quan, L.; Zhao, L.; Tian, Y.; Liu, A. Significance of the microRNA1792 gene cluster expressed in Bcell nonHodgkin's lymphoma. *Mol. Med. Rep.* **2019**, *20*, 2459–2467. [CrossRef] [PubMed]
59. Musilova, K.; Devan, J.; Cerna, K.; Seda, V.; Pavlasova, G.; Sharma, S.; Oppelt, J.; Pytlik, R.; Prochazka, V.; Prouzova, Z.; et al. miR-150 downregulation contributes to the high-grade transformation of follicular lymphoma by upregulating FOXP1 levels. *Blood* **2018**, *132*, 2389–2400. [CrossRef]
60. Jardin, F.; Jais, J.P.; Molina, T.J.; Parmentier, F.; Picquenot, J.M.; Ruminy, P.; Tilly, H.; Bastard, C.; Salles, G.A.; Feugier, P.; et al. Diffuse large B-cell lymphomas with CDKN2A deletion have a distinct gene expression signature and a poor prognosis under R-CHOP treatment: A GELA study. *Blood* **2010**, *116*, 1092–1104. [CrossRef]
61. Yu, T.; Ma, P.; Wu, D.; Shu, Y.; Gao, W. Functions and mechanisms of microRNA-31 in human cancers. *Biomed. Pharmacother.* **2018**, *108*, 1162–1169. [CrossRef] [PubMed]
62. Thompson, M.A.; Edmonds, M.D.; Liang, S.; McClintock-Treep, S.; Wang, X.; Li, S.; Eischen, C.M. miR-31 and miR-17-5p levels change during transformation of follicular lymphoma. *Hum. Pathol.* **2016**, *50*, 118–126. [CrossRef] [PubMed]
63. Wang, W.; Corrigan-Cummins, M.; Hudson, J.; Maric, I.; Simakova, O.; Neelapu, S.S.; Kwak, L.W.; Janik, J.E.; Gause, B.; Jaffe, E.S.; et al. MicroRNA profiling of follicular lymphoma identifies microRNAs related to cell proliferation and tumor response. *Haematologica* **2012**, *97*, 586–594. [CrossRef] [PubMed]
64. Hoffman, A.E.; Liu, R.; Fu, A.; Zheng, T.; Slack, F.; Zhu, Y. Targetome profiling, pathway analysis and genetic association study implicate miR-202 in lymphomagenesis. *Cancer Epidemiol. Biomark. Prev.* **2013**, *22*, 327–336. [CrossRef] [PubMed]
65. Fu, A.; Hoffman, A.E.; Liu, R.; Jacobs, D.I.; Zheng, T.; Zhu, Y. Targetome profiling and functional genetics implicate miR-618 in lymphomagenesis. *Epigenetics* **2014**, *9*, 730–737. [CrossRef]
66. Lin, J.; Lwin, T.; Zhao, J.J.; Tam, W.; Choi, Y.S.; Moscinski, L.C.; Dalton, W.S.; Sotomayor, E.M.; Wright, K.L.; Tao, J. Follicular dendritic cell-induced microRNA-mediated upregulation of PRDM1 and downregulation of BCL-6 in non-Hodgkin's B-cell lymphomas. *Leukemia* **2011**, *25*, 145–152. [CrossRef]
67. Raju, S.; Kretzmer, L.Z.; Koues, O.I.; Payton, J.E.; Oltz, E.M.; Cashen, A.; Polic, B.; Schreiber, R.D.; Shaw, A.S.; Markiewicz, M.A. NKG2D-NKG2D Ligand Interaction Inhibits the Outgrowth of Naturally Arising Low-Grade B Cell Lymphoma In Vivo. *J. Immunol.* **2016**, *196*, 4805–4813. [CrossRef]
68. Craig, V.J.; Cogliatti, S.B.; Imig, J.; Renner, C.; Neuenschwander, S.; Rehrauer, H.; Schlapbach, R.; Dirnhofer, S.; Tzankov, A.; Muller, A. Myc-mediated repression of microRNA-34a promotes high-grade transformation of B-cell lymphoma by dysregulation of FoxP1. *Blood* **2011**, *117*, 6227–6236. [CrossRef]

69. Chang, T.C.; Yu, D.; Lee, Y.S.; Wentzel, E.A.; Arking, D.E.; West, K.M.; Dang, C.V.; Thomas-Tikhonenko, A.; Mendell, J.T. Widespread microRNA repression by Myc contributes to tumorigenesis. *Nat. Genet.* **2008**, *40*, 43–50. [CrossRef] [PubMed]
70. Sagaert, X.; de Paepe, P.; Libbrecht, L.; Vanhentenrijk, V.; Verhoef, G.; Thomas, J.; Wlodarska, I.; de Wolf-Peeters, C. Forkhead box protein P1 expression in mucosa-associated lymphoid tissue lymphomas predicts poor prognosis and transformation to diffuse large B-cell lymphoma. *J. Clin. Oncol.* **2006**, *24*, 2490–2497. [CrossRef]
71. He, M.; Gao, L.; Zhang, S.; Tao, L.; Wang, J.; Yang, J.; Zhu, M. Prognostic significance of miR-34a and its target proteins of FOXP1, p53, and BCL2 in gastric MALT lymphoma and DLBCL. *Gastric Cancer* **2014**, *17*, 431–441. [CrossRef] [PubMed]
72. Song, Y.; Jiang, K.; Su, S.; Wang, B.; Chen, G. Clinical manifestations and epigenetic mechanisms of gastric mucosa associated lymphoid tissue lymphoma and long-term follow-up following Helicobacter pylori eradication. *Exp. Ther. Med.* **2018**, *15*, 553–561. [CrossRef] [PubMed]
73. Craig, V.J.; Cogliatti, S.B.; Rehrauer, H.; Wundisch, T.; Muller, A. Epigenetic silencing of microRNA-203 dysregulates ABL1 expression and drives Helicobacter-associated gastric lymphomagenesis. *Cancer Res.* **2011**, *71*, 3616–3624. [CrossRef]
74. Fernandez, C.; Bellosillo, B.; Ferraro, M.; Seoane, A.; Sanchez-Gonzalez, B.; Pairet, S.; Pons, A.; Barranco, L.; Vela, M.C.; Gimeno, E.; et al. MicroRNAs 142-3p, miR-155 and miR-203 are Deregulated in Gastric MALT Lymphomas Compared to Chronic Gastritis. *Cancer Genom. Proteom.* **2017**, *14*, 75–82. [CrossRef]
75. Saito, Y.; Suzuki, H.; Tsugawa, H.; Imaeda, H.; Matsuzaki, J.; Hirata, K.; Hosoe, N.; Nakamura, M.; Mukai, M.; Saito, H.; et al. Overexpression of miR-142-5p and miR-155 in gastric mucosa-associated lymphoid tissue (MALT) lymphoma resistant to Helicobacter pylori eradication. *PLoS ONE* **2012**, *7*, e47396. [CrossRef]
76. Blosse, A.; Levy, M.; Robe, C.; Staedel, C.; Copie-Bergman, C.; Lehours, P. Deregulation of miRNA in Helicobacter pylori-Induced Gastric MALT Lymphoma: From Mice to Human. *J. Clin. Med.* **2019**, *8*, 845. [CrossRef]
77. Peveling-Oberhag, J.; Crisman, G.; Schmidt, A.; Doring, C.; Lucioni, M.; Arcaini, L.; Rattotti, S.; Hartmann, S.; Piiper, A.; Hofmann, W.P.; et al. Dysregulation of global microRNA expression in splenic marginal zone lymphoma and influence of chronic hepatitis C virus infection. *Leukemia* **2012**, *26*, 1654–1662. [CrossRef]
78. Fry, A.M.; O'Regan, L.; Sabir, S.R.; Bayliss, R. Cell cycle regulation by the NEK family of protein kinases. *J. Cell Sci.* **2012**, *125*, 4423–4433. [CrossRef] [PubMed]
79. Roccaro, A.M.; Sacco, A.; Jia, X.; Azab, A.K.; Maiso, P.; Ngo, H.T.; Azab, F.; Runnels, J.; Quang, P.; Ghobrial, I.M. microRNA-dependent modulation of histone acetylation in Waldenstrom macroglobulinemia. *Blood* **2010**, *116*, 1506–1514. [CrossRef] [PubMed]
80. Fulciniti, M.; Amodio, N.; Bandi, R.L.; Cagnetta, A.; Samur, M.K.; Acharya, C.; Prabhala, R.; D'Aquila, P.; Bellizzi, D.; Passarino, G.; et al. miR-23b/SP1/c-myc forms a feed-forward loop supporting multiple myeloma cell growth. *Blood Cancer J.* **2016**, *6*, e380. [CrossRef]
81. Roccaro, A.M.; Sacco, A.; Chen, C.; Runnels, J.; Leleu, X.; Azab, F.; Azab, A.K.; Jia, X.; Ngo, H.T.; Melhem, M.R.; et al. microRNA expression in the biology, prognosis, and therapy of Waldenstrom macroglobulinemia. *Blood* **2009**, *113*, 4391–4402. [CrossRef]
82. Roccaro, A.M.; Sacco, A.; Jia, X.; Banwait, R.; Maiso, P.; Azab, F.; Flores, L.; Manier, S.; Azab, A.K.; Ghobrial, I.M. Mechanisms of activity of the TORC1 inhibitor everolimus in Waldenstrom macroglobulinemia. *Clin. Cancer Res.* **2012**, *18*, 6609–6622. [CrossRef]
83. Gaudette, B.T.; Dwivedi, B.; Chitta, K.S.; Poulain, S.; Powell, D.; Vertino, P.; Leleu, X.; Lonial, S.; Chanan-Khan, A.A.; Kowalski, J.; et al. Low expression of pro-apoptotic Bcl-2 family proteins sets the apoptotic threshold in Waldenstrom macroglobulinemia. *Oncogene* **2016**, *35*, 479–490. [CrossRef] [PubMed]
84. Bradshaw, G.; Sutherland, H.G.; Haupt, L.M.; Griffiths, L.R. Dysregulated MicroRNA Expression Profiles and Potential Cellular, Circulating and Polymorphic Biomarkers in Non-Hodgkin Lymphoma. *Genes* **2016**, *7*, 130. [CrossRef] [PubMed]
85. Capuano, C.; Pighi, C.; Maggio, R.; Battella, S.; Morrone, S.; Palmieri, G.; Santoni, A.; Klein, C.; Galandrini, R. CD16 pre-ligation by defucosylated tumor-targeting mAb sensitizes human NK cells to gammac cytokine stimulation via PI3K/mTOR axis. *Cancer Immunol. Immunother.* **2020**, *69*, 501–512. [CrossRef]
86. Battella, S.; Cox, M.C.; Santoni, A.; Palmieri, G. Natural killer (NK) cells and anti-tumor therapeutic mAb: Unexplored interactions. *J. Leukoc. Biol.* **2016**, *99*, 87–96. [CrossRef]
87. Culpin, R.E.; Proctor, S.J.; Angus, B.; Crosier, S.; Anderson, J.J.; Mainou-Fowler, T. A 9 series microRNA signature differentiates between germinal centre and activated B-cell-like diffuse large B-cell lymphoma cell lines. *Int. J. Oncol.* **2010**, *37*, 367–376. [CrossRef]
88. Malpeli, G.; Barbi, S.; Greco, C.; Zupo, S.; Bertolaso, A.; Scupoli, M.T.; Krampera, M.; Kamga, P.T.; Croce, C.M.; Scarpa, A.; et al. MicroRNA signatures and Foxp3(+) cell count correlate with relapse occurrence in follicular lymphoma. *Oncotarget* **2018**, *9*, 19961–19979. [CrossRef]
89. Sinkovics, J.G. Molecular biology of oncogenic inflammatory processes. I. Non-oncogenic and oncogenic pathogens, intrinsic inflammatory reactions without pathogens, and microRNA/DNA interactions (Review). *Int. J. Oncol.* **2012**, *40*, 305–349. [CrossRef]
90. Ruskone-Fourmestraux, A.; Lavergne, A.; Aegerter, P.H.; Megraud, F.; Palazzo, L.; de Mascarel, A.; Molina, T.; Rambaud, J.L. Predictive factors for regression of gastric MALT lymphoma after anti-Helicobacter pylori treatment. *Gut* **2001**, *48*, 297–303. [CrossRef]
91. Tzioufas, A.G.; Kapsogeorgou, E.K.; Moutsopoulos, H.M. Pathogenesis of Sjogren's syndrome: What we know and what we should learn. *J. Autoimmun.* **2012**, *39*, 4–8. [CrossRef] [PubMed]

92. Steinbach, G.; Ford, R.; Glober, G.; Sample, D.; Hagemeister, F.B.; Lynch, P.M.; McLaughlin, P.W.; Rodriguez, M.A.; Romaguera, J.E.; Sarris, A.H.; et al. Antibiotic treatment of gastric lymphoma of mucosa-associated lymphoid tissue. An uncontrolled trial. *Ann. Intern. Med.* **1999**, *131*, 88–95. [CrossRef]
93. Morgner, A.; Miehlke, S.; Fischbach, W.; Schmitt, W.; Muller-Hermelink, H.; Greiner, A.; Thiede, C.; Schetelig, J.; Neubauer, A.; Stolte, M.; et al. Complete remission of primary high-grade B-cell gastric lymphoma after cure of Helicobacter pylori infection. *J. Clin. Oncol.* **2001**, *19*, 2041–2048. [CrossRef] [PubMed]
94. Liu, H.; Ye, H.; Ruskone-Fourmestraux, A.; de Jong, D.; Pileri, S.; Thiede, C.; Lavergne, A.; Boot, H.; Caletti, G.; Wundisch, T.; et al. T(11;18) is a marker for all stage gastric MALT lymphomas that will not respond to H. pylori eradication. *Gastroenterology* **2002**, *122*, 1286–1294. [CrossRef] [PubMed]
95. Akagi, T.; Motegi, M.; Tamura, A.; Hosokawa, Y.; Suzuki, H.; Ota, H.; Nakamura, S.; Morishima, Y.; Taniwaki, M.; Seto, M. A novel gene, MALT1 at 18q21, is involved in t(11;18) (q21;q21) found in low-grade B-cell lymphoma of mucosa-associated lymphoid tissue. *Oncogene* **1999**, *18*, 5785–5794. [CrossRef]
96. Hosokawa, Y.; Suzuki, H.; Suzuki, Y.; Takahashi, R.; Seto, M. Antiapoptotic function of apoptosis inhibitor 2-MALT1 fusion protein involved in t(11;18)(q21;q21) mucosa-associated lymphoid tissue lymphoma. *Cancer Res.* **2004**, *64*, 3452–3457. [CrossRef]
97. Rosebeck, S.; Lucas, P.C.; McAllister-Lucas, L.M. Protease activity of the API2-MALT1 fusion oncoprotein in MALT lymphoma development and treatment. *Future Oncol.* **2011**, *7*, 613–617. [CrossRef] [PubMed]
98. Tomasini, R.; Seux, M.; Nowak, J.; Bontemps, C.; Carrier, A.; Dagorn, J.C.; Pebusque, M.J.; Iovanna, J.L.; Dusetti, N.J. TP53INP1 is a novel p73 target gene that induces cell cycle arrest and cell death by modulating p73 transcriptional activity. *Oncogene* **2005**, *24*, 8093–8104. [CrossRef]
99. Dai, Y.H.; Tang, Y.P.; Zhu, H.Y.; Lv, L.; Chu, Y.; Zhou, Y.Q.; Huo, J.R. ZEB2 promotes the metastasis of gastric cancer and modulates epithelial mesenchymal transition of gastric cancer cells. *Dig. Dis. Sci.* **2012**, *57*, 1253–1260. [CrossRef] [PubMed]
100. Gebauer, N.; Kuba, J.; Senft, A.; Schillert, A.; Bernard, V.; Thorns, C. MicroRNA-150 is up-regulated in extranodal marginal zone lymphoma of MALT type. *Cancer Genom. Proteom.* **2014**, *11*, 51–56.
101. Wu, Q.; Jin, H.; Yang, Z.; Luo, G.; Lu, Y.; Li, K.; Ren, G.; Su, T.; Pan, Y.; Feng, B.; et al. MiR-150 promotes gastric cancer proliferation by negatively regulating the pro-apoptotic gene EGR2. *Biochem. Biophys. Res. Commun.* **2010**, *392*, 340–345. [CrossRef] [PubMed]
102. Young, E.; Noerenberg, D.; Mansouri, L.; Ljungstrom, V.; Frick, M.; Sutton, L.A.; Blakemore, S.J.; Galan-Sousa, J.; Plevova, K.; Baliakas, P.; et al. EGR2 mutations define a new clinically aggressive subgroup of chronic lymphocytic leukemia. *Leukemia* **2017**, *31*, 1547–1554. [CrossRef] [PubMed]
103. Krysiak, K.; Gomez, F.; White, B.S.; Matlock, M.; Miller, C.A.; Trani, L.; Fronick, C.C.; Fulton, R.S.; Kreisel, F.; Cashen, A.F.; et al. Recurrent somatic mutations affecting B-cell receptor signaling pathway genes in follicular lymphoma. *Blood* **2017**, *129*, 473–483. [CrossRef]
104. Sun, M.; Liu, X.H.; Li, J.H.; Yang, J.S.; Zhang, E.B.; Yin, D.D.; Liu, Z.L.; Zhou, J.; Ding, Y.; Li, S.Q.; et al. MiR-196a is upregulated in gastric cancer and promotes cell proliferation by downregulating p27(kip1). *Mol. Cancer Ther.* **2012**, *11*, 842–852. [CrossRef]
105. Yuan, Y.; Du, W.; Wang, Y.; Xu, C.; Wang, J.; Zhang, Y.; Wang, H.; Ju, J.; Zhao, L.; Wang, Z.; et al. Suppression of AKT expression by miR-153 produced anti-tumor activity in lung cancer. *Int. J. Cancer* **2015**, *136*, 1333–1340. [CrossRef]
106. Xie, J.; Chen, M.; Zhou, J.; Mo, M.S.; Zhu, L.H.; Liu, Y.P.; Gui, Q.J.; Zhang, L.; Li, G.Q. miR-7 inhibits the invasion and metastasis of gastric cancer cells by suppressing epidermal growth factor receptor expression. *Oncol. Rep.* **2014**, *31*, 1715–1722. [CrossRef] [PubMed]
107. Zhao, X.; Dou, W.; He, L.; Liang, S.; Tie, J.; Liu, C.; Li, T.; Lu, Y.; Mo, P.; Shi, Y.; et al. MicroRNA-7 functions as an anti-metastatic microRNA in gastric cancer by targeting insulin-like growth factor-1 receptor. *Oncogene* **2013**, *32*, 1363–1372. [CrossRef] [PubMed]
108. Pang, L.; Li, B.; Zheng, B.; Niu, L.; Ge, L. miR-138 inhibits gastric cancer growth by suppressing SOX4. *Oncol. Rep.* **2017**, *38*, 1295–1302. [CrossRef] [PubMed]
109. Yao, Y.; Suo, A.L.; Li, Z.F.; Liu, L.Y.; Tian, T.; Ni, L.; Zhang, W.G.; Nan, K.J.; Song, T.S.; Huang, C. MicroRNA profiling of human gastric cancer. *Mol. Med. Rep.* **2009**, *2*, 963–970. [CrossRef] [PubMed]
110. Rao, D.S.; O'Connell, R.M.; Chaudhuri, A.A.; Garcia-Flores, Y.; Geiger, T.L.; Baltimore, D. MicroRNA-34a perturbs B lymphocyte development by repressing the forkhead box transcription factor Foxp1. *Immunity* **2010**, *33*, 48–59. [CrossRef]
111. Streubel, B.; Vinatzer, U.; Lamprecht, A.; Raderer, M.; Chott, A. T(3;14)(p14.1;q32) involving IGH and FOXP1 is a novel recurrent chromosomal aberration in MALT lymphoma. *Leukemia* **2005**, *19*, 652–658. [CrossRef] [PubMed]
112. Goatly, A.; Bacon, C.M.; Nakamura, S.; Ye, H.; Kim, I.; Brown, P.J.; Ruskone-Fourmestraux, A.; Cervera, P.; Streubel, B.; Banham, A.H.; et al. FOXP1 abnormalities in lymphoma: Translocation breakpoint mapping reveals insights into deregulated transcriptional control. *Mod. Pathol.* **2008**, *21*, 902–911. [CrossRef] [PubMed]
113. Gu, S.; Wang, X.; Feng, L.; Xie, Y.; Ping, L.; Liu, W.; Li, J.; Liu, Y.; Ding, N.; Song, Y.; et al. Prognostic role of serum miRNA-16 in primary gastric lymphoma. *Int. J. Clin. Exp. Pathol.* **2018**, *11*, 5430–5435. [PubMed]
114. Cimmino, A.; Calin, G.A.; Fabbri, M.; Iorio, M.V.; Ferracin, M.; Shimizu, M.; Wojcik, S.E.; Aqeilan, R.I.; Zupo, S.; Dono, M.; et al. miR-15 and miR-16 induce apoptosis by targeting BCL2. *Proc. Natl. Acad. Sci. USA* **2005**, *102*, 13944–13949. [CrossRef] [PubMed]
115. Cerroni, L.; Zochling, N.; Putz, B.; Kerl, H. Infection by Borrelia burgdorferi and cutaneous B-cell lymphoma. *J. Cutan. Pathol.* **1997**, *24*, 457–461. [CrossRef]
116. Bogle, M.A.; Riddle, C.C.; Triana, E.M.; Jones, D.; Duvic, M. Primary cutaneous B-cell lymphoma. *J. Am. Acad. Dermatol.* **2005**, *53*, 479–484. [CrossRef] [PubMed]

117. Monsalvez, V.; Montes-Moreno, S.; Artiga, M.J.; Rodriguez, M.E.; Sanchez-Espiridion, B.; Lozano, M.; Fernandez-de-Misa, R.; Rodriguez-Peralto, J.L.; Piris, M.A.; Ortiz-Romero, P.L. MicroRNAs as prognostic markers in indolent primary cutaneous B-cell lymphoma. *Mod. Pathol.* **2013**, *26*, 171–181. [CrossRef]
118. Hother, C.; Rasmussen, P.K.; Joshi, T.; Reker, D.; Ralfkiaer, U.; Workman, C.T.; Heegaard, S.; Ralfkiaer, E.; Gronbaek, K. MicroRNA profiling in ocular adnexal lymphoma: A role for MYC and NFKB1 mediated dysregulation of microRNA expression in aggressive disease. *Investig. Ophthalmol. Vis. Sci.* **2013**, *54*, 5169–5175. [CrossRef] [PubMed]
119. Kwon, J.J.; Factora, T.D.; Dey, S.; Kota, J. A Systematic Review of miR-29 in Cancer. *Mol. Ther. Oncolytics* **2019**, *12*, 173–194. [CrossRef] [PubMed]
120. Wang, Q.; Ye, B.; Wang, P.; Yao, F.; Zhang, C.; Yu, G. Overview of microRNA-199a Regulation in Cancer. *Cancer Manag. Res.* **2019**, *11*, 10327–10335. [CrossRef]
121. Troppan, K.; Wenzl, K.; Pichler, M.; Pursche, B.; Schwarzenbacher, D.; Feichtinger, J.; Thallinger, G.G.; Beham-Schmid, C.; Neumeister, P.; Deutsch, A. miR-199a and miR-497 are Associated with Better Overall Survival due to Increased Chemosensitivity in Diffuse Large B-Cell Lymphoma Patients. *Int. J. Mol. Sci.* **2015**, *16*, 18077–18095. [CrossRef]
122. Alunno, A.; Leone, M.C.; Giacomelli, R.; Gerli, R.; Carubbi, F. Lymphoma and Lymphomagenesis in Primary Sjogren's Syndrome. *Front. Med.* **2018**, *5*, 102. [CrossRef]
123. Gourzi, V.C.; Kapsogeorgou, E.K.; Kyriakidis, N.C.; Tzioufas, A.G. Study of microRNAs (miRNAs) that are predicted to target the autoantigens Ro/SSA and La/SSB in primary Sjogren's Syndrome. *Clin. Exp. Immunol.* **2015**, *182*, 14–22. [CrossRef] [PubMed]
124. Kapsogeorgou, E.K.; Papageorgiou, A.; Protogerou, A.D.; Voulgarelis, M.; Tzioufas, A.G. Low miR200b-5p levels in minor salivary glands: A novel molecular marker predicting lymphoma development in patients with Sjogren's syndrome. *Ann. Rheum. Dis.* **2018**, *77*, 1200–1207. [CrossRef]
125. Rhodes, L.V.; Martin, E.C.; Segar, H.C.; Miller, D.F.; Buechlein, A.; Rusch, D.B.; Nephew, K.P.; Burow, M.E.; Collins-Burow, B.M. Dual regulation by microRNA-200b-3p and microRNA-200b-5p in the inhibition of epithelial-to-mesenchymal transition in triple-negative breast cancer. *Oncotarget* **2015**, *6*, 16638–16652. [CrossRef] [PubMed]
126. Robinson, J.E.; Cutucache, C.E. Deciphering splenic marginal zone lymphoma pathogenesis: The proposed role of microRNA. *Oncotarget* **2018**, *9*, 30005–30022. [CrossRef]
127. Mateo, M.; Mollejo, M.; Villuendas, R.; Algara, P.; Sanchez-Beato, M.; Martínez, P.; Piris, M.A. 7q31-32 Allelic Loss is a Frequent Finding in Splenic Marginal Zone Lymphoma. *Am. J. Pathol.* **1999**, *154*, 1583–1589. [CrossRef]
128. Watkins, A.J.; Hamoudi, R.A.; Zeng, N.; Yan, Q.; Huang, Y.; Liu, H.; Zhang, J.; Braggio, E.; Fonseca, R.; de Leval, L.; et al. An integrated genomic and expression analysis of 7q deletion in splenic marginal zone lymphoma. *PLoS ONE* **2012**, *7*, e44997. [CrossRef]
129. Ruiz-Ballesteros, E.; Mollejo, M.; Rodriguez, A.; Camacho, F.I.; Algara, P.; Martinez, N.; Pollan, M.; Sanchez-Aguilera, A.; Menarguez, J.; Campo, E.; et al. Splenic marginal zone lymphoma: Proposal of new diagnostic and prognostic markers identified after tissue and cDNA microarray analysis. *Blood* **2005**, *106*, 1831–1838. [CrossRef] [PubMed]
130. Pekarsky, Y.; Santanam, U.; Cimmino, A.; Palamarchuk, A.; Efanov, A.; Maximov, V.; Volinia, S.; Alder, H.; Liu, C.G.; Rassenti, L.; et al. Tcl1 expression in chronic lymphocytic leukemia is regulated by miR-29 and miR-181. *Cancer Res.* **2006**, *66*, 11590–11593. [CrossRef] [PubMed]
131. Karaayvaz, M.; Zhai, H.; Ju, J. miR-129 promotes apoptosis and enhances chemosensitivity to 5-fluorouracil in colorectal cancer. *Cell Death Dis.* **2013**, *4*, e659. [CrossRef]
132. Suzuki, R.; Amatya, V.J.; Kushitani, K.; Kai, Y.; Kambara, T.; Takeshima, Y. miR-182 and miR-183 Promote Cell Proliferation and Invasion by Targeting FOXO1 in Mesothelioma. *Front. Oncol.* **2018**, *8*, 446. [CrossRef]
133. Bouteloup, M.; Verney, A.; Rachinel, N.; Callet-Bauchu, E.; Ffrench, M.; Coiffier, B.; Magaud, J.P.; Berger, F.; Salles, G.A.; Traverse-Glehen, A. MicroRNA expression profile in splenic marginal zone lymphoma. *Br. J. Haematol.* **2012**, *156*, 279–281. [CrossRef]
134. Fuertes, T.; Ramiro, A.R.; de Yebenes, V.G. miRNA-Based Therapies in B Cell Non-Hodgkin Lymphoma. *Trends Immunol.* **2020**, *41*, 932–947. [CrossRef]
135. Xiong, W.; Lv, R.; Li, H.; Li, Z.; Wang, H.; Liu, W.; Zou, D.; Qiu, L.; Yi, S. Prevalence of hepatitis B and hepatitis C viral infections in various subtypes of B-cell non-Hodgkin lymphoma: Confirmation of the association with splenic marginal zone lymphoma. *Blood Cancer J.* **2017**, *7*, e548. [CrossRef] [PubMed]
136. Hermine, O.; Lefrere, F.; Bronowicki, J.P.; Mariette, X.; Jondeau, K.; Eclache-Saudreau, V.; Delmas, B.; Valensi, F.; Cacoub, P.; Brechot, C.; et al. Regression of splenic lymphoma with villous lymphocytes after treatment of hepatitis C virus infection. *N. Engl. J. Med.* **2002**, *347*, 89–94. [CrossRef]
137. Arcaini, L.; Vallisa, D.; Rattotti, S.; Ferretti, V.V.; Ferreri, A.J.M.; Bernuzzi, P.; Merli, M.; Varettoni, M.; Chiappella, A.; Ambrosetti, A.; et al. Antiviral treatment in patients with indolent B-cell lymphomas associated with HCV infection: A study of the Fondazione Italiana Linfomi. *Ann. Oncol.* **2014**, *25*, 1404–1410. [CrossRef] [PubMed]
138. Jeon, Y.J.; Lee, K.Y.; Cho, Y.Y.; Pugliese, A.; Kim, H.G.; Jeong, C.H.; Bode, A.M.; Dong, Z. Role of NEK6 in tumor promoter-induced transformation in JB6 C141 mouse skin epidermal cells. *J. Biol. Chem.* **2010**, *285*, 28126–28133. [CrossRef] [PubMed]
139. Arribas, A.J.; Campos-Martin, Y.; Gomez-Abad, C.; Algara, P.; Sanchez-Beato, M.; Rodriguez-Pinilla, M.S.; Montes-Moreno, S.; Martinez, N.; Alves-Ferreira, J.; Piris, M.A.; et al. Nodal marginal zone lymphoma: Gene expression and miRNA profiling identify diagnostic markers and potential therapeutic targets. *Blood* **2012**, *119*, e9–e21. [CrossRef] [PubMed]

140. Jardin, F.; Figeac, M. MicroRNAs in lymphoma, from diagnosis to targeted therapy. *Curr. Opin. Oncol.* **2013**, *25*, 480–486. [CrossRef] [PubMed]
141. Van den Brand, M.; van Krieken, J.H. Recognizing nodal marginal zone lymphoma: Recent advances and pitfalls. A systematic review. *Haematologica* **2013**, *98*, 1003–1013. [CrossRef]
142. Barh, D.; Malhotra, R.; Ravi, B.; Sindhurani, P. MicroRNA let-7: An emerging next-generation cancer therapeutic. *Curr. Oncol.* **2010**, *17*, 70–80. [CrossRef]
143. Gebauer, N.; Thorns, C.; Bernard, V.; Senft, A.; Schillert, A.; Merz, H.; Feller, A.C.; Bernd, H.W. MicroRNA profiling of low-grade and transformed nodal marginal zone lymphoma reveals a similar signature pattern distinct from diffuse large B cell lymphoma. *Acta Haematol.* **2015**, *133*, 214–220. [CrossRef] [PubMed]
144. Angelopoulou, M.K. Transformed nodal marginal zone lymphoma versus diffuse large B cell lymphoma: The microRNA aspect. *Acta Haematol.* **2015**, *133*, 212–213. [CrossRef] [PubMed]
145. Braggio, E.; Keats, J.J.; Leleu, X.; van Wier, S.; Jimenez-Zepeda, V.H.; Valdez, R.; Schop, R.F.; Price-Troska, T.; Henderson, K.; Sacco, A.; et al. Identification of copy number abnormalities and inactivating mutations in two negative regulators of nuclear factor-kappaB signaling pathways in Waldenstrom's macroglobulinemia. *Cancer Res.* **2009**, *69*, 3579–3588. [CrossRef]
146. Moussay, E.; Wang, K.; Cho, J.H.; van Moer, K.; Pierson, S.; Paggetti, J.; Nazarov, P.V.; Palissot, V.; Hood, L.E.; Berchem, G.; et al. MicroRNA as biomarkers and regulators in B-cell chronic lymphocytic leukemia. *Proc. Natl. Acad. Sci. USA* **2011**, *108*, 6573–6578. [CrossRef] [PubMed]
147. Kitagawa, Y.; Brahmachary, M.; Tiacci, E.; Dalla-Favera, R.; Falini, B.; Basso, K. A microRNA signature specific for hairy cell leukemia and associated with modulation of the MAPK-JNK pathways. *Leukemia* **2012**, *26*, 2564–2567. [CrossRef]
148. Koens, L.; Qin, Y.; Leung, W.Y.; Corver, W.E.; Jansen, P.M.; Willemze, R.; Vermeer, M.H.; Tensen, C.P. MicroRNA profiling of primary cutaneous large B-cell lymphomas. *PLoS ONE* **2013**, *8*, e82471. [CrossRef]
149. Hodge, L.S.; Elsawa, S.F.; Grote, D.M.; Price-Troska, T.L.; Asmann, Y.W.; Fonseca, R.; Gertz, M.A.; Witzig, T.E.; Novak, A.J.; Ansell, S.M. MicroRNA expression in tumor cells from Waldenstrom's macroglobulinemia reflects both their normal and malignant cell counterparts. *Blood Cancer J.* **2011**, *1*, e24. [CrossRef] [PubMed]
150. Qi, L.; Bart, J.; Tan, L.P.; Platteel, I.; Sluis, T.; Huitema, S.; Harms, G.; Fu, L.; Hollema, H.; Berg, A. Expression of miR-21 and its targets (PTEN, PDCD4, TM1) in flat epithelial atypia of the breast in relation to ductal carcinoma in situ and invasive carcinoma. *BMC Cancer* **2009**, *9*, 163. [CrossRef]
151. Kubiczkova Besse, L.; Sedlarikova, L.; Kryukov, F.; Nekvindova, J.; Radova, L.; Almasi, M.; Pelcova, J.; Minarik, J.; Pika, T.; Pikalova, Z.; et al. Combination of serum microRNA-320a and microRNA-320b as a marker for Waldenstrom macroglobulinemia. *Am. J. Hematol.* **2015**, *90*, E51–E52. [CrossRef]
152. Bouyssou, J.M.; Liu, C.J.; Bustoros, M.; Sklavenitis-Pistofidis, R.; Aljawai, Y.; Manier, S.; Yosef, A.; Sacco, A.; Kokubun, K.; Tsukamoto, S.; et al. Profiling of circulating exosomal miRNAs in patients with Waldenstrom Macroglobulinemia. *PLoS ONE* **2018**, *13*, e0204589. [CrossRef] [PubMed]
153. Vazquez, M.I.; Catalan-Dibene, J.; Zlotnik, A. B cells responses and cytokine production are regulated by their immune microenvironment. *Cytokine* **2015**, *74*, 318–326. [CrossRef] [PubMed]
154. De Silva, N.S.; Klein, U. Dynamics of B cells in germinal centres. *Nat. Rev. Immunol.* **2015**, *15*, 137–148. [CrossRef]
155. Garavelli, S.; de Rosa, V.; de Candia, P. The Multifaceted Interface between Cytokines and microRNAs: An Ancient Mechanism to Regulate the Good and the Bad of Inflammation. *Front. Immunol.* **2018**, *9*, 3012. [CrossRef]
156. Barnes, N.A.; Stephenson, S.; Cocco, M.; Tooze, R.M.; Doody, G.M. BLIMP-1 and STAT3 counterregulate microRNA-21 during plasma cell differentiation. *J. Immunol.* **2012**, *189*, 253–260. [CrossRef]
157. Loffler, D.; Brocke-Heidrich, K.; Pfeifer, G.; Stocsits, C.; Hackermuller, J.; Kretzschmar, A.K.; Burger, R.; Gramatzki, M.; Blumert, C.; Bauer, K.; et al. Interleukin-6 dependent survival of multiple myeloma cells involves the Stat3-mediated induction of microRNA-21 through a highly conserved enhancer. *Blood* **2007**, *110*, 1330–1333. [CrossRef] [PubMed]
158. Papanota, A.M.; Karousi, P.; Kontos, C.K.; Ntanasis-Stathopoulos, I.; Scorilas, A.; Terpos, E. Multiple Myeloma Bone Disease: Implication of MicroRNAs in Its Molecular Background. *Int. J. Mol. Sci.* **2021**, *22*, 2375. [CrossRef]
159. Pedersen, I.M.; Otero, D.; Kao, E.; Miletic, A.V.; Hother, C.; Ralfkiaer, E.; Rickert, R.C.; Gronbaek, K.; David, M. Onco-miR-155 targets SHIP1 to promote TNFalpha-dependent growth of B cell lymphomas. *EMBO Mol. Med.* **2009**, *1*, 288–295. [CrossRef] [PubMed]
160. Cheng, Y.; Ma, Z.; Kim, B.H.; Wu, W.; Cayting, P.; Boyle, A.P.; Sundaram, V.; Xing, X.; Dogan, N.; Li, J.; et al. Principles of regulatory information conservation between mouse and human. *Nature* **2014**, *515*, 371–375. [CrossRef] [PubMed]
161. Bose, D.; Neumann, A.; Timmermann, B.; Meinke, S.; Heyd, F. Differential Interleukin-2 Transcription Kinetics Render Mouse but Not Human T Cells Vulnerable to Splicing Inhibition Early after Activation. *Mol. Cell. Biol.* **2019**, *39*. [CrossRef] [PubMed]
162. Heykers, A.; Leemans, A.; van der Gucht, W.; de Schryver, M.; Cos, P.; Delputte, P. Differences in Susceptibility of Human and Mouse Macrophage Cell Lines to Respiratory Syncytial Virus Infection. *Intervirology* **2019**, *62*, 134–144. [CrossRef] [PubMed]
163. LeBien, T.W.; Tedder, T.F. B lymphocytes: How they develop and function. *Blood* **2008**, *112*, 1570–1580. [CrossRef] [PubMed]
164. Vijg, J. Somatic mutations, genome mosaicism, cancer and aging. *Curr. Opin. Genet. Dev.* **2014**, *26*, 141–149. [CrossRef]
165. Fernandez, L.C.; Torres, M.; Real, F.X. Somatic mosaicism: On the road to cancer. *Nat. Rev. Cancer* **2016**, *16*, 43–55. [CrossRef]
166. Sato, F.; Tsuchiya, S.; Terasawa, K.; Tsujimoto, G. Intra-platform repeatability and inter-platform comparability of microRNA microarray technology. *PLoS ONE* **2009**, *4*, e5540. [CrossRef]

167. De Tullio, G.; de Fazio, V.; Sgherza, N.; Minoia, C.; Serrati, S.; Merchionne, F.; Loseto, G.; Iacobazzi, A.; Rana, A.; Petrillo, P.; et al. Challenges and opportunities of microRNAs in lymphomas. *Molecules* **2014**, *19*, 14723–14781. [CrossRef] [PubMed]
168. Rupaimoole, R.; Slack, F.J. MicroRNA therapeutics: Towards a new era for the management of cancer and other diseases. *Nat. Rev. Drug Discov.* **2017**, *16*, 203–222. [CrossRef] [PubMed]
169. Hayes, J.; Peruzzi, P.P.; Lawler, S. MicroRNAs in cancer: Biomarkers, functions and therapy. *Trends Mol. Med.* **2014**, *20*, 460–469. [CrossRef]
170. Yao, Q.; Chen, Y.; Zhou, X. The roles of microRNAs in epigenetic regulation. *Curr. Opin. Chem. Biol.* **2019**, *51*, 11–17. [CrossRef]
171. Sohel, M.M.H. Circulating microRNAs as biomarkers in cancer diagnosis. *Life Sci.* **2020**, *248*, 117473. [CrossRef] [PubMed]

biomedicines

MDPI

Review

MiR-7 in Cancer Development

Petra Korać [1], Mariastefania Antica [2] and Maja Matulić [1,*]

1 Department of Biology, Division of Molecular Biology, Faculty of Science, University of Zagreb, Horvatovac 102, 10000 Zagreb, Croatia; petra.korac@biol.pmf.hr
2 Division of Molecular Biology, Rudjer Bosković Institute, Bijenička 54, 10000 Zagreb, Croatia; antica@irb.hr
* Correspondence: mmatulic@biol.pmf.hr

Abstract: MicroRNAs (miRNAs) are short non-coding RNA involved in the regulation of specific mRNA translation. They participate in cellular signaling circuits and can act as oncogenes in tumor development, so-called oncomirs, as well as tumor suppressors. miR-7 is an ancient miRNA involved in the fine-tuning of several signaling pathways, acting mainly as tumor suppressor. Through downregulation of PI3K and MAPK pathways, its dominant role is the suppression of proliferation and survival, stimulation of apoptosis and inhibition of migration. Besides these functions, it has numerous additional roles in the differentiation process of different cell types, protection from stress and chromatin remodulation. One of the most investigated tissues is the brain, where its downregulation is linked with glioblastoma cell proliferation. Its deregulation is found also in other tumor types, such as in liver, lung and pancreas. In some types of lung and oral carcinoma, it can act as oncomir. miR-7 roles in cell fate determination and maintenance of cell homeostasis are still to be discovered, as well as the possibilities of its use as a specific biotherapeutic.

Keywords: microRNAs; miR-7; gene expression; tumor suppressor; cancer cell

Citation: Korać, P.; Antica, M.; Matulić, M. MiR-7 in Cancer Development. *Biomedicines* **2021**, *9*, 325. https://doi.org/10.3390/biomedicines9030325

Academic Editor: Francesca Lovat

Received: 26 February 2021
Accepted: 22 March 2021
Published: 23 March 2021

Publisher's Note: MDPI stays neutral with regard to jurisdictional claims in published maps and institutional affiliations.

1. Introduction

MicroRNAs are short non-coding RNAs involved in the regulation of specific mRNA translation. Through this process, they regulate numerous cellular functions, participate in signaling circuits and fine-tune cellular differentiation.

miRNAs (miRs) have a complex pathway of biogenesis and regulation of their function. While final miRNAs are short single-stranded noncoding RNAs of 20–23 nt, they start as pri-miRNAs, several hundred base pairs long with a complex formation pathway. These primary miRNAs are first processed by a microprocessor containing Drosha, an enzyme that cleaves the stem of a hairpin structure formed by future miR sequence and producing pre-miRNA. After nuclear export, further processing is done by Dicer in the cytoplasm, which removes the loop region and produces miRNA duplex. Only one strand of the duplex is chosen to become the mature miRNA, loaded on an RNA-induced silencing complex (RISC) containing the Argonaute protein. RISC complex with specific miR targets complementary mRNAs and fully complementary mRNA are degraded. Since mature miRNAs in higher eukaryotic cells most often are not fully complementary to their target mRNA, they can lead to translation inhibition [1].

Usually, one mRNA can be targeted by several miRNAs on its 3′UTR. It is supposed that the target site spacing can influence cooperative repression. Although a great number of genes can be influenced by a single miR, in general, miRs act according to the cellular program in a specific cell type and target only a subset of transcripts [2]. However, the regulation of these processes is still poorly understood.

One of the first known, and also most investigated miRNAs is miR-7. The seed sequence GGAAGA is evolutionarily conserved and is found in Nematodes, Insects and Vertebrates [3]. In Mammals miR-7 dominantly acts as a tumor suppressor and regulates several basic cellular processes, which include proliferation, differentiation, apoptosis, migration and expression of stem cell features. It was also one of the miRs used for the classification of the regulatory clusters. Most discoveries are in regard to its role in the brain and sensory cell differentiation in man and Drosophila, respectively. Li X, and his collaborators (2009) suggest that miRs, in general, may stabilize different regulatory networks depending on the conditions of environmental fluctuation during development [4]. This hypothesis was developed on an example of miR-7, participating in Notch and Epidermal growth factor receptor (EGFR) coherent and incoherent feedforward loops during photoreceptor determination in *Drosophila*. On the other hand, miR-7 downregulation is linked to cell proliferation in many tumors, and its regulation is tightly connected with differentiation processes in the pancreas, brain and other organs [5,6].

miR-7 is encoded in three different sites in the human genome. *MIR7-1* sequence is present inside the last intron of the heterogeneous nuclear ribonucleoprotein K (hnRNPK) gene, on chromosome 9 (9q21.32) and is considered to be the dominant gene responsible for miR-7 expression. *MIR7-2* sequence is present in the intergenic region on chromosome 15, and *MIR7-3* in the intron of pituitary gland specific factor 1 gene (*PGSF1*) or *MIR-7* host gene on chromosome 19 [7].

2. Regulation of MiR-7 Expression

miRNA genes, like the protein-coding genes, have a regulated promoter and their products are members of signaling circuits of different cellular processes. miRs are also regulated at several steps during processing into their active form by means of binding to different proteins [8]. miRs can bind different long non-coding RNAs and circular RNAs either to be degraded or to be "preserved" for later function. Different RNA classes can function as miR "sponges" and bind miRs to keep them out of function: 3'UTR mRNA [9], long non-coding RNAs (lncRNAs) and circular RNAs. Different proteins can also regulate pri-miR degradation [1].

miR-7 is considered to be a network stabilizer, connecting different signaling pathways through feedback and feedforward loops [4]. Its function in buffering gene expression and providing robustness in cell response was demonstrated. Caygill and Brand (2017) showed on the *Drosophila* model where miR-7 targeted the Notch pathway, that miR-7 buffers the differentiation of the neuroepithelial cells into neuroblasts. Its role was to enable precision in the process despite conditions of environmental stress [10].

As a tumor suppressor, miR-7 expression is often downregulated in different cancer cells (i.e., in brain, lung and colon cancer cells [11–13]). Interestingly, it is also involved in signaling circuits directing differentiation in different tissues and it is regulated by specific transcription factors [6,14–16]. miR-7 promoter was found to be silenced by DNA methylation in cancer stem cells [17]. In breast carcinoma, its expression is estrogen-dependent [18]. Duex et al. found miR-7 to be in a signaling loop with EGFR through Usp18 (Ubp43), a ubiquitin-specific peptidase, whose downregulation elevates miR-7 levels [19]. It was also found that Hepatitis B virus (HBV) protein HBx can upregulate miR-7 expression through EGFR [20] and in breast cells by hepatocyte growth factor (HGF) [21]. However, miR-7 inhibition promotes breast cancer metastasis [22].

miR-7 was found to belong to a p53-dependent non-coding RNA network [23,24], as well as the Myc signaling circuit [25]. Hansen et al. described the existence of circular RNAs, which can pair with complementary miRNAs [26]. Circular RNAs have a structure of covalently closed single-stranded RNA molecules, produced by a specific type of splicing. These molecules are more stable than linear. Some of them can act as miR sponges: RNA molecules, which contain multiple target sites complementary to a specific miR and influence its activities by binding to it. The first such molecule was detected in neurons and it was Cdr1as (ciRS-7) regulating miR-7. It contained miR-7 sequences transcribed in the antisense orientation from the *CDR1* gene, forming circular RNA (circRNA) Cdr1as with more than 70 binding sites for miR-7 and one perfectly complementary site for miR-671 [26–28]. It seems that Cdr1as binds miR-7s and serves as their reservoir, and their release is regulated with miR-671, which causes cleavage of Cdr1as and liberation of miR-7s to exert their activities. Furthermore, Kleaveland et al. (found miR-7 to be a member of a regulatory network consisting of four ncRNAs: one long ncRNA, one circular and two microRNAs, in the mouse cerebellum [29]. Cyrano is a long ncRNA, which pairs to miR-7 and triggers its destruction. At the same time, this long ncRNA enables upregulation of circular Cdr1as, otherwise downregulated by miR-7. miR-671 was found to be involved in Cdr1as destruction.

Numerous long noncoding RNAs were found to bind to miR-7 and downregulate its activities: LINC00115 and XIST in breast cancer [22,30], LINC00240 in lung cancer [31], RSC1-AS1 in hepatocellular carcinoma, TINCR and Zing Finger Antisense 1 (ZFAS1) in breast and colorectal cancer [32–34], LPP-AS2 in glioma cells [35], etc. LncRNA SOX21-AS1 influenced cervical cancer progression by inhibiting miR-7/VDAC1 (voltage-dependent anion channel 1) [36]. lncRNA KCNQ1OT1 modulated cell resistance to chemotherapy [37], and lncRNA FOXD2-AS1 was found to bind miR-7 in thyroid cancer, upregulating the expression of hTERT [38]. lncRNA UCA1 downregulated miR-7, influencing the EGFR axis in gastric cancer cells resistant to hypoxia [39]. Upregulation of long noncoding RNA ANRIL caused by hypoxia modulated miR-7/SIRT1 axis and protected cells from cell death [40]. lncRNA CASC21 influenced miR-7/YAP1 signaling in colorectal cancer [41], and lncRNA Oip5-as1 in stem cells was found to modulate NANOG expression [42].

Several circular RNAs, besides ciRS, also regulate miR-7 and its downstream targets: circHIPK3 in colorectal cancer, circ-0015756 in hepatocellular carcinoma influencing downstream FAK [43,44], hsa_circRNA_0006528 in breast cancer influencing proliferation through MAPK/ERK pathway [45], circ-U2AF1 in gliomas influencing the expression of NOVA 2 [46], circ-TFCP2L1 decreasing mir-7-PAK1 signaling [47], circAkap17b regulating FSH secretion in pituitary gland [48]. circSNCA, *SNCA* and miR-7 were found to be regulated by endogenous competition and could influence the progression of Parkinson's disease [49].

Other types of RNA can also modulate miR-7 activity: 3′UTR Ube3a-1 mRNA [9] and Small Nucleolar RNA Host Gene 15 (SNHG15) regulating Klf4 through miR-7 [50].

There are also several proteins, which influence miR-7 maturation. Wang et al. described miR-7 regulation by protein quaking isoforms (QKI) [51]. The QKI proteins have heteronuclear ribonucleoprotein particle K (hnRNPK) homology KH and belong to RNA binding proteins. These proteins interact with a QKI response element sequence in introns and mature mRNAs, and it was shown that nuclear isoforms QKI-5 and QKI-6 associated with pri-miR-7-1 to prevent its processing. They were also found to interact with Ago2, during stress conditions. Similarly, it was shown that miR-7 can be downregulated by NF90-NF45 complex, through the binding of this double-stranded RNA-binding protein complex to primary miR-7 [52]. miR-7, in turn, targeted the coding region of NF90 mRNA. Nerve cells have a posttranscriptional regulation of miR-7 through the expression of Musashi homolog 2 (MSI2) and Hu antigen R (HuR), miR processing inhibitors and tissue-specific factors, regulating miR-7 expression and activity during neural differentiation [8]. A similar regulation was found in human lung cancer cells as a response to TLR9 signaling [53]. In addition, mechanisms of miR-7 targeted degradation linked with its 3′ end modifications were recently discovered [54].

On the other side, SF2/ASF increases Drosha cleavage of primary miR-7 transcript and promotes miR-7 maturation, and miR-7 in the feedback loop can decrease SF2/ASF expression. This molecule does not only affect miR-7, but also other miRs, coordinating their splicing regulation and gene repression [55] (Figure 1).

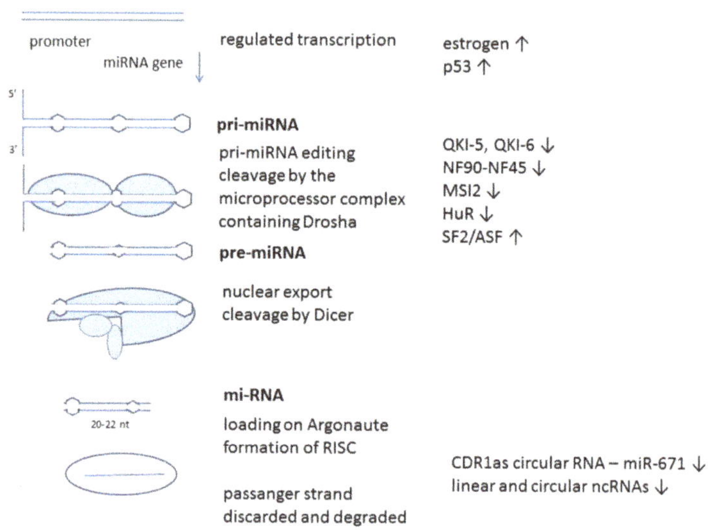

Figure 1. Biogenesis of miRNA. miRNA are transcribed from their genes regulated by promoters. Primary or pri-miRNA is several hundred base pairs long and has a form of a hairpin. It is processed by a microprocessor, a complex containing Drosha enzyme which removes the stem of the structure. Such pre-miRNA is exported from the nucleus and further cleaved by Dicer. miRNA duplex of 20–22 22 nt is produced. One strand of the duplex becomes the mature miRNA loaded on RNA-induced silencing complex (RISC), containing Argonaute protein. miRNA targets mRNA complementary to its sequence and directs it to degradation or inhibits translation, depending on the level of complementarity. Some of the known signaling molecules regulating miR-7 expression are shown [1,8,18,23,26,31–51,55].

3. MiR-7 and Chromatin Regulation

miR-7 was found to regulate a number of genes involved in chromatin modulation. It can downregulate histone methyl-transferase gene, *SETDB1* in different types of cancer cells [56,57], as well as *TET2* and *SMARCD1* [58,59]. It can also influence global cellular expression through the regulation of master transcription factors, such as *KLF4*, and thus impact the fate of cancer stem cells and human embryonic stem cells [60]. miR-7 is also found in extracellular vesicles and besides the possibility to influence the fate of the cell where it is expressed, it could also interfere with the biology of the cells to which it is delivered [61].

4. MiR-7 in Nerve Cells and Glioblastoma

In brain development, a fine regulation of cell proliferation, cell differentiation and regulation of symmetric and asymmetric division, as well as cell migration is necessary. It seems that miR-7 has a role in fine-tuning of these processes, in general as a suppressor of proliferation (Tables 1–4, Figure 2).

Table 1. miR-7 target genes/proteins in proliferation.

Gene/Protein	Cell Type	Pathway	Function	Reference
EGFR	glioblastoma schwanoma lung cancer melanoma	EGFR signaling	inhibition of proliferation	[11,12,62–64]
PIK3R3 PIK3CD PI3K	glioblastoma lung cancer	PI3K/Akt pathway TLR9 pathway	proliferation inhibition	[65–68]
IRS-1, IRS-2	glioblastoma tong squamous cell carcinoma melanoma	PI3K/Akt pathway	inhibition of proliferation viability	[67,69–71]
Raf1	glioblastoma	EGFR signaling	inhibition of proliferation	[63,67,72,73]
FOS	gastric cancer	MAPK signaling	inhibition of proliferation	[74]
ARF4	glioblastoma	MAPK signaling	inhibition of proliferation	[63]
mTOR p70S6K eIF4E Mapkap1 Mknk1 Mknk2	glioblastoma pancreas	PI3K/Akt pathway	proliferation inhibition	[66,75]
MAP3K9	pancreatic cancer	MAPK pathway	inhibition of proliferation and migration	[76]
IGF-1R	gastric metastasis glioma tong squamous carcinoma osteosarcoma	PI3K/Akt pathway IGF1-Snail EMT	inhibition of migration and survival	[71,77–79]
TAL1	T acute lymphoblastic leukaemia	proliferation	inhibition of proliferation	[80]
RELA	gastric cancer melanoma	NFkappa B signaling	inhibition of proliferation	[74,81]
IKK eta	gastric cancer	NF kappa B	regulation of proliferation	[74]
Skp2 Psme	ovary cells	cell cycle regulation	cell cycle arrest	[82]
CCNE	liver hepatocellular carcinoma	cell cycle	inhibition of proliferation	[83]
KLF4	epithelial brain metastasis endothelial cells lung	stem cell regulation	proliferation migration angiogenesis	[84–86]
RECK reversion inducing cysteine-rich protein with kazal motifs	oral cancer	metalloproteinase inhibitor	increase in proliferation	[87]
ERF Ets2 transcriptional repressor	lung cancer	MAPK pathway	increase in proliferation	[88]
CKS2 cyclin-dependent kinase regulatory subunit 2	thyroid cancer	cell cycle	inhibition of proliferation	[89]
TRIP6 thyroid receptor interactor protein	colorectal cancer	proliferation	inhibition of proliferation and metastasis	[90]
ALDJ1A3	breast cancer	stem cell survival	decrease in stem cell survival	[91]
CUL5 cullin5	hepatocellular	ubiquitination and protein degradation	inhibition of proliferation cell cycle arrest	[92]
TYRO3	colorectal cancer	PI3K/Akt/mTOR	inhibition of proliferation	[93]

Table 2. miR-7 target genes/proteins involved in migration.

Gene/Protein	Cell Type	Pathway	Function	Reference
PAK2 PAK1	lung carcinoma thyroid cancer gliomas tong squamous cell carcinomaschwanoma	Rho kinase effector	inhibition of proliferation, motility, regulation of cytoskeleton apoptosis	[11,63,67,69,71,94,95]
TBX2 T-Box2	glioblastoma	differentiation, EMT	inhibition of invasiveness	[96]
trefoil factor 3	glioblastoma	PI3K/Akt pathway	inhibition of migration	[97]
cdc42	brain damaged	Rho kinase	inhibition of migration and proliferation	[98]
Ack1 associated cdc42 kinase 1	schwannoma	Rho pathways citoskeleton regulation	inhibition of migration	[11]
FAK FAK (PTK2)	glioblastoma breast cancer colon cancer lung cancer	citoskeleton regulation	inhibition of migration and proliferation	[99–102]
NOVA2	lung carcinoma		inhibition of migration	[103]
LASP1	breast cancer			[104]
SATB1 special AT rich sequence binding protein	glioblastoma		inhibition of migration and invasion	[105]
Slug	breast cancer	EMT	decrease in migration	[30]
NFAT	pancreas	EMT	inhibition of migration	[106]
VE cadherin Notch4	hepatocellular carcinoma		inhibition of migration	[107]
KLF4 YY1	Non-Hodgin lymphoma		inhibition of migration and chemosensitivity	[108]

Table 3. miR-7 target genes/proteins involved in apoptosis and protection from stress.

Gene/Protein	Cell Type	Pathway	Function	Reference
BCL-2	lung, liver	apoptosis	apoptosis	[109]
XIAP	glioblastoma cervical cancer hepatocellular carcinoma lung	apoptosis	apoptosis	[68,72,110]
SLC25A37 TIMM50	rhabdomyosarcoma	mitochondria	induction of cell death ??	[111]
REGγ proteasome activator subunit	breast cancer	proteasome	inhibition of proliferation increase of apoptosis	[112]
NEIL Nei endonuclease VIII-like 1	colorectal cancer	inhibition of apoptosis, proliferation	inhibition of proliferation and survival	[113]
UBE2A	brain	ubiquitination and protein degradation	amyloid peptide proteolysis	[114]
YY1	glioblastoma colon cancer	p53 pathway cell cycle arrest wnt signaling	resistance to alkylation	[13,115]

Table 3. *Cont.*

Gene/Protein	Cell Type	Pathway	Function	Reference
1BRCA1	breast cancer	DNA repair	decrease in survival	[104]
Sirtuin /Sirt1	neuroblastoma	regulation of oxygen-glucose deprivation	protection from damage	[116]
VDAC	neuroblastoma hepatocellular carcinoma	ion channel on mitochondria; ROS defense	protection from oxidative stress	[117,118]
KEAP1	neuroblastoma	ROS defense	protection from oxidative stress	[119]
HOXB3	breast cancer retinal epithelial cells	glucose metabolism PI3K/Akt/mTOR	reduction of high glucose damage	[120]
PARP1	lung cancer cells	DNA repair	decreased DNA repair and survival	[121]
REDD1 regulated in development and DNA damage response 1	cervical carcinoma cells under hypoxia	DNA damage response	hypoxamir proliferation modulation	[122]
SMARCD1	lung cancer cells	chromatin regulator p53 pathway	increased chemoresistance	[59]
XRCC2	colorectal cancer cells	DNA repair	proliferation inhibition, induction of apoptosis	[123]
Rad54L	ovary cells	DNA damage repair	survival under cell cycle arrest conditions	[82]
REG1 regenerating islet-derived protein	pancreas	response to glucose starvation	inhibition of proliferation, apoptosis, differentiation	[124]
MRP1/ABCC1	lung carcinoma	multidrug resistance	decreased survival	[125]
NF90	tumor	DNA repair	DNA damage repair inhibition	[52]

Table 4. miR-7 target genes/proteins involved in differentiation and metabolic processes.

Gene/Protein	CELL TYPE	Pathway	Function	Reference
TLR4	brain	inflammation	downregulation of inflammation	[126]
FAM177A	macrophages	inflammation	inhibition of cytokine production	[127]
NLRP3 Nod like receptor	brain	inflammation	downregulation of inflammation	[128]
TET2	hematopoietic malignancies	chromatin modification		[58]
SETDB1 SETD8	pancreas	chromatin regulation		[56,57]
PAX6	brain lung colon pancreas embryonic stem cells		differentiation	[5,129]
Gli3	brain bladder cancer	hedgehog	differentiation	[130,131]
FGFR4	liver	protection from injury	stem cell proliferation	[132]

Table 4. *Cont.*

Gene/Protein	CELL TYPE	Pathway	Function	Reference
HoxD family	brain		differentiation	[133]
TCF4 and TCF12	brain	wnt pathway	differentiation	[134]
TCF7L2	brain	wnt pathway		[134]
SHANK3	brain		differentiation	[135]
ihog Hedgehog receptor	drosophila eye	hedgehog pathway	differentiation	[136]
CRY2	osteoblast	CLOCK/BMAL/p300 pathway	differentiation	[137]
Yorkie	drosophila wings	Hippo pathway	organ size	[138]
G protein signalling 5 RGSS	eye	signaling		[139]
PA28 gamma	lung carcinoma	proteasome	inhibition of protein degradation	[140]
insulin receptor INSR insulin receptor substrate 2 IRS-2 insulin-degrading enzyme IDE	brain	regulation of glucose metabolism	insulin sensitivity	[141]
TfR1 transferrin receptor 1	pancreatic and colon cells	iron transport and storage	iron transport and storage	[142]
beta arrestin 1	pancreatic beta cells	regulation of insulin secretion	metabolism	[143]
Sepp1b selenoprotein P	brain	synaptic function		[144]
Prostaglandin F2 receptor negative regulator PTGFR Golgi glycoprotein 1	pituitary gland	hormone regulation	gonad development	[145,146]
OGT, O-GlcNAcyl Transferase	lung cancer	O-GlcNAcylation	metabolic reprogramming migration	[147]
CAMK2D calponin	smooth muscle cell	calcification	vascular calcification in pulmonary hypertension	[148]
enolase ENO2	nasopharyngeal carcinoma	glycolysis	metabolism radioresistance	[149]
Lactat dehydrogenase A	gastric cancer	glycolysis	metabolism	[150]
Raf1	pituitary gland	production of prolactin	development	[151]
KLF4	myoblasts	differentiation and proliferation	inhibition of differentiation and proliferation	[152]
Follicle stimulating hormone FSH	pituitary gland	metabolism	inhibition of production	[48]
alpha-Synuclein	brain; Parkinson disease	neuron function and survival	inhibition of production	[153]

It has spatiotemporal-dependent expression and regulation [28], and it is found in discrete brain regions [3]. It can also have specific subcellular localization, different in the cell body and neurites. One example is miR-7 role in dopaminergic neuron differentiation by fine-tuning *Pax6* expression [5]. miR-7 also regulates other neural fate markers, elements of the Wnt pathway, interferes with Hedgehog and Notch signaling and takes part in the differentiation process [134]. miR-7 regulates both, specific nerve functions (such as synaptic [144]) and master regulators (such as HoxD family members). It is detected as one of the miRNAs forming "miR signature" in neural stem and neural cancer stem cells [154], which is in accordance with its role in differentiation and proliferation. Interestingly, miR-7 is 40 times more abundant in neurons than in astrocytes (Table 1).

Besides influencing cell differentiation, and thus indirectly interfering with it, miR-7 can also directly inhibit cell proliferation (Figure 2). In glioblastoma and neuroblastoma miR-7 was found to be downregulated compared to normal tissue, indicating its role as a tumor suppressor [11,62]. The functions of miR-7 in glioblastomas are mainly linked to its influence on cell proliferation, differentiation, apoptosis and migration. Although some glioblastoma cells can be refractory to miR-7 expression, its downregulation is often found in nerve cell tumors. Saydam et al. found its downregulation to be the typical miR schwannoma characteristic signature [11].

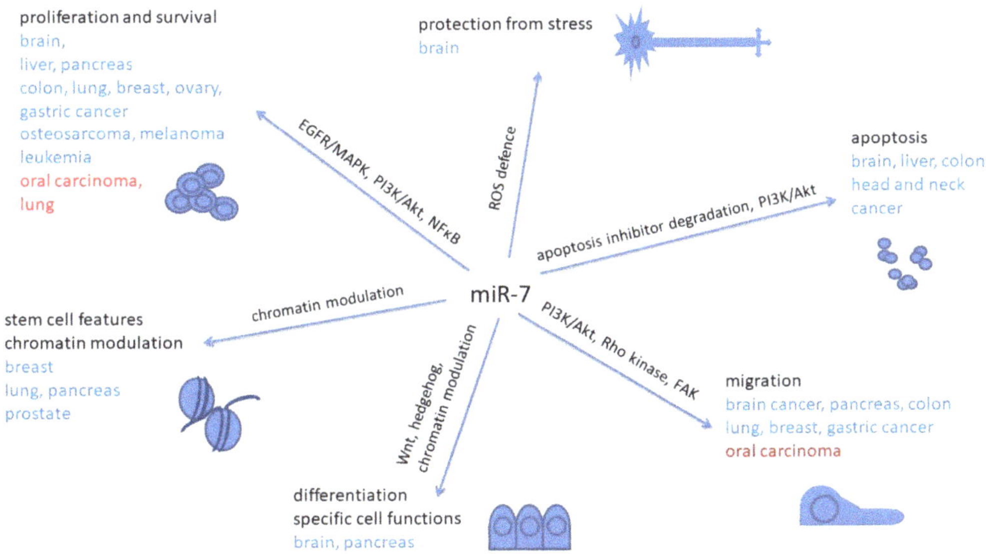

Figure 2. Effects of miR-7 on the process of carcinogenesis in different types of cancer. blue: tumor suppressor's activities; red: activities as oncomirs.

One of the first detected and most investigated targets of miR-7 is Epidermal growth factor receptor, EGFR, whose protein expression is decreased by miR activity. EGFR is linked to several important proliferation-inducing pathways, such as PI3K/Akt and MAPK and their downregulation leads to decreased activation of the Akt and ERK1/2. Kefas found that miR-7 directly regulates EGFR expression [62].

miR-7 targets are also several other proteins involved in downstream signaling. In the PI3K/Akt pathway, these are Akt pathway regulators IRS-1 and IRS-2, PI3K subunits (PIK3R3 and PIK3CD), mTOR [66], and PAK1 (p21/Cdc42/Rac1-activated kinase) [67,155]. The latter is potentially involved not only in oncogenic signaling through EGFR/Akt, but also in motility, regulation of cytoskeleton and apoptosis [63]. On the MAPK pathway, miR-7 influences Raf1 and ARF4 (ADP-ribosylation factor 4) expression, which modulates activation of phospholipase D2 (PLD2) and downstream activation of AP-1 [67,155]. Webster et al. found its influence on JNK and CAMK pathways [63]. In addition, Duex et al. found miR-7 to be involved in the signaling loop with EGFR through Usp18 (Ubp43), a ubiquitin-specific peptidase, whose downregulation elevates miR-7 levels [19].

miR-7 is also involved in the regulation of cell survival [67,84] as it downregulates pro-survival proteins IRS-1, IGF-1R, PAK1, and Raf-1 and leads to the reduction in cell viability. Zhang X et al. found that the expression pattern of miR-7 correlates with the glioblastoma cells' sensitivity to apoptosis induced by TRAIL, a TNF family member [110]. XIAP, an apoptosis inhibitor, was detected as a direct miR-7 target (Tables 1–3).

In another experimental setting, Kabaria et al. found that miR-7 targeted 3'UTR of Keap1 in human neuroblastoma cells [119]. Keap1 takes part in the regulation of Nrf2, a transcription factor involved in the expression of many antioxidant and detoxifying genes in reactive oxygen species (ROS) defense. miR-7, therefore, participated in cellular protection from oxidative stress. In neuroblastoma cells, Sirtuin (Sirt 1) was found to be a direct target of miR-7, and a link to the regulation of oxygen-glucose deprivation and cerebral injury [116]. It was found that miR-7 can target VDAC1, voltage-dependent anion channel, a part of the mitochondrial permeability transition pore, leading to the decrease in the intracellular ROS and protection against mitochondrial dysfunction and cytotoxicity [117]. Jia et al. compared RNA expression in glioblastoma cell lines differently sensitive on alkylation DNA damage and found miR-7 to be downregulated in the resistant cells [115]. They showed that miR-7 upregulation increased the cell sensitivity to alkylation. As a direct target, transcription factor YY1 was identified. However, it is also possible that in glioblastoma cell lines cell-specific regulation exists and that not all cell lines are responsive to miR-7 expression [8,156].

miR-7 also targets the expression of proteins involved in migration and metastasis [99]. Increased expression of miR-7 inhibited migration and invasion through downregulation of MMP-2, MMP-9 and FAK, a kinase involved in motility. Different targets were found to link miR-7 to actin cytoskeleton: Rho GTPases, Ack1 and PAK. In addition, in glioblastoma its target was a special AT-rich sequence binding protein 1 (SATB1), a protein able to promote migration and invasion [105].

Pan CM et al. found miR-7 to target TBX2 mRNA, and due to miR-7 downregulation in glioblastomas, TBX2 is increased [96]. Its high expression correlated with poor prognosis and higher invasivity of glioblastoma cells, EMT features and pulmonary metastasis. TBX2 is involved in the developmental processes and morphogenesis of different organs. It represses E-cadherin and increases the invasiveness of breast cancer cells. miR-7 also influenced TFF3, a signaling molecule downstream of PI3K/Akt pathway. Its downregulation decreased migration and invasion. This process can be reversed by a glioblastoma cell treatment with a glycolytic inhibitor which reduces the expression of miR-7 [97].

5. MiR-7 Role in Gastrointestinal Tumours

In gastric cancer (GC) patients, miR-7 deregulation consequently leads to increased cell proliferation, tumorigenesis and poor survival. In gastric cancer cells, besides targeting the EGFR pathway, miR-7 targets the IGF1R and downstream RELA and FOS [77]. miR-7 indirectly influences RELA activation, through targeting IKKeta. Through the feedback circuit, the NF-kappaB pathway regulates the miR-7 expression. In addition, miR-7 can downregulate the IGF1R-Snail pathway, which is involved in epithelial-mesenchymal transition [74,77]. Similar pathways were influenced in tongue squamous cell carcinoma [71]. Recently it was found that miR-7 could target lactate dehydrogenase A (LDH-A) in gastric cancer cells, so its downregulation can influence glycolysis, cell proliferation and sensitivity to chemotherapy [150].

In oral squamous cell carcinoma cells miR-7 regulated the expression of RECK, which acts as a metalloproteinase inhibitor and can suppress cell proliferation and migration. Therefore, miR-7 acted as an oncogene, and RECK inhibition was associated with poor prognosis and aggressiveness of tumors [87] (Figure 2).

miR-7 has also been reported to target a specific set of genes in the liver. Some of them code for proteins involved in cell cycle and apoptosis regulation, such as CCNE1 [83], Bcl-2 and XIAP. In hepatocellular carcinoma cells miR-7 directly regulates *CUL5*, influencing cell proliferation and inducing cell cycle arrest [92]. As miR-7 targets Notch3, its downregulation leads to Notch signaling activation in the same type of cancer cells [157]. Besides Notch3, Notch4 and VE cadherin were also found to be miR-7 targets [107]. miR-7 also downregulates *VDAC1* in hepatocellular carcinoma and influences proliferation and migration [118], as well as the fibroblast growth factor receptor FGFR4, a key molecule for liver protection from chronic injury. In the conditions of increased fibrosis miR-7 was found to be upregulated and promoted HSC proliferation and activation as a consequence of *FGFR4* downregulation [132].

miR-7 is involved in the differentiation of pancreatic endocrine cells [6]. In pancreatic carcinoma, miR-7 can suppress NFAT. This transcription factor can regulate epithelial-mesenchymal transition and act as an oncogene in pancreatic carcinoma cells [106]. Downing et al. found miR-7 to directly target *REG1*, a protein that increases proliferation and influences apoptosis and differentiation of pancreatic cells [124]. miR-7 was found to suppress SOX18 and to influence the gp130/JAK2/STAT3 pathway. Wang et al. found miR-7 to target members of the mTOR signaling pathway (p70S6K, eIF4E, Mapkap1, Mknk1 and MknK2) [75], influencing cell proliferation, as well as MAP3K9 [76]. In addition, miR-7 targets also SET8, a histone methyltransferase, thus potentially influencing the expression of a number of downstream genes [56] (Table 4).

In colon cancer cells miR-7 also suppresses proliferation, increases apoptosis and causes cell-cycle arrest, by targeting YY1 and by influencing downstream p53, caspases and c-jun, as well as wnt signaling (through beta-catenin, survivin and FGF4) [13]. Neil, an endonuclease that inhibits apoptosis and increases cell survival and proliferation was found to be regulated by miR7 [113].

Other targets are *TYRO3*, influencing PI3K/Akt/mTOR pathway [93], *TRIP6* which regulates proliferation and metastases [90], *FAK* [101] and *XRCC2*, a gene involved in homologous recombination repair pathway [123].

6. MiR-7 Roles in Lung Cancer

Promoter mutation of miR-7 was found to be associated with a poor prognosis of lung cancer [12]. The main targets released from miR-7 downregulation are those of EGFR and PIK3/Akt pathways, apoptosis inhibitors [109], and proteins involved in migration, FAK, PAK2 and NOVA2 [12,72,94,102,103]. PIK3/Akt signaling also connects TLR9 and miR-7 regulation [65]. However, Chou found that miR-7 could act as an oncomir in lung tumorigenesis [88]. EGFR, through the Ras/ERK/Myc pathway, increased the production of miR-7-1, which targets ERF, a transcriptional repressor. Therefore, in carcinoma samples, a positive correlation between EGFR and miR-7 expression was found, and miR-7 increased cell proliferation and tumor volume. Another example of oncogene activity was miR-7 modulation of the MYC pathway, in a positive feedback loop. The miRNA target is *AMBRA1*, an important regulator of early autophagy and a mediator in MYC dephosphorylation [158].

Hong et al. identified *SMARCD1*, a chromatin remodeling protein, to be a direct target of miR-7 in lung cancer cells [59]. They concluded that miR-7 influences the coupling of SMARCD1 with p53, which leads to an increased chemoresistance of lung cancer cells. miR-7 also downregulates PARP1, thus influencing DNA homologous recombination repair and survival after Adriamycin treatment of small cell lung cancer cells [121]. Furthermore, miR-7 modulates chemoresistance by targeting the multidrug resistance-associated protein MRP1/ABCC1 [125].

In addition to that, miR-7 was found to target several proteins linked to protein degradation as PA28gamma, a proteasome activator, targeted in non-small cell lung carcinoma [140]. O-GlcNAcyl Transferase (OGT), an enzyme involved in O-linked *N*-acetylglucosaminylation and contributing to cancer phenotype, is regulated by miR-7 [147]. In nasopharyngeal carcinoma cells, miR-7 was found to regulate the expression of enolase, ENO2, and therefore its downregulation can influence cell glycolysis [149].

7. MiR-7 Roles in Melanoma and Skin Cancer

Similarly to its role in other tissues, in melanoma cells, miR-7 takes part in the suppression of proliferation. However, as melanoma cells are not typically EGFR-driven, Giles et al. found miR-7 to target RelA and thus inhibit NF-κB activity and its downstream genes, such as *IL-1β*, *IL-6* and *IL-8* [70]. The analysis of melanoma patient samples revealed a correlation between RelA expression and poor survival.

On the contrary, Meza-Sosa et al. found *KLF4* to be a miR-7 direct target in epithelial cells, and miR-7 overexpression in lung and skin epithelial cells enhanced cell proliferation, migration and tumorigenesis [84]. Tumors with an increased miR-7 had a decreased p21 and cyclin D. In thyroid papillary cancer, miR-7 targeted *CKS2*, a cyclin-dependent kinase regulator, and downstream cyclin B1 and cdk1 [89]. As a target, also *PAK1* was detected [95].

It has been shown that in the cancer-associated fibroblasts of head and neck cancers, overexpression of miR-7 downregulates *RSSF2*, a proapoptotic molecule influencing proliferation and migration, and decreases the secretion of a tumor suppressor PAR-4 (prostate apoptosis response 4) [159]. In the human ocular tissue, miR-7 targets *RGS5*, a regulator of G protein signaling [139].

8. MiR-7 Roles in Breast, Prostate and Ovarian Cancer

In breast cancer miR-7 inhibits the metastases and influences epithelial-mesenchymal transition by targeting FAK, a kinase that acts as a mediator in ECM-integrin signaling [100]. Overexpression of miR-7 induces an increase in E cadherin and downregulation of mesenchymal proteins, suppresses proliferation, anchorage-independent growth, migration and invasion, as well as anchorage-independent growth in matrigel. The level of miR-7 is associated with the aggressiveness of estrogen receptor-positive breast tumors [160]. It also targets proteasome activator subunit 3 (*REG*γ) and contributes to the decrease in the cancer stem cell population survival, proliferation and migration [30,91,112,161]. Several miR-7 targets influence chemotherapy resistance, such as members of EGFR/PI3K signaling, *BRCA1, LASP1, BCL-2* and *MRP1* [104,162,163]. Okuda et al. (2013) found that miR-7 suppresses the ability of breast cancer stem cells to metastasize to the brain [85]. The correlation was found with miR-7 modulation of KLF4 expression, involved in stem cell biology. In addition, in a breast cancer cell line miR-7 was also found to be in regulation circuit with HOXD10, and, together with miR-218, to downregulate HoxB3 [69,120]. These changes were further connected with increased activity of other tumor suppressors, RASSF1A and Claudin-6 through epigenetic regulation, leading to cell cycle inhibition. Seong et al. found miR-7 to target *REDD1*, a negative regulator of mTOR signaling in the stress conditions [122]. miR-7 was therefore assigned to so-called hypoxamirs, miRNAs involved in hypoxic response. In HeLa cells, hypoxia caused downregulation of miR-7, in order to increase REDD1 level and inhibit mTOR signaling. In prostate cancer, it was found that miR-7 can regulate the expression of AXL, a receptor tyrosine kinase, associated with tumorigenesis, inhibition of apoptosis and EMT, often deregulated in different types of carcinomas [164]. miR-7 also inhibited the stemness of prostate stem cancer cells through repression of *KLF4* and PI3K/Akt/p21 downstream pathway [165].

miR-7 overexpression in hamster ovary cell line CHO decreased the cell proliferation, without influencing viability. Transient transfection of CHO led to upregulation of nearly 200 genes and downregulation of around 350 genes. The pathways involved included translation, RNA and DNA processing, secretion and protein folding. miR-7 has been found to target regulators of G1-S transition, *Skp2* and *Psme*, to upregulate p27KIP and arrest the cells in the G1 phase. Furthermore, it was found that miR-7 coordinately changes the levels of many genes in order to maintain homeostasis under the arrest conditions. It regulates *Rad54L*, a DNA repair protein, and influences the proapoptotic regulator p53 and the antiapoptotic Akt pathway to insure cell survival [82,166].

In testicular germ cell tumors, miR-7 was found to be one of the four hub miRNAs in regulatory networks of nonseminoma tumors [167].

In addition, expression of miR-7 was found increased in renal cell carcinomas in comparison with normal tissue, suggesting its activities as an oncogene [168].

9. MiR-7 Roles in Mesenchymal Tissue and Tumours

In osteosarcoma miR-7 influences *IGF1R*, and in paediatric rhabdomyosarcomas targets *SLC25A37* and *TIMM50*, two mitochondrial proteins, important for the induction of cell death [79,111]. In osteosarcomas, miR-7 is supposed to be a regulating link between Linc00852 lncRNA, and AXL, a tyrosine kinase involved in tumor growth [169].

10. MiR-7 Roles in Leukaemia

miR-7 had a low expression in haematopoietic cells and in B-chronic lymphocytic leukaemia (Antica et al. unpublished results). In chronic myeloid leukaemia, it was found to interfere with Bcr/Abl signaling [170]. A higher expression of miR-7 was found in acute lymphocytic leukaemia (ALL) patients with CNS relapse compared to those without [171]. In B cell lymphoma it was found to be regulated by c-Myc [172]. In T-cell acute lymphocytic leukaemia (T-ALL), upregulation of long noncoding RNA ANRIL caused miR-7 sponging, binding multiple tandem miRNAs through response elements binding seed sequences, in order to sequester them from their target sequences. Consequently, *TCF4*, a miR-7 direct target, is upregulated and is involved in the disease progression [173]. In T-ALL, miR-7 was found to bind to *TAL1*, coding for T-cell acute lymphocytic leukaemia protein. In T-ALL, expression of miR-7 is often attenuated, while TAL1 expression is increased and solicitates cell proliferation [80]. In Non-Hodgkin lymphoma cells, miR-7 regulates migration and chemoresistance through KLF4 and YY1 [108] and miR-7 downregulation can increase the aggressiveness of follicular lymphoma by FasL upregulation in macrophages which modulate immunosuppressive stroma [174].

11. Conclusions

miR-7 is one of the most conserved and oldest miRs, and is engaged in numerous signaling circuits involved in differentiation, regulation of proliferation, apoptosis and migration. It targets numerous mRNAs depending on the intracellular milieu and is also regulated by different transcription factors and molecules involved in its processing and degradation. It was suggested that its role could be to buffer cellular processes under stress conditions and to coordinate cell proliferation with other functions. This could be the reason for its involvement in numerous diseases. In most tumors its expression is downregulated, as its dominant activity is tumor suppression by inhibition of cell proliferation and survival. In some cancer types, it acts as an oncomir, stressing the importance of nuances of signaling circuits in which it is involved. We believe that numerous functions in the maintenance of cell homeostasis and cell fate determination are still to be discovered.

Author Contributions: Conceptualization: P.K., M.A. and M.M.; writing—original draft preparation, M.M. and P.K.; writing—review and editing, M.A., P.K. and M.M. All authors have read and agreed to the published version of the manuscript.

Funding: This work was supported by the University of ZagrebUniversity Support 20285121 for 2020 was given to M. Matulić.

Institutional Review Board Statement: Not applicable.

Informed Consent Statement: Not applicable.

Data Availability Statement: Not applicable.

Acknowledgments: We thank all the colleagues from the Department of Molecular Biology, Faculty of Science, University of Zagreb.

Conflicts of Interest: The authors declare no conflict of interest. The funders had no role in the design of the study; in the collection, analyses, or interpretation of data; in the writing of the manuscript, or in the decision to publish the results.

References

1. Treiber, T.; Treiber, N.; Meister, G. Regulation of microRNA biogenesis and its crosstalk with other cellular pathways. *Nat. Rev. Mol. Cell Biol.* **2019**, *20*, 5–20. [CrossRef]
2. Oliveira, A.C.; Bovolenta, L.A.; Alves, L.; Figueiredo, L.; Ribeiro, A.O.; Campos, V.F.; Lemke, N.; Pinhal, D. Understanding the Modus Operandi of MicroRNA Regulatory Clusters. *Cells* **2019**, *8*, 1103. [CrossRef] [PubMed]
3. Zhao, J.; Zhou, Y.; Guo, M.; Yue, D.; Chen, C.; Liang, G.; Xu, L. MicroRNA-7: Expression and function in brain physiological and pathological processes. *Cell Biosci.* **2020**, *10*, 1–12. [CrossRef] [PubMed]
4. Li, X.; Cassidy, J.J.; Reinke, C.A.; Fischboeck, S.; Carthew, R.W. A MicroRNA Imparts Robustness against Environmental Fluctuation during Development. *Cell* **2009**, *137*, 273–282. [CrossRef]

5. De Chevigny, A.; Coré, N.; Follert, P.; Gaudin, M.; Barbry, P.; Beclin, C.; Cremer, H. miR-7a regulation of Pax6 controls spatial origin of forebrain dopaminergic neurons. *Nat. Neurosci.* **2012**, *15*, 1120–1126. [CrossRef]
6. Kredo-Russo, S.; Mandelbaum, A.D.; Ness, A.; Alon, I.; Lennox, K.A.; Behlke, M.A.; Hornstein, E. Pancreas-enriched miRNA refines endocrine cell differentiation. *Development* **2012**, *139*, 3021–3031. [CrossRef]
7. Horsham, J.L.; Ganda, C.; Kalinowski, F.C.; Brown, R.A.; Epis, M.R.; Leedman, P.J. MicroRNA-7: A miRNA with expanding roles in development and disease. *Int. J. Biochem. Cell Biol.* **2015**, *69*, 215–224. [CrossRef] [PubMed]
8. Choudhury, N.R.; Alves, F.D.L.; De Andrés-Aguayo, L.; Graf, T.; Cáceres, J.F.; Rappsilber, J.; Michlewski, G. Tissue-specific control of brain-enriched miR-7 biogenesis. *Genes Dev.* **2013**, *27*, 24–38. [CrossRef] [PubMed]
9. Valluy, J.; Bicker, S.; Aksoy-Aksel, A.; Lackinger, M.; Sumer, S.; Fiore, R.; Wüst, T.; Seffer, D.; Metge, F.; Dieterich, C.; et al. A coding-independent function of an alternative Ube3a transcript during neuronal development. *Nat. Neurosci.* **2015**, *18*, 666–673. [CrossRef]
10. Caygill, E.E.; Brand, A.H. miR-7 Buffers Differentiation in the Developing Drosophila Visual System. *Cell Rep.* **2017**, *20*, 1255–1261. [CrossRef] [PubMed]
11. Saydam, O.; Senol, O.; Würdinger, T.; Mizrak, A.; Ozdener, G.B.; Stemmer-Rachamimov, A.O.; Yi, M.; Stephens, R.M.; Krichevsky, A.M.; Saydam, N.; et al. miRNA-7 Attenuation in Schwannoma Tumors Stimulates Growth by Upregulating Three Oncogenic Signaling Pathways. *Cancer Res.* **2011**, *71*, 852–861. [CrossRef] [PubMed]
12. Zhao, J.; Wang, K.; Liao, Z.; Li, Y.; Yang, H.; Chen, C.; Zhou, Y.; Tao, Y.; Guo, M.; Ren, T.; et al. Promoter mutation of tumor suppressor microRNA-7 is associated with poor prognosis of lung cancer. *Mol. Clin. Oncol.* **2015**, *3*, 1329–1336. [CrossRef]
13. Zhang, N.; Li, X.; Wu, C.W.; Dong, Y.; Cai, M.; Mok, M.T.S.; Wang, H.; Chen, J.; Ng, S.S.M.; Chen, M.; et al. microRNA-7 is a novel inhibitor of YY1 contributing to colorectal tumorigenesis. *Oncogene* **2012**, *32*, 5078–5088. [CrossRef] [PubMed]
14. Midgley, A.; Morris, G.; Phillips, A.O.; Steadman, R. 17β-estradiol ameliorates age-associated loss of fibroblast function by attenuating IFN-γ/STAT1-dependent miR-7 upregulation. *Aging Cell* **2016**, *15*, 531–541. [CrossRef] [PubMed]
15. Marzioni, M.; Agostinelli, L.; Candelaresi, C.; Saccomanno, S.; De Minicis, S.; Maroni, L.; Mingarelli, E.; Rychlicki, C.; Trozzi, L.; Banales, J.M.; et al. Activation of the developmental pathway neurogenin-3/microRNA-7a regulates cholangiocyte proliferation in response to injury. *Hepatology* **2014**, *60*, 1324–1335. [CrossRef]
16. Chen, H.; Shalom-Feuerstein, R.; Riley, J.; Zhang, S.-D.; Tucci, P.; Agostini, M.; Aberdam, D.; Knight, R.A.; Genchi, G.; Nicotera, P.; et al. miR-7 and miR-214 are specifically expressed during neuroblastoma differentiation, cortical development and embryonic stem cells differentiation, and control neurite outgrowth in vitro. *Biochem. Biophys. Res. Commun.* **2010**, *394*, 921–927. [CrossRef]
17. Xin, L.; Liu, L.; Liu, C.; Zhou, L.; Zhou, Q.; Yuan, Y.; Li, S.; Zhang, H. DNA-methylation-mediated silencing of miR-7-5p promotes gastric cancer stem cell invasion via increasing Smo and Hes1. *J. Cell. Physiol.* **2020**, *235*, 2643–2654. [CrossRef] [PubMed]
18. Masuda, M.; Miki, Y.; Hata, S.; Takagi, K.; Sakurai, M.; Ono, K.; Suzuki, K.; Yang, Y.; Abe, E.; Hirakawa, H.; et al. An induction of microRNA, miR-7 through estrogen treatment in breast carcinoma. *J. Transl. Med.* **2012**, *10*, S2. [CrossRef] [PubMed]
19. Duex, J.E.; Comeau, L.; Sorkin, A.; Purow, B.; Kefas, B. Usp18 Regulates Epidermal Growth Factor (EGF) Receptor Expression and Cancer Cell Survival via MicroRNA-7. *J. Biol. Chem.* **2011**, *286*, 25377–25386. [CrossRef] [PubMed]
20. Chen, Y.-J.; Chien, P.-H.; Chen, W.-S.; Chien, Y.-F.; Hsu, Y.-Y.; Wang, L.-Y.; Chen, J.-Y.; Lin, C.-W.; Huang, T.-C.; Yu, Y.-L.; et al. Hepatitis B Virus-Encoded X Protein Downregulates EGFR Expression via Inducing MicroRNA-7 in Hepatocellular Carcinoma Cells. *Evidence-Based Complement. Altern. Med.* **2013**, *2013*, 1–10. [CrossRef] [PubMed]
21. Jeong, D.; Ham, J.; Park, S.; Lee, S.; Lee, H.; Kang, H.-S.; Kim, S.J. MicroRNA-7-5p mediates the signaling of hepatocyte growth factor to suppress oncogenes in the MCF-10A mammary epithelial cell. *Sci. Rep.* **2017**, *7*, 15425. [CrossRef] [PubMed]
22. Yuan, C.; Luo, X.; Duan, S.; Guo, L. Long noncoding RNA LINC00115 promotes breast cancer metastasis by inhibiting miR-7. *FEBS Open Bio* **2020**, *10*, 1230–1237. [CrossRef] [PubMed]
23. Blume, C.J.; Hotzwagenblatt, A.; Hullein, J.; Sellner, L.; Jethwa, A.; Stolz, T.; Slabicki, M.; Lee, K.; Sharathchandra, A.; Benner, A.; et al. p53-dependent non-coding RNA networks in chronic lymphocytic leukemia. *Leukemia* **2015**, *29*, 2015–2023. [CrossRef]
24. Pollock, A.; Bian, S.; Zhang, C.; Chen, Z.; Sun, T. Growth of the Developing Cerebral Cortex Is Controlled by MicroRNA-7 through the p53 Pathway. *Cell Rep.* **2014**, *7*, 1184–1196. [CrossRef] [PubMed]
25. Gou, Q.; Wu, K.; Zhou, J.-K.; Xie, Y.; Liu, L.; Peng, Y. Profiling and bioinformatic analysis of circular RNA expression regulated by c-Myc. *Oncotarget* **2017**, *8*, 71587–71596. [CrossRef]
26. Hansen, T.B.; Jensen, T.I.; Clausen, B.H.; Bramsen, J.B.; Finsen, B.; Damgaard, C.K.; Kjems, J. Natural RNA circles function as efficient microRNA sponges. *Nat. Cell Biol.* **2013**, *495*, 384–388. [CrossRef] [PubMed]
27. Memczak, S.; Jens, M.; Elefsinioti, A.; Torti, F.; Krueger, J.; Rybak, A.; Maier, L.; Mackowiak, S.D.; Gregersen, L.H.; Munschauer, M.; et al. Circular RNAs are a large class of animal RNAs with regulatory potency. *Nat. Cell Biol.* **2013**, *495*, 333–338. [CrossRef] [PubMed]
28. Rajman, M.; Schratt, G. MicroRNAs in neural development: From master regulators to fine-tuners. *Development* **2017**, *144*, 2310–2322. [CrossRef] [PubMed]
29. Kleaveland, B.; Shi, C.Y.; Stefano, J.; Bartel, D.P. A Network of Noncoding Regulatory RNAs Acts in the Mammalian Brain. *Cell* **2018**, *174*, 350–362. [CrossRef]
30. Li, M.; Pan, M.; You, C.; Zhao, F.; Wu, D.; Guo, M.; Xu, H.; Shi, F.; Zheng, D.; Dou, J. MiR-7 reduces the BCSC subset by inhibiting XIST to modulate the miR-92b/Slug/ESA axis and inhibit tumor growth. *Breast Cancer Res.* **2020**, *22*, 1–14. [CrossRef]

31. Ku, G.W.; Kang, Y.; Yu, S.-L.; Park, J.; Park, S.; Jeong, I.B.; Kang, M.W.; Son, J.W.; Kang, J. LncRNA LINC00240 suppresses invasion and migration in non-small cell lung cancer by sponging miR-7-5p. *BMC Cancer* **2021**, *21*, 1–13. [CrossRef]
32. Liu, Y.; Du, Y.; Hu, X.; Zhao, L.; Xia, W. Up-regulation of ceRNA TINCR by SP1 contributes to tumorigenesis in breast cancer. *BMC Cancer* **2018**, *18*, 367. [CrossRef] [PubMed]
33. Yu, S.; Wang, D.; Shao, Y.; Zhang, T.; Xie, H.; Jiang, X.; Deng, Q.; Jiao, Y.; Yang, J.; Cai, C.; et al. SP1-induced lncRNA TINCR overexpression contributes to colorectal cancer progression by sponging miR-7-5p. *Aging* **2019**, *11*, 1389–1403. [CrossRef] [PubMed]
34. Mo, D.; Liu, W.; Li, Y.; Cui, W. Long Non-coding RNA Zinc Finger Antisense 1 (ZFAS1) Regulates Proliferation, Migration, Invasion, and Apoptosis by Targeting MiR-7-5p in Colorectal Cancer. *Med. Sci. Monit.* **2019**, *25*, 5150–5158. [CrossRef]
35. Zhang, X.; Niu, W.; Mu, M.; Hu, S.; Niu, C. Long non-coding RNA LPP-AS2 promotes glioma tumorigenesis via miR-7-5p/EGFR/PI3K/AKT/c-MYC feedback loop. *J. Exp. Clin. Cancer Res.* **2020**, *39*, 1–20. [CrossRef]
36. Zhang, X.; Zhao, X.; Li, Y.; Zhou, Y.; Zhang, Z. Long noncoding RNA SOX21-AS1 promotes cervical cancer progression by competitively sponging miR-7/VDAC1. *J. Cell. Physiol.* **2019**, *234*, 17494–17504. [CrossRef] [PubMed]
37. Hu, H.; Yang, L.; Li, L.; Zeng, C. Long non-coding RNA KCNQ1OT1 modulates oxaliplatin resistance in hepatocellular carcinoma through miR-7-5p/ ABCC1 axis. *Biochem. Biophys. Res. Commun.* **2018**, *503*, 2400–2406. [CrossRef] [PubMed]
38. Liu, X.; Fu, Q.; Li, S.; Liang, N.; Li, F.; Li, C.; Sui, C.; Dionigi, G.; Sun, H. LncRNA FOXD2-AS1 Functions as a Competing Endogenous RNA to Regulate TERT Expression by Sponging miR-7-5p in Thyroid Cancer. *Front. Endocrinol.* **2019**, *10*. [CrossRef]
39. Yang, Z.; Shi, X.; Li, C.; Wang, X.; Hou, K.; Li, Z.; Zhang, X.; Fan, Y.; Qu, X.; Che, X.; et al. Long non-coding RNA UCA1 upregulation promotes the migration of hypoxia-resistant gastric cancer cells through the miR-7-5p/EGFR axis. *Exp. Cell Res.* **2018**, *368*, 194–201. [CrossRef]
40. Shu, L.; Zhang, W.; Huang, C.; Huang, G.; Su, G.; Xu, J. lncRNA ANRIL protects H9c2 cells against hypoxia-induced injury through targeting the miR-7-5p/SIRT1 axis. *J. Cell. Physiol.* **2020**, *235*, 1175–1183. [CrossRef]
41. Zheng, Y.; Nie, P.; Xu, S. Long noncoding RNA CASC21 exerts an oncogenic role in colorectal cancer through regulating miR-7-5p/YAP1 axis. *Biomed. Pharmacother.* **2020**, *121*, 109628. [CrossRef]
42. Arunkumar, G.; Anand, S.; Raksha, P.; Dhamodharan, S.; Rao, H.P.S.; Subbiah, S.; Murugan, A.K.; Munirajan, A.K. LncRNA OIP5-AS1 is overexpressed in undifferentiated oral tumors and integrated analysis identifies as a downstream effector of stemness-associated transcription factors. *Sci. Rep.* **2018**, *8*, 1–13. [CrossRef] [PubMed]
43. Liu, L.; Yang, X.; Li, N.-F.; Lin, L.; Luo, H. Circ_0015756 promotes proliferation, invasion and migration by microRNA-7-dependent inhibition of FAK in hepatocellular carcinoma. *Cell Cycle* **2019**, *18*, 2939–2953. [CrossRef] [PubMed]
44. Zeng, K.; Chen, X.; Xu, M.; Liu, X.; Hu, X.; Xu, T.; Sun, H.; Pan, Y.; He, B.; Wang, S. CircHIPK3 promotes colorectal cancer growth and metastasis by sponging miR-7. *Cell Death Dis.* **2018**, *9*, 1–15. [CrossRef] [PubMed]
45. Gao, D.; Qi, X.; Zhang, X.; Fang, K.; Guo, Z.; Li, L. hsa_circRNA_0006528 as a competing endogenous RNA promotes human breast cancer progression by sponging miR-7-5p and activating the MAPK/ERK signaling pathway. *Mol. Carcinog.* **2019**, *58*, 554–564. [CrossRef] [PubMed]
46. Li, G.; Huang, M.; Cai, Y.; Yang, Y.; Sun, X.; Ke, Y. Circ-U2AF1 promotes human glioma via derepressing neuro-oncological ventral antigen 2 by sponging hsa-miR-7-5p. *J. Cell. Physiol.* **2019**, *234*, 9144–9155. [CrossRef]
47. Wang, Q.; Li, Z.; Hu, Y.; Zheng, W.; Tang, W.; Zhai, C.; Gu, Z.; Tao, J.; Wang, H. RETRACTED ARTICLE: Circ-TFCP2L1 Promotes the Proliferation and Migration of Triple Negative Breast Cancer through Sponging miR-7 by Inhibiting PAK1. *J. Mammary Gland. Biol. Neoplasia* **2019**, *24*, 323–331. [CrossRef]
48. Wang, C.-J.; Gao, F.; Huang, Y.-J.; Han, D.-X.; Zheng, Y.; Wang, W.-H.; Jiang, H.; Gao, Y.; Yuan, B.; Zhang, J.-B. circAkap17b acts as a miR-7 family molecular sponge to regulate FSH secretion in rat pituitary cells. *J. Mol. Endocrinol.* **2020**, *65*, 135–148. [CrossRef] [PubMed]
49. Sang, Q.; Liu, X.; Wang, L.; Qi, L.; Sun, W.; Wang, W.; Sun, Y.; Zhang, H. CircSNCA downregulation by pramipexole treatment mediates cell apoptosis and autophagy in Parkinson's disease by targeting miR-7. *Aging* **2018**, *10*, 1281–1293. [CrossRef]
50. Chen, Y.; Guo, H.; Li, L.; Bao, D.; Gao, F.; Li, Q.; Huang, Q.; Duan, X.; Xiang, Z. Long Non-Coding RNA (lncRNA) Small Nucleolar RNA Host Gene 15 (SNHG15) Alleviates Osteoarthritis Progression by Regulation of Extracellular Matrix Homeostasis. *Med. Sci. Monit.* **2020**, *26*. [CrossRef]
51. Wang, Y.; Vogel, G.; Yu, Z.; Richard, S. The QKI-5 and QKI-6 RNA Binding Proteins Regulate the Expression of MicroRNA 7 in Glial Cells. *Mol. Cell. Biol.* **2013**, *33*, 1233–1243. [CrossRef] [PubMed]
52. Higuchi, T.; Todaka, H.; Sugiyama, Y.; Ono, M.; Tamaki, N.; Hatano, E.; Takezaki, Y.; Hanazaki, K.; Miwa, T.; Lai, S.; et al. Suppression of MicroRNA-7 (miR-7) Biogenesis by Nuclear Factor 90-Nuclear Factor 45 Complex (NF90-NF45) Controls Cell Proliferation in Hepatocellular Carcinoma. *J. Biol. Chem.* **2016**, *291*, 21074–21084. [CrossRef] [PubMed]
53. Li, Y.-J.; Wang, C.-H.; Zhou, Y.; Liao, Z.-Y.; Zhu, S.-F.; Hu, Y.; Chen, C.; Luo, J.-M.; Wen, Z.-K.; Xu, L. TLR9 signaling repressed tumor suppressor miR-7 expression through up-regulation of HuR in human lung cancer cells. *Cancer Cell Int.* **2013**, *13*, 90. [CrossRef] [PubMed]
54. Yang, A.; Shao, T.-J.; Ros, X.B.-D.; Lian, C.; Villanueva, P.; Dai, L.; Gu, S. AGO-bound mature miRNAs are oligouridylated by TUTs and subsequently degraded by DIS3L2. *Nat. Commun.* **2020**, *11*, 1–13. [CrossRef] [PubMed]
55. Wu, H.; Sun, S.; Tu, K.; Gao, Y.; Xie, B.; Krainer, A.R.; Zhu, J. A Splicing-Independent Function of SF2/ASF in MicroRNA Processing. *Mol. Cell* **2010**, *38*, 67–77. [CrossRef]

56. Yu, N.; Huangyang, P.; Yang, X.; Han, X.; Yan, R.; Jia, H.; Shang, Y.; Sun, L. microRNA-7 Suppresses the Invasive Potential of Breast Cancer Cells and Sensitizes Cells to DNA Damages by Targeting Histone Methyltransferase SET8. *J. Biol. Chem.* **2013**, *288*, 19633–19642. [CrossRef] [PubMed]
57. Zhang, H.; Cai, K.; Wang, J.; Wang, X.; Cheng, K.; Shi, F.; Jiang, L.; Zhang, Y.; Dou, J. MiR-7, Inhibited Indirectly by LincRNA HOTAIR, Directly Inhibits SETDB1 and Reverses the EMT of Breast Cancer Stem Cells by Downregulating the STAT3 Pathway. *Stem Cells* **2014**, *32*, 2858–2868. [CrossRef]
58. Cheng, J.; Guo, S.; Chen, S.; Mastriano, S.J.; Liu, C.; D'Alessio, A.C.; Hysolli, E.; Guo, Y.; Yao, H.; Megyola, C.M.; et al. An Extensive Network of TET2-Targeting MicroRNAs Regulates Malignant Hematopoiesis. *Cell Rep.* **2013**, *5*, 471–481. [CrossRef] [PubMed]
59. Hong, C.-F.; Lin, S.-Y.; Chou, Y.-T.; Wu, C.-W. MicroRNA-7 Compromises p53 Protein-dependent Apoptosis by Controlling the Expression of the Chromatin Remodeling Factor SMARCD1. *J. Biol. Chem.* **2016**, *291*, 1877–1889. [CrossRef]
60. López-Beas, J.; Capilla-González, V.; Aguilera, Y.; Mellado, N.; Lachaud, C.C.; Martín, F.; Smani, T.; Soria, B.; Hmadcha, A. miR-7 Modulates hESC Differentiation into Insulin-Producing Beta-like Cells and Contributes to Cell Maturation. *Mol. Ther. Nucleic Acids* **2018**, *12*, 463–477. [CrossRef]
61. Titze-De-Almeida, R.; Titze-De-Almeida, S.S. miR-7 Replacement Therapy in Parkinson's Disease. *Curr. Gene Ther.* **2018**, *18*, 143–153. [CrossRef] [PubMed]
62. Kefas, B.; Godlewski, J.; Comeau, L.; Li, Y.; Abounader, R.; Hawkinson, M.; Lee, J.; Fine, H.; Chiocca, E.A.; Lawler, S.; et al. microRNA-7 Inhibits the Epidermal Growth Factor Receptor and the Akt Pathway and Is Down-regulated in Glioblastoma. *Cancer Res.* **2008**, *68*, 3566–3572. [CrossRef]
63. Webster, R.J.; Giles, K.M.; Price, K.J.; Zhang, P.M.; Mattick, J.S.; Leedman, P.J. Regulation of Epidermal Growth Factor Receptor Signaling in Human Cancer Cells by MicroRNA-7. *J. Biol. Chem.* **2009**, *284*, 5731–5741. [CrossRef]
64. Sun, X.; Li, J.; Sun, Y.; Zhang, Y.; Dong, L.; Shen, C.; Yang, L.; Yang, M.; Li, Y.; Shen, G.; et al. miR-7 reverses the resistance to BRAFi in melanoma by targeting EGFR/IGF-1R/CRAF and inhibiting the MAPK and PI3K/AKT signaling pathways. *Oncotarget* **2016**, *7*, 53558–53570. [CrossRef] [PubMed]
65. Xu, L.; Wen, Z.; Zhou, Y.; Liu, Z.; Li, Q.; Fei, G.; Luo, J.; Ren, T. MicroRNA-7–regulated TLR9 signaling–enhanced growth and metastatic potential of human lung cancer cells by altering the phosphoinositide-3-kinase, regulatory subunit 3/Akt pathway. *Mol. Biol. Cell* **2013**, *24*, 42–55. [CrossRef]
66. Fang, Y.; Xue, J.-L.; Shen, Q.; Chen, J.; Tian, L. MicroRNA-7 inhibits tumor growth and metastasis by targeting the phosphoinositide 3-kinase/Akt pathway in hepatocellular carcinoma. *Hepatology* **2012**, *55*, 1852–1862. [CrossRef] [PubMed]
67. Liu, Z.; Jiang, Z.; Huang, J.; Huang, S.; Li, Y.; Yu, S.; Yu, S.; Liu, X. miR-7 inhibits glioblastoma growth by simultaneously interfering with the PI3K/ATK and Raf/MEK/ERK pathways. *Int. J. Oncol.* **2014**, *44*, 1571–1580. [CrossRef] [PubMed]
68. Liu, S.; Zhang, P.; Chen, Z.; Liu, M.; Li, X.; Tang, H. MicroRNA-7 downregulates XIAP expression to suppress cell growth and promote apoptosis in cervical cancer cells. *FEBS Lett.* **2013**, *587*, 2247–2253. [CrossRef]
69. Reddy, S.D.N.; Ohshiro, K.; Rayala, S.K.; Kumar, R. MicroRNA-7, a Homeobox D10 Target, Inhibits p21-Activated Kinase 1 and Regulates Its Functions. *Cancer Res.* **2008**, *68*, 8195–8200. [CrossRef] [PubMed]
70. Giles, K.M.; Brown, R.A.; Epis, M.R.; Kalinowski, F.C.; Leedman, P.J. miRNA-7-5p inhibits melanoma cell migration and invasion. *Biochem. Biophys. Res. Commun.* **2013**, *430*, 706–710. [CrossRef]
71. Jiang, L.; Liu, X.; Chen, Z.; Jin, Y.; Heidbreder, C.E.; Kolokythas, A.; Wang, A.; Dai, Y.; Zhou, X. MicroRNA-7 targets IGF1R (insulin-like growth factor 1 receptor) in tongue squamous cell carcinoma cells. *Biochem. J.* **2010**, *432*, 199–207. [CrossRef] [PubMed]
72. Rai, K.; Takigawa, N.; Ito, S.; Kashihara, H.; Ichihara, E.; Yasuda, T.; Shimizu, K.; Tanimoto, M.; Kiura, K. Liposomal Delivery of MicroRNA-7–Expressing Plasmid Overcomes Epidermal Growth Factor Receptor Tyrosine Kinase Inhibitor-Resistance in Lung Cancer Cells. *Mol. Cancer Ther.* **2011**, *10*, 1720–1727. [CrossRef] [PubMed]
73. Kalinowski, F.C.; Brown, R.A.; Ganda, C.; Giles, K.M.; Epis, M.R.; Horsham, J.; Leedman, P.J. microRNA-7: A tumor suppressor miRNA with therapeutic potential. *Int. J. Biochem. Cell Biol.* **2014**, *54*, 312–317. [CrossRef] [PubMed]
74. Zhao, X.-D.; Lu, Y.-Y.; Guo, H.; Xie, H.-H.; He, L.-J.; Shen, G.-F.; Zhou, J.-F.; Li, T.; Hu, S.-J.; Zhou, L.; et al. MicroRNA-7/NF-κB signaling regulatory feedback circuit regulates gastric carcinogenesis. *J. Cell Biol.* **2015**, *210*, 613–627. [CrossRef]
75. Wang, Y.; Wang, Q.; Song, J. Inhibition of autophagy potentiates the proliferation inhibition activity of microRNA-7 in human hepatocellular carcinoma cells. *Oncol. Lett.* **2017**, *14*, 3566–3572. [CrossRef]
76. Xia, J.; Cao, T.; Ma, C.; Shi, Y.; Sun, Y.; Wang, Z.P.; Ma, J. miR-7 Suppresses Tumor Progression by Directly Targeting MAP3K9 in Pancreatic Cancer. *Mol. Ther. Nucleic Acids* **2018**, *13*, 121–132. [CrossRef]
77. Zhao, X.; Dou, W.; He, L.; Liang, S.; Tie, J.; Liu, C.; Li, T.; Lu, Y.; Mo, P.; Shi, Y.; et al. MicroRNA-7 functions as an anti-metastatic microRNA in gastric cancer by targeting insulin-like growth factor-1 receptor. *Oncogene* **2013**, *32*, 1363–1372. [CrossRef] [PubMed]
78. Wang, B.; Sun, F.; Dong, N.; Sun, Z.; Diao, Y.; Zheng, C.; Sun, J.; Yang, Y.; Jiang, D. MicroRNA-7 directly targets insulin-like growth factor 1 receptor to inhibit cellular growth and glucose metabolism in gliomas. *Diagn. Pathol.* **2014**, *9*, 1–6. [CrossRef] [PubMed]
79. Zhang, Z.; Zhao, M.; Wang, G. Upregulation of microRNA-7 contributes to inhibition of the growth and metastasis of osteosarcoma cells through the inhibition of IGF1R. *J. Cell. Physiol.* **2019**, *234*, 22195–22206. [CrossRef]

80. Sun, H.; Zhang, Z.; Luo, W.; Liu, J.; Lou, Y.; Xia, S. MiR-7 Functions as a Tumor Suppressor by Targeting the Oncogenes TAL1 in T-Cell Acute Lymphoblastic Leukemia. *Technol. Cancer Res. Treat.* **2020**, *19*. [CrossRef]
81. Giles, K.M.; Brown, R.A.; Ganda, C.; Podgorny, M.J.; Candy, P.A.; Wintle, L.C.; Richardson, K.L.; Kalinowski, F.C.; Stuart, L.M.; Epis, M.R.; et al. microRNA-7-5p inhibits melanoma cell proliferation and metastasis by suppressing RelA/NF-κB. *Oncotarget* **2016**, *7*, 31663–31680. [CrossRef]
82. Meleady, P.; Gallagher, M.; Clarke, C.; Henry, M.; Sanchez, N.; Barron, N.; Clynes, M. Impact of miR-7 over-expression on the proteome of Chinese hamster ovary cells. *J. Biotechnol.* **2012**, *160*, 251–262. [CrossRef] [PubMed]
83. Zhang, X.; Hu, S.; Zhang, X.; Wang, L.; Zhang, X.; Yan, B.; Zhao, J.; Yang, A.; Zhang, R. MicroRNA-7 arrests cell cycle in G1 phase by directly targeting CCNE1 in human hepatocellular carcinoma cells. *Biochem. Biophys. Res. Commun.* **2014**, *443*, 1078–1084. [CrossRef]
84. Meza-Sosa, K.F.; Pérez-García, E.I.; Camacho Concha, N.; López-Gutiérrez, O.; Pedraza-Alva, G.; Pérez-Martínez, L. MiR-7 Promotes Epithelial Cell Transformation by Targeting the Tumor Suppressor KLF4. *PLoS ONE* **2014**, *9*, e103987. [CrossRef]
85. Okuda, H.; Xing, F.; Pandey, P.R.; Sharma, S.; Watabe, M.; Pai, S.K.; Mo, Y.-Y.; Iiizumi-Gairani, M.; Hirota, S.; Liu, Y.; et al. miR-7 Suppresses Brain Metastasis of Breast Cancer Stem-Like Cells By Modulating KLF4. *Cancer Res.* **2013**, *73*, 1434–1444. [CrossRef]
86. Li, Y.-Z.; Wen, L.; Wei, X.; Wang, Q.-R.; Xu, L.-W.; Zhang, H.-M.; Liu, W.-C. Inhibition of miR-7 promotes angiogenesis in human umbilical vein endothelial cells by upregulating VEGF via KLF4. *Oncol. Rep.* **2016**, *36*, 1569–1575. [CrossRef] [PubMed]
87. Jung, H.M.; Phillips, B.L.; Patel, R.S.; Cohen, D.M.; Jakymiw, A.; Kong, W.W.; Cheng, J.Q.; Chan, E.K.L. Keratinization-associated miR-7 and miR-21 Regulate Tumor Suppressor Reversion-inducing Cysteine-rich Protein with Kazal Motifs (RECK) in Oral Cancer*. *J. Biol. Chem.* **2012**, *287*, 29261–29272. [CrossRef] [PubMed]
88. Chou, Y.-T.; Lin, H.-H.; Lien, Y.-C.; Wang, Y.-H.; Hong, C.-F.; Kao, Y.-R.; Lin, S.-C.; Chang, Y.-C.; Lin, S.-Y.; Chen, S.-J.; et al. EGFR Promotes Lung Tumorigenesis by Activating miR-7 through a Ras/ERK/Myc Pathway That Targets the Ets2 Transcriptional Repressor ERF. *Cancer Res.* **2010**, *70*, 8822–8831. [CrossRef]
89. Hua, K.; Jin, J.; Zhang, H.; Zhao, B.; Wu, C.; Xu, H.; Fang, L. MicroRNA-7 inhibits proliferation, migration and invasion of thyroid papillary cancer cells via targeting CKS2. *Int. J. Oncol.* **2016**, *49*, 1531–1540. [CrossRef]
90. Ling, Y.; Cao, C.; Li, S.; Qiu, M.; Shen, G.; Chen, Z.; Yao, F.; Chen, W. TRIP6, as a target of miR-7, regulates the proliferation and metastasis of colorectal cancer cells. *Biochem. Biophys. Res. Commun.* **2019**, *514*, 231–238. [CrossRef]
91. Pan, M.; Li, M.; You, C.; Zhao, F.; Guo, M.; Xu, H.; Li, L.; Wang, L.; Dou, J. Inhibition of breast cancer growth via miR-7 suppressing ALDH1A3 activity concomitant with decreasing breast cancer stem cell subpopulation. *J. Cell. Physiol.* **2019**, *235*, 1405–1416. [CrossRef] [PubMed]
92. Ma, C.; Qi, Y.; Shao, L.; Liu, M.; Li, X.; Tang, H. Downregulation of miR-7 upregulates Cullin 5 (CUL5) to facilitate G1/S transition in human hepatocellular carcinoma cells. *IUBMB Life* **2013**, *65*, 1026–1034. [CrossRef] [PubMed]
93. Qin, A.; Qian, W. MicroRNA-7 inhibits colorectal cancer cell proliferation, migration and invasion via TYRO3 and phosphoinositide 3-kinase/protein B kinase/mammalian target of rapamycin pathway suppression. *Int. J. Mol. Med.* **2018**, *42*, 2503–2514. [CrossRef]
94. Li, Q.; Wu, X.; Guo, L.; Shi, J.; Li, J. MicroRNA-7-5p induces cell growth inhibition, cell cycle arrest and apoptosis by targeting PAK2 in non-small cell lung cancer. *FEBS Open Bio* **2019**, *9*, 1983–1993. [CrossRef]
95. Yue, K.; Wang, X.; Wu, Y.; Zhou, X.; He, Q.; Duan, Y. microRNA-7 regulates cell growth, migration and invasion via direct targeting of PAK1 in thyroid cancer. *Mol. Med. Rep.* **2016**, *14*, 2127–2134. [CrossRef]
96. Pan, C.-M.; Chan, K.-H.; Chen, C.-H.; Jan, C.-I.; Liu, M.-C.; Lin, C.-M.; Cho, D.-Y.; Tsai, W.-C.; Chu, Y.-T.; Cheng, C.-H.; et al. MicroRNA-7 targets T-Box 2 to inhibit epithelial-mesenchymal transition and invasiveness in glioblastoma multiforme. *Cancer Lett.* **2020**, *493*, 133–142. [CrossRef]
97. Shukla, A.; Gupta, P.; Singh, R.; Mishra, D.P. Glycolytic inhibitor 2-Deoxy-d-Glucose activates migration and invasion in glioblastoma cells through modulation of the miR-7-5p/TFF3 signaling pathway. *Biochem. Biophys. Res. Commun.* **2018**, *499*, 829–835. [CrossRef] [PubMed]
98. Zhou, N.; Hao, S.; Huang, Z.; Wang, W.; Yan, P.; Zhou, W.; Zhu, Q.; Liu, X. MiR-7 inhibited peripheral nerve injury repair by affecting neural stem cells migration and proliferation through cdc42. *Mol. Pain* **2018**, *14*. [CrossRef] [PubMed]
99. Wu, D.-G.; Wang, Y.-Y.; Fan, L.-G.; Luo, H.; Han, B.; Sun, L.-H.; Wang, X.-F.; Zhang, J.-X.; Cao, L.; Wang, X.-R.; et al. MicroRNA-7 regulates glioblastoma cell invasion via targeting focal adhesion kinase expression. *Chin. Med. J.* **2011**, *124*, 2616–2621. [PubMed]
100. Kong, X.; Li, G.; Yuan, Y.; He, Y.; Wu, X.; Zhang, W.; Wu, Z.; Chen, T.; Wu, W.; Lobie, P.E.; et al. MicroRNA-7 Inhibits Epithelial-to-Mesenchymal Transition and Metastasis of Breast Cancer Cells via Targeting FAK Expression. *PLoS ONE* **2012**, *7*, e41523. [CrossRef] [PubMed]
101. Zeng, C.-Y.; Zhan, Y.-S.; Huang, J.; Chen, Y.-X. MicroRNA-7 suppresses human colon cancer invasion and proliferation by targeting the expression of focal adhesion kinase. *Mol. Med. Rep.* **2015**, *13*, 1297–1303. [CrossRef]
102. Cao, Q.; Mao, Z.-D.; Shi, Y.-J.; Chen, Y.; Sun, Y.; Zhang, Q.; Song, L.; Peng, L.-P. MicroRNA-7 inhibits cell proliferation, migration and invasion in human non-small cell lung cancer cells by targeting FAK through ERK/MAPK signaling pathway. *Oncotarget* **2016**, *7*, 77468–77481. [CrossRef]
103. Xiao, H. MiR-7-5p suppresses tumor metastasis of non-small cell lung cancer by targeting NOVA2. *Cell. Mol. Biol. Lett.* **2019**, *24*, 1–13. [CrossRef]

104. Moazzeni, H.; Najafi, A.; Khani, M. Identification of direct target genes of miR-7, miR-9, miR-96, and miR-182 in the human breast cancer cell lines MCF-7 and MDA-MB-231. *Mol. Cell. Probes* **2017**, *34*, 45–52. [CrossRef] [PubMed]
105. Yin, C.; Kong, W.; Jiang, J.; Xu, H.; Zhao, W. miR-7-5p inhibits cell migration and invasion in glioblastoma through targeting SATB1. *Oncol. Lett.* **2018**, *17*, 1819–1825. [CrossRef] [PubMed]
106. Bi, Y.; Shen, W.; Min, M.; Liu, Y. MicroRNA-7 functions as a tumor-suppressor gene by regulating ILF2 in pancreatic carcinoma. *Int. J. Mol. Med.* **2017**, *39*, 900–906. [CrossRef] [PubMed]
107. Bao, S.; Jin, S.; Wang, C.; Tu, P.; Hu, K.; Lu, J. Androgen receptor suppresses vasculogenic mimicry in hepatocellular carcinoma via circRNA7/miRNA7-5p/VE-cadherin/Notch4 signalling. *J. Cell. Mol. Med.* **2020**, *24*, 14110–14120. [CrossRef]
108. Morales-Martinez, M.; Vega, G.G.; Neri, N.; Nambo, M.J.; Alvarado, I.; Cuadra, I.; Duran-Padilla, M.A.; Huerta-Yepez, S.; Vega, M.I. MicroRNA-7 Regulates Migration and Chemoresistance in Non-Hodgkin Lymphoma Cells Through Regulation of KLF4 and YY1. *Front. Oncol.* **2020**, *10*, 588893. [CrossRef]
109. Xiong, S.; Zheng, Y.; Jiang, P.; Liu, R.; Liu, X.; Chu, Y. MicroRNA-7 Inhibits the Growth of Human Non-Small Cell Lung Cancer A549 Cells through Targeting BCL-2. *Int. J. Biol. Sci.* **2011**, *7*, 805–814. [CrossRef]
110. Zhang, X.; Zhang, X.; Hu, S.; Zheng, M.; Zhang, J.; Zhao, J.; Zhang, X.; Yan, B.; Jia, L.; Zhao, J.; et al. Identification of miRNA-7 by genome-wide analysis as a critical sensitizer for TRAIL-induced apoptosis in glioblastoma cells. *Nucleic Acids Res.* **2017**, *45*, 5930–5944. [CrossRef] [PubMed]
111. Yang, L.; Kong, D.; He, M.; Gong, J.; Nie, Y.; Tai, S.; Teng, C.-B. MiR-7 mediates mitochondrial impairment to trigger apoptosis and necroptosis in Rhabdomyosarcoma. *Biochim. Biophys. Acta (BBA) Bioenerg.* **2020**, *1867*, 118826. [CrossRef] [PubMed]
112. Shi, Y.; Luo, X.; Li, P.; Tan, J.; Wang, X.; Xiang, T.; Ren, G. miR-7-5p suppresses cell proliferation and induces apoptosis of breast cancer cells mainly by targeting REGγ. *Cancer Lett.* **2015**, *358*, 27–36. [CrossRef] [PubMed]
113. Xue, W.; Liu, Y.; Xin, N.; Miao, J.; Du, J.; Wang, Y.; Shi, H.; Wei, Y.; Zhang, H.; Chen, Y.; et al. Nei Endonuclease VIII-Like1 (NEIL1) Inhibits Apoptosis of Human Colorectal Cancer Cells. *BioMed Res. Int.* **2020**, *2020*, 1–11. [CrossRef] [PubMed]
114. Zhao, Y.; Alexandrov, P.N.; Jaber, V.; Lukiw, W.J. Deficiency in the Ubiquitin Conjugating Enzyme UBE2A in Alzheimer's Disease (AD) is Linked to Deficits in a Natural Circular miRNA-7 Sponge (circRNA; ciRS-7). *Genes* **2016**, *7*, 116. [CrossRef]
115. Jia, B.; Liu, W.; Gu, J.; Wang, J.; Lv, W.; Zhang, W.; Hao, Q.; Pang, Z.; Mu, N.; Zhang, W.; et al. MiR-7-5p suppresses stemness and enhances temozolomide sensitivity of drug-resistant glioblastoma cells by targeting Yin Yang 1. *Exp. Cell Res.* **2019**, *375*, 73–81. [CrossRef] [PubMed]
116. Zhao, J.; Wang, B. MiR-7-5p enhances cerebral ischemia-reperfusion injury by degrading sirt1 mRNA. *J. Cardiovasc. Pharmacol.* **2020**, *76*, 227–236. [CrossRef] [PubMed]
117. Chaudhuri, A.D.; Choi, D.C.; Kabaria, S.; Tran, A.; Junn, E. MicroRNA-7 Regulates the Function of Mitochondrial Permeability Transition Pore by Targeting VDAC1 Expression. *J. Biol. Chem.* **2016**, *291*, 6483–6493. [CrossRef]
118. Wang, F.; Qiang, Y.; Zhu, L.; Jiang, Y.; Wang, Y.; Shao, X.; Yin, L.; Chen, J.; Chen, Z. MicroRNA-7 downregulates the oncogene VDAC1 to influence hepatocellular carcinoma proliferation and metastasis. *Tumor Biol.* **2016**, *37*, 10235–10246. [CrossRef] [PubMed]
119. Kabaria, S.; Choi, D.C.; Chaudhuri, A.D.; Jain, M.R.; Li, H.; Junn, E. MicroRNA-7 activates Nrf2 pathway by targeting Keap1 expression. *Free Radic. Biol. Med.* **2015**, *89*, 548–556. [CrossRef]
120. Li, Q.; Zhu, F.; Chen, P. miR-7 and miR-218 epigenetically control tumor suppressor genes RASSF1A and Claudin-6 by targeting HoxB3 in breast cancer. *Biochem. Biophys. Res. Commun.* **2012**, *424*, 28–33. [CrossRef]
121. Lai, J.; Yang, H.; Zhu, Y.; Ruan, M.; Huang, Y.; Zhang, Q. MiR-7-5p-mediated downregulation of PARP1 impacts DNA homologous recombination repair and resistance to doxorubicin in small cell lung cancer. *BMC Cancer* **2019**, *19*, 1–9. [CrossRef]
122. Seong, M.; Lee, J.; Kang, H. Hypoxia-induced regulation of mTOR signaling by miR-7 targeting REDD1. *J. Cell. Biochem.* **2019**, *120*, 4523–4532. [CrossRef]
123. Xu, K.; Song, X.; Chen, Z.; Qin, C. miR-7 inhibits colorectal cancer cell proliferation and induces apoptosis by targeting XRCC2. *OncoTargets Ther.* **2014**, *7*, 325–332. [CrossRef]
124. Downing, S.; Zhang, F.; Chen, Z.; Tzanakakis, E.S. MicroRNA-7 directly targets Reg1 in pancreatic cells. *Am. J. Physiol. Physiol.* **2019**, *317*, C366–C374. [CrossRef] [PubMed]
125. Liu, H.; Wu, X.; Huang, J.; Peng, J.; Guo, L. miR-7 modulates chemoresistance of small cell lung cancer by repressing MRP1/ABCC1. *Int. J. Exp. Pathol.* **2015**, *96*, 240–247. [CrossRef] [PubMed]
126. Zhang, X.D.; Fan, Q.Y.; Qiu, Z.; Chen, S. MiR-7 alleviates secondary inflammatory response of microglia caused by cerebral hem-orrhage through inhibiting TLR4 expression. *Eur. Rev. Med. Pharmacol. Sci.* **2018**, *22*, 5597–5604. [CrossRef]
127. Chen, H.; Guo, M.; Yue, D.; Zhao, J.; Zhou, Y.; Chen, C.; Liang, G.; Xu, L. MicroRNA-7 negatively regulates Toll-like receptor 4 signaling pathway through FAM177A. *Immunology* **2021**, *162*, 44–57. [CrossRef]
128. Fan, Z.; Lu, M.; Qiao, C.; Zhou, Y.; Ding, J.-H.; Hu, G. MicroRNA-7 Enhances Subventricular Zone Neurogenesis by Inhibiting NLRP3/Caspase-1 Axis in Adult Neural Stem Cells. *Mol. Neurobiol.* **2016**, *53*, 7057–7069. [CrossRef]
129. Luo, J.; Li, H.; Zhang, C. MicroRNA-7 inhibits the malignant phenotypes of non-small cell lung cancer in vitro by targeting Pax6. *Mol. Med. Rep.* **2015**, *12*, 5443–5448. [CrossRef] [PubMed]
130. Zhang, L.; Mubarak, T.; Chen, Y.; Lee, T.; Pollock, A.; Sun, T. Counter-Balance Between Gli3 and miR-7 Is Required for Proper Morphogenesis and Size Control of the Mouse Brain. *Front. Cell. Neurosci.* **2018**, *12*, 259. [CrossRef] [PubMed]

131. Li, J.; Qiu, M.; An, Y.; Huang, J.; Gong, C. miR-7-5p acts as a tumor suppressor in bladder cancer by regulating the hedgehog pathway factor Gli3. *Biochem. Biophys. Res. Commun.* **2018**, *503*, 2101–2107. [CrossRef] [PubMed]
132. Tian, S.; Chen, M.; Wang, B.; Han, Y.; Shang, H.; Chen, J. miR-7-5p Promotes Hepatic Stellate Cell Activation by Targeting Fibroblast Growth Factor Receptor 4. *Gastroenterol. Res. Pr.* **2020**, *2020*, 1–9. [CrossRef]
133. Kim, T.; Mehta, S.L.; Morris-Blanco, K.C.; Chokkalla, A.K.; Chelluboina, B.; Lopez, M.; Sullivan, R.; Kim, H.T.; Cook, T.D.; Kim, J.Y.; et al. The microRNA miR-7a-5p ameliorates ischemic brain damage by repressing α-synuclein. *Sci. Signal.* **2018**, *11*, eaat4285. [CrossRef] [PubMed]
134. Adusumilli, L.; Facchinello, N.; Teh, C.; Busolin, G.; Le, M.T.; Yang, H.; Beffagna, G.; Campanaro, S.; Tam, W.L.; Argenton, F.; et al. miR-7 Controls the Dopaminergic/Oligodendroglial Fate through Wnt/β-catenin Signaling Regulation. *Cells* **2020**, *9*, 711. [CrossRef] [PubMed]
135. Choi, S.-Y.; Pang, K.; Kim, J.Y.; Ryu, J.R.; Kang, H.; Liu, Z.; Kim, W.-K.; Sun, W.; Kim, H.; Han, K. Post-transcriptional regulation of SHANK3 expression by microRNAs related to multiple neuropsychiatric disorders. *Mol. Brain* **2015**, *8*, 74. [CrossRef]
136. Da Ros, V.G.; Gutierrez-Perez, I.; Ferres-Marco, L.; Dominguez, M. Dampening the Signals Transduced through Hedgehog via MicroRNA miR-7 Facilitates Notch-Induced Tumourigenesis. *PLoS Biol.* **2013**, *11*, e1001554. [CrossRef]
137. Tang, Z.; Xu, T.; Li, Y.; Fei, W.; Yang, G.; Hong, Y. Inhibition of CRY2 by STAT3/miRNA-7-5p Promotes Osteoblast Differentiation through Upregulation of CLOCK/BMAL1/P300 Expression. *Mol. Ther. Nucleic Acids* **2020**, *19*, 865–876. [CrossRef]
138. Yu, X.; Li, M.; Cui, M.; Sun, B.; Zhou, Z. Silence of yki by miR-7 regulates the Hippo pathway. *Biochem. Biophys. Res. Commun.* **2020**, *532*, 446–452. [CrossRef]
139. Banaei-Esfahani, A.; Moazzeni, H.; Nosar, P.N.; Amin, S.; Arefian, E.; Soleimani, M.; Yazdani, S.; Elahi, E. MicroRNAs that target RGS5. *Iran. J. Basic Med. Sci.* **2015**, *18*, 108–114.
140. Xiong, S.; Zheng, Y.; Jiang, P.; Liu, R.; Liu, X.; Qian, J.; Gu, J.; Chang, L.; Ge, D.; Chu, Y. PA28gamma emerges as a novel functional target of tumour suppressor microRNA-7 in non-small-cell lung cancer. *Br. J. Cancer* **2013**, *110*, 353–362. [CrossRef] [PubMed]
141. Frutos, M.F.-D.; Galán-Chilet, I.; Goedeke, L.; Kim, B.; Pardo-Marqués, V.; Pérez-García, A.; Herrero, J.I.; Fernández-Hernando, C.; Kim, J.; Ramírez, C.M. MicroRNA 7 Impairs Insulin Signaling and Regulates Aβ Levels through Posttranscriptional Regulation of the Insulin Receptor Substrate 2, Insulin Receptor, Insulin-Degrading Enzyme, and Liver X Receptor Pathway. *Mol. Cell. Biol.* **2019**, *39*. [CrossRef]
142. Miyazawa, M.; Bogdan, A.R.; Hashimoto, K.; Tsuji, Y. Regulation of transferrin receptor-1 mRNA by the interplay between IRE-binding proteins and miR-7/miR-141 in the 3′-IRE stem–loops. *RNA* **2018**, *24*, 468–479. [CrossRef]
143. Matarese, A.; Gambardella, J.; Lombardi, A.; Wang, X.; Santulli, G. miR-7 Regulates GLP-1-Mediated Insulin Release by Targeting β-Arrestin 1. *Cells* **2020**, *9*, 1621. [CrossRef]
144. Dewing, A.S.; Rueli, R.H.; Robles, M.J.; Nguyen-Wu, E.D.; Zeyda, T.; Berry, M.J.; Bellinger, F.P. Expression and regulation of mouse selenoprotein P transcript variants differing in non-coding RNA. *RNA Biol.* **2012**, *9*, 1361–1369. [CrossRef]
145. Yuan, B.; Sun, G.; Zhang, G.; Wu, J.; Xu, C.; Dai, L.; Chen, J.; Yu, X.; Zhao, Z.; Zhang, J. Identification of target genes for adenohypophysis-prefer miR-7 and miR-375 in cattle. *Genet. Mol. Res.* **2015**, *14*, 9753–9763. [CrossRef] [PubMed]
146. Ahmed, K.; Lapierre, M.P.; Gasser, E.; Denzler, R.; Yang, Y.; Ruelicke, T.; Kero, J.; Latreille, M.; Stoffel, M. Loss of microRNA-7a2 induces hypogonadotropic hypogonadism and infertility. *J. Clin. Investig.* **2017**, *127*, 1061–1074. [CrossRef] [PubMed]
147. Woo, S.Y.; Lee, S.Y.; Yu, S.-L.; Park, S.J.; Kang, D.; Kim, J.S.; Jeong, I.B.; Kwon, S.J.; Hwang, W.J.; Park, C.R.; et al. MicroRNA-7-5p′s role in the O-GlcNAcylation and cancer metabolism. *Non-coding RNA Res.* **2020**, *5*, 201–207. [CrossRef]
148. Ma, C.; Gu, R.; Wang, X.; He, S.; Bai, J.; Zhang, L.; Zhang, J.; Li, Q.; Qu, L.; Xin, W.; et al. circRNA CDR1as Promotes Pulmonary Artery Smooth Muscle Cell Calcification by Upregulating CAMK2D and CNN3 via Sponging miR-7-5p. *Mol. Ther. Nucleic Acids* **2020**, *22*, 530–541. [CrossRef]
149. Peng, J.; Liu, F.; Zheng, H.; Wu, Q.; Liu, S. IncRNA ZFAS1 contributes to the radioresistance of nasopharyngeal carcinoma cells by sponging hsa-miR-7-5p to upregulate ENO2. *Cell Cycle* **2021**, *20*, 126–141. [CrossRef]
150. Jin, H.-F.; Wang, J.-F.; Shao, M.; Zhou, K.; Ma, X.; Lv, X.-P. Down-Regulation of miR-7 in Gastric Cancer Is Associated With Elevated LDH-A Expression and Chemoresistance to Cisplatin. *Front. Cell Dev. Biol.* **2020**, *8*. [CrossRef] [PubMed]
151. Lapierre, M.P.; Godbersen, S.; Esteban, M.T.; Schad, A.N.; Treier, M.; Ghoshdastider, U.; Stoffel, M. MicroRNA-7a2 Regulates Prolactin in Developing Lactotrophs and Prolactinoma Cells. *Endocrinology* **2021**, *162*. [CrossRef]
152. Zhang, G.; Chen, F.; Wu, P.; Li, T.; He, M.; Yin, X.; Shi, H.; Duan, Y.; Zhang, T.; Wang, J.; et al. MicroRNA-7 Targets the KLF4 Gene to Regulate the Proliferation and Differentiation of Chicken Primary Myoblasts. *Front. Genet.* **2020**, *11*, 842. [CrossRef] [PubMed]
153. Junn, E.; Lee, K.-W.; Jeong, B.S.; Chan, T.W.; Im, J.-Y.; Mouradian, M.M. Repression of -synuclein expression and toxicity by microRNA-7. *Proc. Natl. Acad. Sci. USA* **2009**, *106*, 13052–13057. [CrossRef] [PubMed]
154. Diana, A.; Gaido, G.; Murtas, D. MicroRNA Signature in Human Normal and Tumoral Neural Stem Cells. *Int. J. Mol. Sci.* **2019**, *20*, 4123. [CrossRef] [PubMed]
155. Liu, Z.; Liu, Y.; Li, L.; Xu, Z.; Bi, B.; Wang, Y.; Li, J.Y. MiR-7-5p is frequently downregulated in glioblastoma microvasculature and inhibits vascular endothelial cell proliferation by targeting RAF1. *Tumor Biol.* **2014**, *35*, 10177–10184. [CrossRef]
156. Kolic, D.; Horvat, L.; Setinc, M.; Antica, M.; Matulic, M. miR-7 AND miR-34a sequence cloning and expression in a1235 glioblastoma cell line. *Mol. Exp. Boil. Med.* **2020**, *3*, 31–38. [CrossRef]
157. Chen, Y.A.; Cheng, L.; Zhang, Y.; Peng, L.; Yang, H.G. LncRNA RUSC1-AS1 promotes the proliferation of hepatocellular carcinoma cells through modulating NOTCH signaling. *Neoplasma* **2021**, *67*, 1204–1213. [CrossRef]

158. Capizzi, M.; Strappazzon, F.; Cianfanelli, V.; Papaleo, E.; Cecconi, F. MIR7–3HG, a MYC-dependent modulator of cell proliferation inhibits autophagy by a regulatory loop involving AMBRA1. *Autophagy* **2017**, *13*, 554–566. [CrossRef]
159. Shen, Z.; Qin, X.; Yan, M.; Li, R.; Chen, G.; Zhang, J.; Chen, W. Cancer-associated fibroblasts promote cancer cell growth through a miR-7-RASSF2-PAR-4 axis in the tumor microenvironment. *Oncotarget* **2016**, *8*, 1290–1303. [CrossRef]
160. Foekens, J.A.; Sieuwerts, A.M.; Smid, M.; Look, M.P.; De Weerd, V.; Boersma, A.W.M.; Klijn, J.G.M.; Wiemer, E.A.C.; Martens, J.W. Four miRNAs associated with aggressiveness of lymph node-negative, estrogen receptor-positive human breast cancer. *Proc Natl. Acad. Sci. USA* **2008**, *105*, 13021–13026. [CrossRef]
161. Li, M.; Pan, M.; Wang, J.; You, C.; Zhao, F.; Zheng, D.; Guo, M.; Xu, H.; Wu, D.; Wang, L.; et al. miR-7 Reduces Breast Cancer Stem Cell Metastasis via Inhibiting RELA to Decrease ESAM Expression. *Mol. Ther. Oncolytics* **2020**, *18*, 70–82. [CrossRef] [PubMed]
162. Huang, Q.; Wu, Y.-Y.; Xing, S.-J.; Yu, Z.-W. Effect of miR-7 on resistance of breast cancer cells to adriamycin via regulating EGFR/PI3K signaling pathway. *Eur. Rev. Med. Pharmacol. Sci.* **2019**, *23*, 5285–5292. [PubMed]
163. Hong, T.; Ding, J.; Li, W. miR-7 Reverses Breast Cancer Resistance To Chemotherapy By Targeting MRP1 And BCL2. *OncoTargets Ther.* **2019**, *12*, 11097–11105. [CrossRef]
164. Paccez, J.D.; Duncan, K.; Sekar, D.; Correa, R.G.; Wang, Y.; Gu, X.; Bashin, M.; Chibale, K.; Libermann, T.A.; Zerbini, L.F. Dihydroartemisinin inhibits prostate cancer via JARID2/miR-7/miR-34a-dependent downregulation of Axl. *Oncogenesis* **2019**, *8* 1–14. [CrossRef]
165. Chang, Y.-L.; Zhou, P.-J.; Wei, L.; Li, W.; Ji, Z.; Fang, Y.-X.; Gao, W.-Q. MicroRNA-7 inhibits the stemness of prostate cancer stem-like cells and tumorigenesis by repressing KLF4/PI3K/Akt/p21 pathway. *Oncotarget* **2015**, *6*, 24017–24031. [CrossRef]
166. Sanchez, N.; Gallagher, M.; Lao, N.; Gallagher, C.; Clarke, C.; Doolan, P.; Aherne, S.; Blanco, A.; Meleady, P.; Clynes, M.; et al. MiR-7 Triggers Cell Cycle Arrest at the G1/S Transition by Targeting Multiple Genes Including Skp2 and Psme3. *PLoS ONE* **2013** *8*, e65671. [CrossRef]
167. Mallik, S.; Qin, G.; Jia, P.; Zhao, Z. Molecular signatures identified by integrating gene expression and methylation in non-seminoma and seminoma of testicular germ cell tumours. *Epigenetics* **2021**, *16*, 162–176. [CrossRef]
168. Yu, Z.; Ni, L.; Chen, D.; Zhang, Q.; Su, Z.; Wang, Y.; Yu, W.; Wu, X.; Ye, J.; Yang, S.; et al. Identification of miR-7 as an oncogene in renal cell carcinoma. *J. Mol. Histol.* **2013**, *44*, 669–677. [CrossRef]
169. Li, Q.; Wang, X.; Jiang, N.; Xie, X.; Liu, N.; Liu, J.; Shen, J.; Peng, T. Exosome-transmitted linc00852 associated with receptor tyrosine kinase AXL dysregulates the proliferation and invasion of osteosarcoma. *Cancer Med.* **2020**, *9*, 6354–6366. [CrossRef]
170. Jiang, M.-J.; Dai, J.-J.; Gu, D.-N.; Huang, Q.; Tian, L. MicroRNA-7 inhibits cell proliferation of chronic myeloid leukemia and sensitizes it to imatinib in vitro. *Biochem. Biophys. Res. Commun.* **2017**, *494*, 372–378. [CrossRef]
171. Zhang, H.; Luo, X.-Q.; Zhang, P.; Huang, L.-B.; Zheng, Y.-S.; Wu, J.; Zhou, H.; Qu, L.-H.; Xu, L.; Chen, Y.-Q. MicroRNA Patterns Associated with Clinical Prognostic Parameters and CNS Relapse Prediction in Pediatric Acute Leukemia. *PLoS ONE* **2009**, *4*, e7826. [CrossRef]
172. O'Donnell, K.A.; Wentzel, E.A.; Zeller, K.I.; Dang, C.V.; Mendell, J.T. c-Myc-regulated microRNAs modulate E2F1 expression. *Nat. Cell Biol.* **2005**, *435*, 839–843. [CrossRef] [PubMed]
173. Li, G.; Gao, L.; Zhao, J.; Liu, D.; Li, H.; Hu, M. LncRNA ANRIL/miR-7-5p/TCF4 axis contributes to the progression of T cell acute lymphoblastic leukemia. *Cancer Cell Int.* **2020**, *20*, 1–12. [CrossRef] [PubMed]
174. Lou, X.; Fu, J.; Zhao, X.; Zhuansun, X.; Rong, C.; Sun, M.; Niu, H.; Wu, L.; Zhang, Y.; An, L.; et al. MiR-7e-5p downregulation promotes transformation of low-grade follicular lymphoma to aggressive lymphoma by modulating an immunosuppressive stroma through the upregulation of FasL in M1 macrophages. *J. Exp. Clin. Cancer Res.* **2020**, *39*, 1–15. [CrossRef] [PubMed]

biomedicines

MDPI

Review

MicroRNAs as Predictive Biomarkers of Resistance to Targeted Therapies in Gastrointestinal Tumors

Valentina Angerilli [1,†], Francesca Galuppini [1,†], Gianluca Businello [1], Luca Dal Santo [1], Edoardo Savarino [2], Stefano Realdon [3], Vincenza Guzzardo [1], Lorenzo Nicolè [1], Vanni Lazzarin [1], Sara Lonardi [3], Fotios Loupakis [3] and Matteo Fassan [1,3,*]

[1] Surgical Pathology & Cytopathology Unit, Department of Medicine (DIMED), University of Padua, 35100 Padua, Italy; valentina.angerilli@gmail.com (V.A.); francesca.galuppini@unipd.it (F.G.); glc.businello@gmail.com (G.B.); lucas1186dalsanto@gmail.com (L.D.S.); vincenza.guzzardo@unipd.it (V.G.); lorenzo.nic86@gmail.com (L.N.); vanni.lazzarin@unipd.it (V.L.)

[2] Division of Gastroenterology, Department of Surgical, Oncological and Gastroenterological Sciences, University of Padua, 35100 Padua, Italy; edoardo.savarino@unipd.it

[3] Istituto Oncologico Veneto (IOV-IRCCS), 35100 Padua, Italy; stefano.realdon@iov.veneto.it (S.R.); sara.lonardi@iov.veneto.it (S.L.); fotios.loupakis@iov.veneto.it (F.L.)

* Correspondence: matteo.fassan@unipd.it; Tel.: +39-049-821-1312

† These authors contributed equally to this article.

Abstract: The advent of precision therapies against specific gene alterations characterizing different neoplasms is revolutionizing the oncology field, opening novel treatment scenarios. However, the onset of resistance mechanisms put in place by the tumor is increasingly emerging, making the use of these drugs ineffective over time. Therefore, the search for indicators that can monitor the development of resistance mechanisms and above all ways to overcome it, is increasingly important. In this scenario, microRNAs are ideal candidate biomarkers, being crucial post-transcriptional regulators of gene expression with a well-known role in mediating mechanisms of drug resistance. Moreover, as microRNAs are stable molecules, easily detectable in tissues and biofluids, they are the ideal candidate biomarker to identify patients with primary resistance to a specific targeted therapy and those who have developed acquired resistance. The aim of this review is to summarize the major studies that have investigated the role of microRNAs as mediators of resistance to targeted therapies currently in use in gastro-intestinal neoplasms, namely anti-EGFR, anti-HER2 and anti-VEGF antibodies, small-molecule tyrosine kinase inhibitors and immune checkpoint inhibitors. For every microRNA and microRNA signature analyzed, the putative mechanisms underlying drug resistance were outlined and the potential to be translated in clinical practice was evaluated.

Keywords: microRNAs; GI cancers; targeted therapy; drug resistance

Citation: Angerilli, V.; Galuppini, F.; Businello, G.; Dal Santo, L.; Savarino, E.; Realdon, S.; Guzzardo, V.; Nicolè, L.; Lazzarin, V.; Lonardi, S.; et al. MicroRNAs as Predictive Biomarkers of Resistance to Targeted Therapies in Gastrointestinal Tumors. *Biomedicines* **2021**, *9*, 318. https://doi.org/10.3390/biomedicines9030318

Academic Editor: Francesca Lovat

Received: 7 February 2021
Accepted: 18 March 2021
Published: 21 March 2021

1. Introduction

The number of druggable tumor-specific molecular alterations has grown substantially in the past decade and a great survival benefit has been obtained from genomic-driven therapies across many cancer types. However, while molecularly targeted drugs offer a significantly higher response rate than traditional chemotherapy in diseases such as melanoma and non-small cell lung cancer, they have shown relatively modest clinical benefits in gastrointestinal (GI) malignancies [1]. The "one gene, one drug" approach of precision oncology clashes with the reality of the extremely complex molecular landscape of most solid tumors. A crucial aspect towards a successful development and application of targeted therapies is the understanding of resistance mechanisms that limit their effectiveness [1].

Primary or intrinsic resistance is defined as radiographic or clinical disease progression as the best response to an anticancer therapy and implies the pre-existence of resistance-mediating factors within the tumor [2].

Secondary or acquired resistance is defined as therapeutic resistance after an initial period of disease stabilization or response, and be can caused by mutations arising during treatment, as well as through various adaptative mechanisms [3].

A wide range of biological determinants have been associated with drug resistance to targeted therapies; these include the presence of undruggable genomic drivers, the mutation of drug targets, the activation of survival signaling pathways, and the inactivation of downstream death signaling pathways. The tumor microenvironment may also mediate resistance by numerous mechanisms, including: promoting immune evasion of cancer cells, hampering drug absorption and stimulating paracrine cancer cell growth factors. The epithelial–mesenchymal transition (EMT) process and the presence of tumor stem cells have also been identified as causes of drug resistance. Furthermore, the coexistence of genetically and molecularly heterogeneous subclones within the bulk solid tumor may substantially fuel resistance under therapeutic selective pressure [2,4,5].

The purpose of this review is to elucidate the role of microRNAs (miRNAs) as mediators of resistance to targeted therapies in GI tumors.

2. MicroRNAs and Drug Resistance

MiRNAs are a group of short non-coding RNA transcripts (18–22 nucleotides) that control gene expression at the post-transcriptional level, thus regulating several pathways involved in the maintenance of cell homeostasis. In most cases, they act by binding to the 3′-untranslated region (3′-UTR) of their target messenger RNAs (mRNAs), which results in mRNA silencing or degradation and subsequent post-translational degradation or down-modulation of proteins [6,7].

Numerous studies have demonstrated that miRNAs are heavily dysregulated in cancer and in certain conditions they may function as oncogenes or tumor suppressor genes. Hence, microRNAs have been identified as potential diagnostic, prognostic and predictive biomarkers for patient stratification in oncology [8,9]. Particularly, miRNAs have been found to act as mediators of drug resistance through various cellular and molecular mechanisms, which are related to: apoptosis, cell cycle modification, alteration in drug targets, regulation of drug efflux transporters, epithelial-mesenchymal transition (EMT) and cancer stem cells [10–12] (Figure 1).

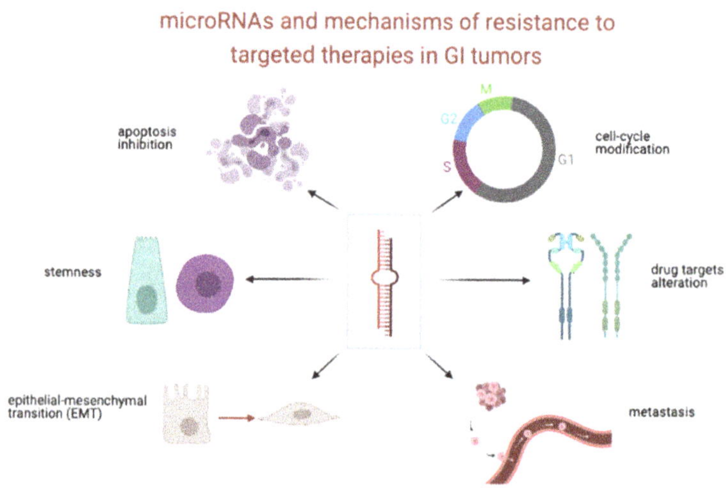

Figure 1. MiRNAs and mechanisms of resistance to targeted therapies in GI tumors. Credits to BioRender.com, accessed date 18 March 2021.

In light of their high stability in tissues and bodily fluids (i.e., plasma and urine), miRNAs are ideal candidates to monitor cancer treatment resistance. In fact, miRNAs are not only present within cells, but are also actively secreted from cells, and included in RNA-binding multiprotein complexes and/or exosomes [13]. In the context of liquid biopsies, analysis of miRNA expression patterns represents a promising tool to perform non-invasive longitudinal tracking of drug sensitivity over time [14,15].

When referring to GI malignancies, the role of miRNAs as predictors of resistance to chemotherapy has been extensively studied [16–18], while further research is needed to gain a better understanding of the complex interplay between miRNA expression and resistance to targeted therapy. Furthermore, many studies have documented the change in miRNA expression profiles following radiotherapy, so that miRNAs could prove useful in modulating radioresponsitivity at the clinical level [19].

With regard to future perspectives, playing such an important role in carcinogenesis, miRNAs have also been proposed and tested as a therapeutic option. In cancer, in order to increase the level of a tumor suppressor miRNA, whose level is downregulated, miRNA mimics can be used to re-establish normal expression and function of a certain miRNA of interest. On the other hand, antagomiRs can be used to suppress the function of a specific oncomiRNA that play a role in the development and progression of the neoplasm. However, while many studies have attempted to evaluate the consequences of the manipulation of these molecules with the purpose to support findings from in vitro or in vivo models, only a few miRNA targeted drugs have entered clinical trials because of a few important limitations, such as dosage and cross-reactions with other miRNAs [20].

3. Anti-EGFR

Epidermal growth factor receptors (EGFRs) are a large family of receptor tyrosine kinases expressed in several types of cancer and are responsible for the activation of a downstream signaling cascade that modulates growth, signaling, differentiation, adhesion, migration and survival of cancer cells [21].

Targeting EGFR has been revealed to be a successful strategy in metastatic colorectal cancer (mCRC). Cetuximab and panitumumab are the two anti-EGFR antibodies currently in use for mCRC [22]. Anti-EGFR antibodies act by binding the extra-cellular domain of EGFR, thereby preventing ligand binding and blocking the downstream signaling pathway. Furthermore, they induce receptor internalization and degradation [23], and are responsible for antibody-dependent cellular cytotoxicity [24].

In patients with metastatic disease, anti-EFGR monoclonal antibodies have shown a clinically significant response when given in monotherapy or in combination with chemotherapy [25,26]. However, drug resistance frequently arises. Many genomic and non-genomic mechanisms underlie resistance to anti-EGFR therapy. Particularly, a plethora of overlapping de novo and acquired genetic alterations have been identified, such as *KRAS*, *NRAS*, *BRAF*, *PIK3C*, and *EFGR* mutations and *MET* and *ERBB2* amplifications [27,28].

Let-7. KRAS mutations are the major predictors of resistance to EGFR inhibitors and are routinely screened in mCRCs [29]. In this regard, members of the Lethal-7 (Let-7) family of miRNAs have been found to display KRAS downregulating activity by binding to specific sites in the 3′-UTR of KRAS mRNA [30]. Ruzzo et al. measured the expression of let-7 in formalin-fixed paraffin-embedded (FFPE) tumor samples of mCRC patients who underwent third-line therapy with cetuximab plus irinotecan. The findings of this study revealed that higher levels of let-7 were significantly associated with better survival outcome in patients, regardless of *KRAS* mutational status. In patients with *KRAS* mutations, let-7 may serve to identify a subpopulation of responders to anti-EGFR therapy [31]. Furthermore, Cappuzzo et al. found that the high-intensity signature of the cluster Let-7c/miR-99a/miR-125b is associated with a favorable response to cetuximab and panitumumab in *KRAS* wild-type mCRC patients [32].

MiR-31-5p/miR-31-3p. The predictive role of miR-31-5p and miR-31-3p has been investigated in several studies. In a study which included an exploratory cohort and a validation cohort, of the 9 miRNAs analyzed in FFPE tumor samples of both cetuximab responders and non-responders, miRNA-31-3p and miRNA-31-5p were found to be strongly associated with time to progress (TTP) in wild-type *RAS* patients treated with cetuximab [33]. In another study it was further discovered that high miR-31-5p expression was associated with shorter progression-free survival (PFS) in mCRC patients treated with anti-EGFR therapy [34]. Furthermore, Manceau et al., after screening 1145 miRNAs on a large cohort of wild-type *RAS* mCRC patients, identified miR-31-3p as a promising predictive biomarker of response to anti-EGFR antibodies [35].

MiR-10/miR-125b. MiR-10 and miR-125b are generated by the long non-coding RNA (lncRNA) MIR100HG, and decrease the expression of five negative regulators of the Wnt/β-catenin pathways, resulting in increased Wnt signaling, which is a hallmark of colorectal carcinogenesis. It was demonstrated that the overexpression of lncRNA MIR100HG and the two embedded miRNAs was associated with cetuximab resistance in tumors derived from CRC patients and in vitro 3-D cell culture models [36].

MiR-181a. MiR-181a has a strong tumor-promoting effect via inhibition of the tumor suppressor factor Wnt Inhibitory factor 1 (WIF-1) and stimulates tumor growth, cell motility and invasion. Pichler et al. analyzed the expression of miR-181a in FFPE tumor samples of a cohort of patients with *KRAS* wild-type mCRC, undergoing cetuximab and panitumumab treatment. The findings of the study highlighted the potential predictive role of miR-181a, as high levels of this miRNA were associated with a longer PFS [37].

MiR-345. According to the results of a study by Schou and colleagues, high levels of miR-345 in whole blood correlated with a lack of response to therapy in patients treated with third-line cetuximab and irinotecan [38]. In the context of liquid biopsy, miR-345 appears to be a valid candidate biomarker of sensitivity to anti-EGFR therapy.

MiR-199a/miR-375. Mussnich et al. performed miRNA expression profiling of human CRC cell lines sensitive to cetuximab and of their resistant counterparts. Among the investigated miRNAs, miR-199a and miR-375 were found to be overexpressed in resistant cells and their upregulation promoted cetuximab resistance. MiR-199a and miR-375 both target PH Domain And Leucine Rich Repeat Protein Phosphatase 1 (PHLPP1), which acts as a tumor suppressor by downregulating the Akt oncogenic pathway [39].

MiR-302. In a recent study, downregulation of miR-302a was observed in cetuximab-resistant CRC cells as well as in patient-derived xenografts. Furthermore, miR-302a overexpression restored cetuximab responsiveness in CRC cell lines both in vitro and in vivo. MiR-302a has been found to act as a tumor suppressor by targeting NFIB (Nuclear factor 1 B-type) and downregulating the NFIB/ITGA6 axis, which is responsible for cell migration and invasion in the metastatic process [40].

MiR-141-3p. The findings of a study by Xing and colleagues indicated that miR-141-3p might be a predictor of response to cetuximab. In CRC cell cultures, miR-141-3p regulated cetuximab sensitivity by directly targeting EGFR and its downstream cascade. Moreover, miR-141-3p improved cetuximab-induced apoptosis in CRC cells, proving to have a tumor-suppressing function [41].

4. Anti-VEGF

Vascular endothelial growth factor (VEGF) promotes tumor angiogenesis through several mechanisms, including enhanced endothelial cell proliferation, survival and migration, chemotaxis of bone-marrow-derived progenitor cells, vascular permeability, and vasodilation. VEGF ligand acts by binding the VEGF receptor (VEGFR) with tyrosine kinase activity, which activates a network of downstream signaling pathways, responsible for mediating numerous changes within the tumor vasculature. Several strategies have been put in place to inhibit the VEGF pathway, including monoclonal antibodies blocking VEGF or VEGFR, soluble VEGF receptors, and tyrosine kinase inhibitors selective for VEGFRs [42].

Bevacizumab is a monoclonal antibody directed against VEGF and has been approved in combination with cytotoxic chemotherapy as first or second-line therapy in mCRC, after several randomized clinical trials had shown improvements in overall survival [43–45]. The VEGF decoy receptor aflibercept has also recently been introduced in clinical practice as second-line therapy in mCRC [46].

In the context of gastro-esophageal malignancies, the anti-VEGFR2 antibody ramucirumab is currently licensed for clinical use in combination with cytotoxic chemotherapy for chemo-refractory metastatic disease [47,48]. Sunitinib and sorafenib are multi-tyrosine kinase inhibitors (TKIs), which target a wide spectrum of pathways involved in tumor growth, including VEGF/VEGFR, and are in use in GI precision oncology.

However, the VEGF pathway inhibitors are failing to provide a durable response in most patients. The activation of alternative angiogenic "escape" pathways is a major contributor to drug resistance and may induce tumor growth, enhancement of invasion and metastasis [49]. A few preliminary studies have been carried out to elucidate the role of miRNAs as mediators of resistance to bevacizumab, however there is still no evidence regarding ramucirumab or aflibercept.

MiR-126. MiR-126 acts by enhancing the proangiogenic actions of VEGF and promotes blood vessel formation and survival [50]. MiR-126-containing micro-vesicles originate from endothelial cells and fuse with the membrane of neighboring cells, thus promoting angiogenesis in a paracrine manner. A negative relationship exists between levels of circulating miR-126 and their response to bevacizumab. Increasing plasma levels of miR-126 are associated with resistance to bevacizumab, while lower levels indicate treatment response. As miR-126 is representative of endothelial cell turn-over, its circulating levels could help perform non-invasive monitoring of bevacizumab resistance.

MiR-664-3p/miR-455-5p. The predictive role of miR-664-3p and miR-455-5p has been evaluated in a study by Boisen et al. Higher miR-664-3p and lower miR-455-5p expression evaluated in FFPE tumor samples was found to be associated with an improved outcome in the cohorts of patients treated with bevacizumab and chemotherapy, in comparison with the cohort treated with chemotherapy only. Therefore, miR-664-3p, with its putative tumor suppressor role, and miR-455-5p, with its putative oncogenic role, could represent potential predictive tissue biomarkers of bevacizumab effectiveness [51].

MiR-20b-5p/miR-29b-3p/miR-155-5p. According to the results of a study by Ulivi and colleagues, higher circulating levels of miR-20b-5p, miR-29b-3p and miR-155-5p are associated with a better outcome in mCRC patients treated with a combination of bevacizumab and chemotherapy. Furthermore, the variation in plasma levels of miR-155-5p could also be indicative of patient survival. In line with these findings, previous studies have found that miR-20b, miR-29b and miR-155-5p play a role in regulating tumor angiogenesis [52]. MiR-29b acts as a tumor suppressor through simultaneously inhibiting angiogenesis and tumorigenesis by targeting Akt3 [53]. Although little information is available on its involvement in the tumor angiogenic process, miR-20 has been found to regulate proliferation and senescence in endothelial cells [54]. MiR-155 contributes to controlling hypoxia-inducible factor (HIF-1α) and promotes angiogenesis under hypoxia condition [55].

5. Anti-HER2

HER2 belongs to the EGFR tyrosine kinase family. It is localized in the cell membrane and when activated initiates intracellular downstream signaling involved in diverse biological processes related to cancer, such as proliferation, migration and apoptosis [56].

Overexpression of HER2 has been detected in 11–20% of gastric and gastroesophageal junction (GEJ) cancers [57–60]. Trastuzumab is an anti-HER2 antibody that acts by blocking the activity of the HER2 receptor and weakening the downstream signaling [61].

In 2010, the results of the Trastuzumab for Gastric cancer (ToGA) trial displayed how a combination of trastuzumab and chemotherapy was able to prolong the overall survival compared to chemotherapy alone in gastric cancer patients with HER2 overexpression. Thus, trastuzumab in combination with chemotherapy has been approved as the standard first-line treatment of advanced gastric or GEJ cancer with HER2 overexpression/*ERBB2* amplification [62]. However, only a fraction of the patients has been found to respond to trastuzumab, and even those who achieved an initial therapeutic response developed resistance within 7 months [63].

MiR-21. It has been found that trastuzumab activates phosphatase and tensin homolog (PTEN) phosphatase, thus decreasing PTEN tyrosine phosphorylation, via inhibition of HER2 receptor bound Src. Reduced PTEN expression could predict trastuzumab resistance [64]. According to the results of a study by Eto et al., overexpression of miR-21 downregulates PTEN and increases the phosphorylation of PTEN downstream target Akt in HER2-positive gastric cancer (GC) cell lines, thus resulting in decreased sensitivity to trastuzumab-induced apoptosis. The opposite effect in PTEN and p-Akt was observed after miR-21 suppression, which instead restored resistance to trastuzumab. These findings suggest that the miR-21/PTEN pathway might play a critical role in regulating trastuzumab resistance of GC cells via modulating apoptosis [65].

MiR-223. F-box and WD repeat domain-containing 7 (FBXW7) is the substrate recognition component of an evolutionarily conserved ubiquitin ligase complex, which appears to have an important role in controlling the stability of several oncoprotein substrates, including cyclin E, c-Myc, Notch, c-Jun, mammalian target of rapamycin (mTOR) and Myeloid Cell Leukemia 1 (MCL1) [66]. Upregulation of miR-223 decreased FBXW7 expression and subsequently reduced the sensitivity of HER2-positive GC cell lines to trastuzumab, thereby acting as an oncomiR and suppressing trastuzumab-induced apoptosis. On the contrary, downregulation of miR-223 restored FBXW7 expression and the sensitivity to trastuzumab [67].

MiR-16. Trastuzumab has the ability to block PI3K/AKT downstream signaling, which results in the inhibition of c-Myc activation and subsequent upregulation of miR-16. A study by Venturutti et al. identified cyclin J and Far Upstream Element Binding Protein 1 (FUBP1) as miR-16 targets. miR-16 acts as a tumor suppressor by exerting an antiproliferative effect via silencing its miRNA targets [68]. Furthermore, it was found that in vitro overexpression of miR-16 and low or null levels of FUBP1 were predictors of trastuzumab sensitivity. These findings suggest that both miR-16 and FUBP1 could represent promising predictive biomarkers of response to trastuzumab [69].

MiR-125b. MiR-125b has been found to be dysregulated in several cancer types. For example, while being downregulated in osteosarcoma, breast cancer, ovarian cancer and hepatocellular carcinoma (HCC), it appears to be upregulated in CRC, prostate cancer, non-small cell lung cancer and GC [70]. In the latter, miR-125b functions as an oncogene by promoting cellular proliferation, migration and invasion by downregulating the expression of Protein Phosphatase 1 Catalytic Subunit Alpha (PPP1CA) and upregulating Rb phosphorylation. Furthermore, it has been recently demonstrated that high miR-125b levels are associated with poor prognosis in patients with HER2-positive GC treated with trastuzumab, thus indicating a speculative role of miR-125b in trastuzumab resistance [71].

MiR-200c. Compelling evidences identify in the EMT one of the mechanisms implicated in drug resistance [72]. In a recent study by Zhou et al. it was observed that trastuzumab-resistant GC cell cultures expressed high levels of EMT markers, as well as TGF-β, which is a master regulator of the EMT. Furthermore, while miR-200c was found to be downregulated in trastuzumab-resistant GC cells, thereby proving to act as a tumor suppressor miRNA, its overexpression restored trastuzumab sensitivity and blocked the EMT by targeting Zinc finger E-box-binding homeobox 1 (ZEB1) and 2 (ZEB2), which are downstream molecules of TGF-β. This study suggested that the TGF-β/ZEB/miR-200c axis is involved in the resistance of trastuzumab in GC by regulating the EMT [73].

MiR-494. Lapatinib is a small molecule which inhibits the tyrosine kinases of HER2 and EGFR1 [74]. MiR-494 has been found to act as a tumor suppressor and to restore lapatinib sensitivity and inhibit formation of cancer-initiating cells (CICs) via reducing expression of FGFR2 in HER2-positive, FGFR2 overexpressing and lapatinib resistant GC cell cultures [75].

6. Small-Molecule Tyrosine Kinase Inhibitors

Receptor tyrosine kinases (RTKs) are widely expressed transmembrane proteins that regulate many fundamental cellular processes, and play a key role in the physiopathology of several diseases. Upon ligand binding, RTKs activate intracellular signaling pathways involved in many functions, such as: differentiation, proliferation, migration, invasion and angiogenesis. Aberrant RTK expression being a well-recognized mechanism of tumorigenesis, a broad variety of inhibitors are currently in clinical use across many cancer types [76]. Particularly, small-molecule tyrosine kinase inhibitors act by blocking the intracellular domain of the receptor or by inhibiting the tyrosine kinase activity of downstream signaling mediators [77].

6.1. Imatinib

Imatinib is a selective inhibitor of certain tyrosine kinases and is highly active in patients with gastrointestinal stromal tumors (GISTs) by blocking the constitutive activity of KIT and platelet-derived growth factor receptor α (PDGFR-α). Imatinib is the treatment of choice for advanced GISTs harboring *KIT* and *PDGFR-α* mutations, and it is also used in adjuvant and neoadjuvant settings. Although imatinib has high response rates, drug resistance remains the main challenge for extending patient survival [78]. The mechanisms underlying imatinib resistance are not completely clear. However, in about half of patients, resistance is caused by secondary mutations in *KIT* (exons 13, 14, 17 or 18) [79].

MiR-125a-5p. In a study on FFPE GIST samples from patients resistant to imatinib in a neoadjuvant setting, it was found that miR-125a-5p can modulate imatinib response in *KIT*-mutated GIST cells by regulating the expression of Protein Tyrosine Phosphatase Non-Receptor Type 18 (PTPN18). Because the targets of miR-125a-5p have been functionally associated with anti-apoptosis, cell cycle progression, signal transduction and protein phosphorylation, miR-125a-5p acts as oncomiR and its overexpression was linked with imatinib resistance [80].

MiR-320a. miR-320 acts as a tumor suppressor by targeting genes involved in Wnt and IGF pathways. In imatinib resistant GIST patients, down-regulation of the tumor suppressor miR-320a is associated with direct up-regulation of β-catenin and subsequent enhanced expression of anti-apoptotic MCL1 [81].

MiR-218. According the findings of a study by Fan and colleagues, the expression of miR-218 was down-regulated in an imatinib-resistant GIST cell line, whereas miR-218 over-expression was able to restore the sensitivity of GIST cells to imatinib, with the PI3K/AKT signaling pathway possibly involved in the mechanism. The PI3K/AKT pathway is down-stream of KIT and it is reactivated when GISTs become resistant to imatinib [82].

MiR-518a-5p. By performing a microarray analysis on GIST samples from patients resistant to imatinib, miR-518a-5p was found to be a potential predictor of drug sensitivity. Downregulation of miR-518a-5p is likely to upregulate PIK3C2A, causing resistance to imatinib in GISTs. PIK3C2A belongs to the phosphoinositide 3-kinase (PI3K) family, whose member proteins have roles in signaling pathways involved in cell proliferation, oncogenic transformation, cell survival, cell migration and intracellular protein trafficking [83].

MiR-92a-3p, miR-99a-5p, and miR-101-3p. MiR-92a-3p, miR-99a-5p and miR-101-3p are three miRNAs implicated in cell cycle regulation. MiR-92a-3p targets Cyclin-dependent kinase inhibitor 1C (CDKN1C), miR-92a-3p regulates mTOR pathway [84], and miR-101-3p also acts as a regulator of the mTOR pathway by mediating AKT activation [85]. The findings of a study, which analyzed miRNA expression profiles across a series of imatinib resistant and sensitive FFPE GIST samples, identified the previously mentioned miRNAs as differentially expressed and therefore possibly implicated in imatinib resistance [86].

MiR-28-5p. A recent study identified miR-28-5p as a potential mediator of imatinib resistance in GISTs. Being overexpressed in imatinib-resistant GIST samples and displaying a significant correlation to imatinib response, miR-28-5p has been proven to function as an oncomiR. However, very little is known about this miRNA and therefore further research is required to confirm these findings [87].

6.2. Sorafenib

Sorafenib is a small molecule which inhibits multiple kinases involved in tumor cell signaling, proliferation, angiogenesis and apoptosis. Sorafenib is currently in clinical use for unresectable HCC. Its approval was based on the successful outcome of the pivotal SHARP and Sorafenib Asia-Pacific trials in Child-Pugh class A patients with advanced HCC. However, only approximately 30% of patients clinically benefit from sorafenib, and this subgroup usually acquires drug resistance within 6 months. Recent studies have highlighted the role of epigenetics, transport processes, regulated cell death, and the tumor microenvironment in the initiation and development of sorafenib resistance in HCC [88].

MiR-122. MiR-122 is the most abundant liver-specific miRNA and it is significantly down-regulated in HCC. miR-122 acts as a tumor suppressor in the liver by inhibiting survival of cancer cells, anchorage-independent growth, migration, invasion, and EMT. The in vitro restoration of expression of miR-122 has been found to sensitize HCC cells to sorafenib. MiR-122 is a negative regulator of A Disintegrin and metalloproteinase domain-containing protein 10 (ADAM10) and serum response factor (SRF), which are both involved in EMT (SFR is also implicated in tumor angiogenesis) and insulin growth factor 1 receptor (IGF1R), which activates the downstream RAS/RAF/ERK pathway, induces proliferation and promotes metastasis [89].

MiR-34. In a study by Yang and colleagues, miR-34 was reported to be downregulated in FFPE HCC samples from patients and HCC cell lines and was associated with poorer survival. MiR-34, which is a direct target of p53, binds the 3′-UTR region of the anti-apoptotic protein B-cell lymphoma 2 (Bcl-2), which was found to be overexpressed in HCC samples and the cells analyzed. The restoration of miR-34a potentiated sorafenib-induced apoptosis, suggesting that miR-34a enhanced the anti-tumor effect of sorafenib in HCC cells [90].

Let-7. To assess the role of miRNAs in HCC, Ohta et al. performed microarray analysis and discovered that let-7 as was downregulated in human HCC cells. Furthermore, it was also found that the upregulation of let-7 was linked to a decreased expression of its putative target, the anti-apoptotic protein B-cell lymphoma xL (Bcl-xL). Ultimately, the expression of let-7c enhanced apoptosis of HCC cells upon exposure to sorafenib, which is responsible for the downregulation of another anti-apoptotic Bcl-2 protein, MCL1. To summarize, let-7 exerts a tumor suppressing function by inducing apoptosis of HCC cells [91].

MiR-338-3p. Hypoxia being one of the main contributors to anti-tumor drug resistance in solid malignancies, hypoxia-inducible factor 1 (HIF-1) has been recognized as one of the key mediators of resistance to sorafenib in HCC. In a study on both patient samples and HCC cell lines, it was discovered that miR-338-3p directly targeted HIF-1α and downregulated the expression of HIF-1α target genes involved in the hypoxia-induced signaling pathway. Via this mechanism, miR-338-3p inhibits HCC tumor growth and sensitizes HCC cells to sorafenib [92].

MiR-216a/217. According to the findings of a study by Xia and colleagues on a series of HCC FFPE samples, chemoresistance against sorafenib is caused by overexpression of the miR-216a/217 cluster. Upregulation of miR-216a/217 is able to induce EMT of cancer cells and decrease expression of SMAD7 and PTEN, consequently activating the transforming growth factor β (TGF-β) and PI3K/AKT pathways [93].

MiR-93. Following a comprehensive miRNA expression profiling using HCC cell lines, miR-93 was identified as a novel target associated with HCC. MiR-93 acts by binding with the 3′-UTR of PTEN and cyclin-dependent kinase inhibitor (CDK1NA), inhibiting their expression and, as a result, activating the oncogenic PI3K/AKT pathway. In light of these findings, miR-93 expression was also found to render HCC cells more sensitive to sorafenib [94].

MiR-21. As previously mentioned, miR-21 dysregulates PTEN, by inhibiting Akt activation. The Akt signaling pathway appears to be highly activated in sorafenib-resistant HCC cells. Moreover, autophagy seems to promote sorafenib sensitivity in sorafenib-resistant HCC cells. Compelling evidence suggests that miR-21 plays a key role in mediating resistance to sorafenib. For example, exposure of HCC cell lines to sorafenib led to an upregulation of miR-21 and a downregulation of PTEN. Furthermore, transfection of miR-21 in HCC cells was able to restore sorafenib resistance by inhibiting autophagy. To summarize, miR-21 contributes to acquired resistance to sorafenib by suppressing autophagy by modulating the Akt/PTEN pathway [95].

MiR-193a. MiR-193a acts a tumor suppressor by negatively regulating the pro-metastatic factor urokinase-type plasminogen activator (uPA) and is downregulated in HCC. Transfection of HCC cell lines with miR-193a was found to decrease proliferation, promote apoptosis and enhance sorafenib anti-tumor activity [96].

MiR-193b. Substantial downregulation of miR-193b and overexpression of its target MCL1 were observed in HBV-positive HCC cells. MCL1 is an anti-apoptotic protein, which is overexpressed in many human malignancies, including HCC, and has been known as an important mediator of chemosensitivity in HCC. Restoration of the expression of miR-193b sensitized HBV-positive HCC cell lines to sorafenib by facilitating sorafenib-induced apoptosis [97]. Similarly, in another study, HCV-positive HCC cells showed decreased expression of miR-193b and upregulation of its target MCL1 [98].

MiR-221. The results of a study by Fornari and colleagues indicated that in both HCC cell lines and xenografts, miR-221 overexpression was linked with sorafenib resistance, by targeting caspase-3 and thereby exerting an anti-apoptotic and oncogenic function. Moreover, when investigating the putative role of miR-221 as a circulating biomarker in HCC patients, lower pre-treatment circulating miR-221 levels were detected in sorafenib responders [99].

MiR-486. In HCC tissues and cell lines, miR-486 appears to function as a tumor suppressor by targeting *CITRON* and *CLDN1*, two genes which are responsible for the regulation of cell proliferation and invasion. In addition to being downregulated in HCC, miR-486 was found to enhance chemosensitivity of HCC cells to sorafenib [100].

MiR-494. In a recent study by Pollutri et al., miR-494 overexpression was reported to enhance sorafenib resistance via mTOR pathway activation both in vitro and in vivo. In fact, p27, PTEN and p53-upregulated-modulator-of-apoptosis (PUMA) were identified as targets of miR-494, contributing to speed up cell cycle progression, survival and invasiveness. Moreover, high miR-494 expression seemed to be associated with stem-like features [101].

MiR-101. As part of the tumor microenvironment, HCC-associated macrophages accelerate tumor progression by releasing growth factors. High TGF-β expression in M2 polarized macrophages is thought to increase tumor growth, metastases and EMT. In HCC cell lines, it was observed that miR-101 targeted dual specificity phosphatase 1 (DUSP1), inhibited TGF-β activation and enhanced the effect of sorafenib in HCC cells by potentiating macrophage modulation of the innate immune responses [102].

6.3. Regorafenib

Regorafenib is small-molecule multiple kinase inhibitor. In mCRC, it is indicated for patients who have been previously treated with, or are not considered candidates for, available therapies, including chemotherapy, an anti-VEGF therapy and, if RAS wild-type, an anti-EGFR therapy [103]. It is also licensed as second-line therapy for advanced HCC patients [104] and as third-line therapy for metastatic or unresectable GIST patients [105].

Despite the observed survival benefits, resistance to regorafenib is fairly common. Moreover, toxicity is not insignificant and the absolute clinical benefit is rather small. Therefore, it is crucial to identify efficient predictive biomarkers in order to optimize the use of regorafenib [106].

MiR-34a. In a recent study on CRC cell lines, Cai et al. showed that regorafenib is able to reduce stemness and tumorigenesis in vitro, by upregulating the tumor suppressor miR-34, which targets the WNT/β-catenin pathway. Thereby it can be inferred that regorafenib is able to suppress the generation of drug-resistant cancer stem-like cells via modulation of miR-34a-associated signaling [107].

MiR-30a-5p. Signal transducer and activator of transcription 3 (STAT3) is a transcriptional factor which contributes to drug resistance in cancer therapies by promoting tumor growth and cancer stemness. miR-30a-5p is a STAT3 downstream miRNA and targets Heat Shock Protein Family A Member 5 (HSPA5), which is a master regulator of unfolded protein response (UPR). A dysregulation of the STAT3-miR-30a-5p-HSPA5 axis, with a downregulation of miR-30a-5p, was observed in regorafenib-resistant in CRC tumorspheres [108].

In the context of liquid biopsies, a few preliminary studies have been carried out to investigate putative circulating predictive biomarkers. In a recent study by Schirripa and colleagues, an miRNA signature involving c-miR-21, c-miR-221 and c-miR-760 was found to be prognostic and predictive of a response to regorafenib in an exploratory cohort of CRC patients. However, the results were not confirmed by the validation cohort [103]. In another study on a large cohort of HCC patients, of the 750 miRNAs analyzed, increased plasma levels of miR-30a, miR-122, miR-125b, miR-200a, and miR-374b decreased levels of miR-15b, miR-107 and miR-320b, and absence of miR-645 were all predictive of survival benefit with regorafenib [109].

7. Immune Checkpoint Inhibitors

The introduction of immune-checkpoint blockades in precision oncology led to a paradigm change in the management of advanced cancers. The rationale behind these novel therapies is that cancer cells are able to evade immunosurveillance through different mechanisms, including activation of immune checkpoint pathways that suppress antitumor immune responses [110].

Immunotherapy has recently been incorporated in treatment regimens of GI malignancies. The immune-checkpoint inhibitors currently in clinical use in GI oncology are nivolumab and pembrolizumab, which target programmed death-1 (PD-1) and ipilimumab, which targets cytotoxic T-lymphocyte antigen-4 (CTLA-4) [111].

Immunotherapy approaches have been extensively studied in CRC. In the metastatic setting, ICI therapies provide clinical benefit to defective mismatch repair and microsatellite instable (dMMR/MSI) tumors. Pembrolizumab has been approved as first-line treatment in dMMR/MSI mCRCs based on the successful outcomes of the KEYNOTE-177 trial [112].

As with gastroesophageal cancers, pembrolizumab is currently licensed for recurrent locally advanced or metastatic cancers with PD-L1 expression [113]. Much like in CRC, MSI gastric cancers greatly benefit from ICIs [114]. Moreover, EBV-positive GCs have shown an even more promising response to pembrolizumab [115]. ICIs have also been recently approved as second-line therapy in HCC, after a single agent checkpoint blockade trials obtained partial success [116].

However, a subset of patients who initially responded to ICIs, later relapsed and acquired therapeutic resistance [117]. The role of miRNAs as predictors of response to ICIs has not been elucidated yet. However, several studies have explored how miRNAs regulate the immune checkpoint signaling pathways, leading the way for the discovery of new predictive biomarkers.

Helicobacter Pylori, which is the most common cause of GC, promotes PD-L1 expression and causes an immune escape by downregulating miR-200b and miR-152. Furthermore, a single nucleotide mutation in 3′-UTR of PD-L1 leads to protein overexpression by disrupting the complementarity between miR-570 and its 3′-UTR binding site. This mutation is linked with high PD-L1 levels in GCs [118]. Moreover, according to the findings of a large population-based study from The Cancer Genome Atlas (TGCA) project, a cluster of EBV-miRNAs is linked with a high expression of PD-1/PD-L1 in solid malignancies, including GC [115]. In the context of hematological malignancies, recent studies have found that EBV-associated lymphomas showed high levels of PD-L1. By restoring the expression of PD-L1 targeting tumor suppressor miRNA miR-34a in vitro, PD-L1 expression was reduced and the tumor immunogenicity was increased [119,120]. Thereby, it could be inferred that miR-34a overexpression could be used to subvert PD-L1 induction in

EBV-associated neoplasms. Moreover, miR-34a has been found to be frequently methylated in GIST [121].

In CRC, the tumor suppressor miR-138-5p acts by downregulating PD-L1, leading to cancer cell growth in vitro and tumorigenesis in vivo [122]. In a recent study on a series of CRC samples, low miR-200 and high *ZEB* oncogene expression, which is a profile compatible with EMT, was associated with upregulation of PD-L1 [123]. Moreover, a comprehensive miRNA screening using the TCGA dataset led to the identification of miR-148a-3p as a potential negative regulator of PD-L1 expression, particularly in dMMR/MSI CRC [124].

MiRNAs involved in drug resistance to targeted therapies in GI cancers are summarized in Table 1. The biological materials and molecular techniques used to investigate miRNAs as predictive biomarkers are outlined in Figure 2.

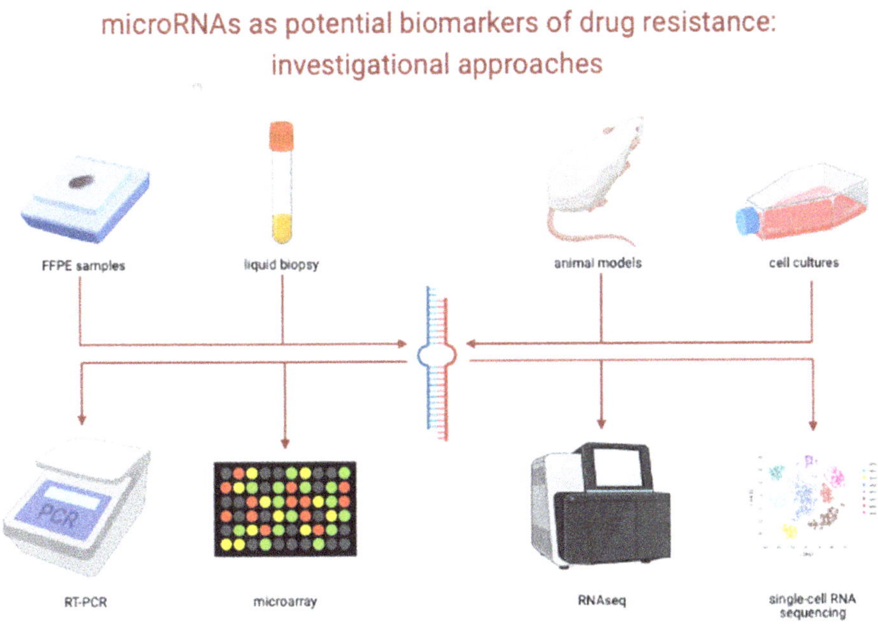

Figure 2. Investigational approaches of miRNAs as potential predictive biomarkers. Credits to BioRender.com, accessed date 18 March 2021.

Table 1. miRNAs involved in resistance to targeted therapies in GI cancers.

Targeted Therapies Class	Type of Cancer	MiRNAs Involved	Gene Mechanism Involved	Action
Anti-EGFR	mCRC	Let-7	KRAS downregulation activity	1. high levels were significantly associated with better survival outcome in patients with KRAS mutation [31,32]
		miR-31-5p/miR-31-3p	BRAF activation and role in the signaling pathway downstream of EGFR	2. high levels are associated with TTP and shorter PFS in wt RAS patients treated with cetuximab [33–35]
		miR-10/miR-125b	Increasing of Wnt signaling	3. high levels are associated with cetuximab resistance [36]
		miR181a	Inibition of WIF-1	4. high levels are correlated with a longer PFS [37]
		miR-345	EGFR pathway dysregulation	5. high levels are associated with lack of response to cetuximab and irinotecan [38]
		miR-199/miR-375	PHLPP1 and downregulation of Akt pathway	6. high levels correlate with cetuximab resistance [39]
		miR-302	Downregulation of NFIB/ITGA6 axis	7. upregulation restored cetuximab responsiveness [40]
		miR 141-3p	EGFR downstream pathway	8. upregulation improves cetuximab activity [41]
Anti-VEGF	mCRC	miR-126	Enhancing the angiogenic effect of VEGF	9. high levels are associated with resistance to bevacizumab [50]
		miR-664-3p/miR-455-5p	Downregulation of the neuroligin and VRGF system	10. potential predictive tissue biomarkers of bevacizumab effectiveness [51]
		miR-20b-5p/miR-29b-3p/miR-155-5p	Inhibition of Akt pathway/Controlling of HIF-1α signalling	11. high levels are associated with a better outcome of mCRC patients treated with a bevacizumab and chemotherapy [52–54]
Anti-HER2	GC	miR-21	PTEN deregulation	12. high levels result in decreased sensitivity to trastuzumab [64]
		miR-223, miR-125b	FBXW7 decrease/PPP1CA downregulation	13. high levels reduce the sensitivity to trastuzumab [67,71]
		miR-16	Akt downregulation via FUBP1 action	14. overexpression is predictors of trastuzumab sensitivity [66]
		miR-200c	EMT block by ZEB1 and ZEB2 targeting	15. downregulated in trastuzumab-resistant GC [70,71]
	GIST	miR-494	FGFR2 reduced expression	16. restores lapatinib sensitivity [75]

Table 1. *Cont.*

Targeted Therapies Class	Type of Cancer	MiRNAs Involved	Gene Mechanism Involved	Action
RTKs inhibitors	GIST	miR-125a-5p	PTPN18 regulation	17. modulates imatinib response [80]
		miR-320a, miR-518a-5p	Enhanced MCL1 expression via B-catenin/PIK3C2A upregulation	18. downregulated in imatinib-resistant GIST [80,83]
		miR-218	Inhibition of PI3K/AKT pathway	19. overexpression is able to restore the sensitivity to imatinib [82]
		miR-28-5p	NA	20. overexpressed in imatinib resistant GIST samples [87]
	HCC	miR-122, miR-34, let-7, miR-338-3p, miR-93, miR-193a/b, miR-486, miR-101	Downregulation of ADAM10/SRF/Bcl2/Bcl-xL/HIF-1α/CIT-RON/CLDN1/DUSP1	21. overexpression sensitizes HCC cells to sorafenib [89–92,94,96–98,100,102]
	HCC	miR-216a/217, miR-21, miR-221, miR-494	TGF-β and PI3K/AKT pathways activation/inhibition of Caspase 3/mTOR activation	22. overexpression causes chemoresistance against sorafenib [93,95,99,101]
	mCRC	miR-34a	WNT/β-catenin pathway downregulation	23. overexpression sensitizes CRC cells to regorafenib [107]
		miR-30a-5p	Dysregulation of STAT3-HSPA5 axis	24. downregulation is observed in regorafenib-resistant CRC [108]
Immune checkpoint inhibitors	GC	miR-200b, miR-152, miR-570		25. downregulation promotes PD-L1 expression [118]
	mCRC	miR-138-5p, miR-148a-3p		26. overexpression acts by downregulating PD-L1 [122,124]
		miR-200		27. downregulation promotes PD-L1 expression [113]

Abbreviation: mCRC, metastatic colorectal cancer; GC, gastric cancer; GIST, gastrointestinal stroma tumor; TTP, time to progress PFS, progression-free survival; HCC, hepatocellular carcinoma.

8. Conclusions

MicroRNAs are well-documented post-transcriptional regulators of gene-expression in physiological and pathological conditions. They are expressed in an organ- and tissue-specific manner and function through different molecular mechanisms. Given that miRNAs can modulate several oncogenic and tumor-suppressing pathways, increasing evidence points out their role as mediators of drug resistance and as therapeutic options. Because miRNAs can be detected rapidly and efficiently in tissues and biofluids, they are the ideal candidate biomarkers to identify patients with primary resistance to a specific targeted therapy and those who have developed acquired resistance. However, regardless of the numerous studies on the matter, only some miRNAs have been extensively proven to act as mediators of resistance to targeted therapies in GI malignancies, by using large patient cohorts for discovery and validation, and in vitro and in vivo modeling for confirmation. In the era of personalized medicine, more investigational studies are needed to translate the use of miRNA to monitor and forecast treatment response and resistance in a clinical setting.

Author Contributions: Conceptualization, V.A., F.G., F.L. and M.F.; methodology, V.A., G.B., L.D.S., V.G., L.N. and V.L.; data curation, V.A., F.G. and M.F.; writing—original draft preparation, V.A., F.G., G.B. and L.D.S.; writing—review and editing, E.S., S.R., S.L. and F.L.; supervision, M.F. All authors have read and agreed to the published version of the manuscript.

Funding: This research was partially funded by a grant from the Italian Health Ministry's research program NET-2016–02363853. The funding agency had no role in the design and performance of the study.

Institutional Review Board Statement: Not applicable.

Informed Consent Statement: Not applicable.

Data Availability Statement: Data available upon request.

Conflicts of Interest: The authors declare no conflict of interest regarding the present work.

References

1. Marquart, J.; Chen, E.Y.; Prasad, V. Estimation of the Percentage of US Patients with Cancer Who Benefit from Genome-Driven Oncology. *JAMA Oncol.* **2018**, *4*, 1093–1098. [CrossRef]
2. Redmond, K.L.; Papafili, A.; Lawler, M.; van Schaeybroeck, S. Overcoming Resistance to Targeted Therapies in Cancer. *Semin. Oncol.* **2015**, *42*, 896–908. [CrossRef] [PubMed]
3. Chen, C.; di Bartolomeo, M.; Corallo, S.; Strickler, J.H.; Goyal, L. Overcoming Resistance to Targeted Therapies in Gastrointestinal Cancers: Progress to Date and Progress to Come. *Am. Soc. Clin. Oncol. Educ. Book* **2020**, *40*, 161–173. [CrossRef] [PubMed]
4. Holohan, C.; van Schaeybroeck, S.; Longley, D.B.; Johnston, P.G. Cancer drug resistance: An evolving paradigm. *Nat. Rev. Cancer* **2013**, *13*, 714–726. [CrossRef] [PubMed]
5. Vasan, N.; Baselga, J.; Hyman, D.M. A view on drug resistance in cancer. *Nature* **2019**, *575*, 299–309. [CrossRef] [PubMed]
6. He, L.; Hannon, G.J. MicroRNAs: Small RNAs with a big role in gene regulation. *Nat. Rev. Genet.* **2004**, *5*, 522–531. [CrossRef]
7. Yates, L.A.; Norbury, C.J.; Gilbert, R.J.C. The long and short of microRNA. *Cell* **2013**, *153*, 516–519. [CrossRef]
8. Drusco, A.; Croce, C.M. MicroRNAs and Cancer: A Long Story for Short RNAs. In *Advances in Cancer Research*, 1st ed.; Elsevier Inc.: Amsterdam, The Netherlands, 2017; Volume 135.
9. Peng, Y.; Croce, C.M. The role of microRNAs in human cancer. *Signal. Transduct. Target. Ther.* **2016**, *1*, 1–9. [CrossRef]
10. Hahne, J.C.; Valeri, N. Non-Coding RNAs and resistance to anticancer drugs in gastrointestinal tumors. *Front. Oncol.* **2018**, *8*. [CrossRef]
11. Gomes, B.C.; Rueff, J.; Rodrigues, A.S. MicroRNAs and cancer drug resistance. *Cancer Drug Resist.* **2016**, *1395*, 137–162. [CrossRef]
12. Lampis, A.; Hahne, J.C.; Hedayat, S.; Valeri, N. MicroRNAs as mediators of drug resistance mechanisms. *Curr. Opin. Pharmacol.* **2020**, *54*, 44–50. [CrossRef]
13. Condrat, C.E.; Thompson, D.C.; Barbu, M.G.; Bugnar, O.L.; Boboc, A.; Cretoiu, D.; Suciu, N.; Cretoiu, S.M.; Voinea, S.C. miRNAs as Biomarkers in Disease: Latest Findings Regarding Their Role in Diagnosis and Prognosis. *Cells* **2020**, *9*, 276. [CrossRef] [PubMed]
14. Perdas, E.; Stawski, R.; Nowak, D.; Zubrzycka, M. Potential of Liquid Biopsy in Papillary Thyroid Carcinoma in Context of miRNA, BRAF and p53 Mutation. *Curr. Drug Targets* **2018**, *19*, 1721–1729. [CrossRef] [PubMed]
15. Giannopoulou, L.; Zavridou, M.; Kasimir-Bauer, S.; Lianidou, E.S. Liquid biopsy in ovarian cancer: The potential of circulating miRNAs and exosomes. *Transl. Res.* **2019**, *205*, 77–91. [CrossRef] [PubMed]

16. Gasiulė, S.; Dreize, N.; Kaupinis, A.; Ražanskas, R.; Čiupas, L.; Stankevičius, V.; Kapustina, Z.; Laurinavičius, A.; Valius, M.; Vilkaitis, G. Molecular Insights into miRNA-Driven Resistance to 5-Fluorouracil and Oxaliplatin Chemotherapy: MiR-23b Modulates the Epithelial–Mesenchymal Transition of Colorectal Cancer Cells. *J. Clin. Med.* **2019**, *8*, 2115. [CrossRef] [PubMed]
17. Lee, S.D.; Yu, D.; Lee, D.Y.; Shin, H.-S.; Jo, J.-H.; Lee, Y.C. Upregulated microRNA-193a-3p is responsible for cisplatin resistance in CD44(+) gastric cancer cells. *Cancer Sci.* **2019**, *110*, 662–673. [CrossRef] [PubMed]
18. Daoud, A.Z.; Mulholland, E.J.; Cole, G.; McCarthy, H.O. MicroRNAs in Pancreatic Cancer: Biomarkers, prognostic, and therapeutic modulators. *BMC Cancer* **2019**, *19*, 1130. [CrossRef]
19. Metheetrairut, C.; Slack, F.J. MicroRNAs in the ionizing radiation response and in radiotherapy. *Curr. Opin. Genet. Dev.* **2013**, *23*, 12–19. [CrossRef]
20. Bajan, S.; Hutvagner, G. RNA-Based Therapeutics: From Antisense Oligonucleotides to miRNAs. *Cells* **2020**, *9*, 137. [CrossRef] [PubMed]
21. Seshacharyulu, P.; Ponnusamy, M.P.; Haridas, D.; Jain, M.; Ganti, A.K.; Batra, S.K. Targeting the EGFR signaling pathway in cancer therapy. *Expert Opin. Ther. Targets* **2012**, *16*, 15–31. [CrossRef]
22. García-Foncillas, J.; Sunakawa, Y.; Aderka, D.; Wainberg, Z.; Ronga, P.; Witzler, P.; Stintzing, S. Distinguishing Features of Cetuximab and Panitumumab in Colorectal Cancer and Other Solid Tumors. *Front. Oncol.* **2019**, *9*, 849. [CrossRef] [PubMed]
23. Sunada, H.; Magun, B.E.; Mendelsohn, J.; MacLeod, C.L. Monoclonal antibody against epidermal growth factor receptor is internalized without stimulating receptor phosphorylation. *Proc. Natl. Acad. Sci. USA* **1986**, *83*, 3825–3829. [CrossRef] [PubMed]
24. Kurai, J.; Chikumi, H.; Hashimoto, K.; Yamaguchi, K.; Yamasaki, A.; Sako, T.; Touge, H.; Makino, H.; Takata, M.; Miyata, M.; et al. Antibody-dependent cellular cytotoxicity mediated by cetuximab against lung cancer cell lines. *Clin. Cancer Res.* **2007**, *13*, 1552–1561. [CrossRef] [PubMed]
25. Cunningham, D.; Humblet, Y.; Siena, S.; Khayat, D.; Bleiberg, H.; Santoro, A.; Bets, D.; Mueser, M.; Harstrick, A.; Verslype, C.; et al. Cetuximab monotherapy and cetuximab plus irinotecan in irinotecan-refractory metastatic colorectal cancer. *N. Engl. J. Med.* **2004**, *351*, 337–345. [CrossRef] [PubMed]
26. Cremolini, C.; Antoniotti, C.; Lonardi, S.; Aprile, G.; Bergamo, F.; Masi, G.; Grande, R.; Tonini, G.; Mescoli, C.; Cardellino, G.G.; et al. Activity and Safety of Cetuximab Plus Modified FOLFOXIRI Followed by Maintenance with Cetuximab or Bevacizumab for RAS and BRAF Wild-type Metastatic Colorectal Cancer: A Randomized Phase 2 Clinical Trial. *JAMA Oncol.* **2018**, *4*, 529–536. [CrossRef] [PubMed]
27. Misale, S.; Di Nicolantonio, F.; Sartore-Bianchi, A.; Siena, S.; Bardelli, A. Resistance to anti-EGFR therapy in colorectal cancer: From heterogeneity to convergent evolution. *Cancer Discov.* **2014**, *4*, 1269–1280. [CrossRef] [PubMed]
28. Zhao, B.; Wang, L.; Qiu, H.; Zhang, M.; Sun, L.; Peng, P.; Yu, Q.; Yuan, X. Mechanisms of resistance to anti-EGFR therapy in colorectal cancer. *Oncotarget* **2017**, *8*, 3980–4000. [CrossRef]
29. Karapetis, C.S.; Khambata-Ford, S.; Jonker, D.J.; O'Callaghan, C.J.; Tu, D.; Tebbutt, N.C.; Simes, R.J.; Chalchal, H.; Shapiro, J.D.; Robitaille, S.; et al. K-ras mutations and benefit from cetuximab in advanced colorectal cancer. *N. Engl. J. Med.* **2008**, *359*, 1757–1765. [CrossRef]
30. Johnson, S.M.; Grosshans, H.; Shingara, J.; Byrom, M.; Jarvis, R.; Cheng, A.; Labourier, E.; Reinert, K.L.; Brown, D.; Slack, F.J. RAS is regulated by the let-7 microRNA family. *Cell* **2005**, *120*, 635–647. [CrossRef]
31. Ruzzo, A.; Graziano, F.; Vincenzi, B.; Canestrari, E.; Perrone, G.; Galluccio, N.; Catalano, V.; Loupakis, F.; Rabitti, C.; Santini, D.; et al. High Let-7a MicroRNA Levels in KRAS -Mutated Colorectal Carcinomas May Rescue Anti-EGFR Therapy Effects in Patients with Chemotherapy-Refractory Metastatic Disease. *Oncologist* **2012**, *17*, 823–829. [CrossRef]
32. Cappuzzo, F.; Sacconi, A.; Landi, L.; Ludovini, V.; Biagioni, F.; D'Incecco, A.; Capodanno, A.; Salvini, J.; Corgna, E.; Cupini, S.; et al. MicroRNA signature in metastatic colorectal cancer patients treated with anti-EGFR monoclonal antibodies. *Clin. Colorectal Cancer* **2014**, *13*, 37–45. [CrossRef] [PubMed]
33. Mlcochova, J.; Faltejskova-Vychytilova, P.; Ferracin, M.; Zagatti, B.; Radova, L.; Svoboda, M.; Nemecek, R.; John, S.; Kiss, I.; Vyzula, R.; et al. MicroRNA expression profiling identifies miR-31-5p/3p as associated with time to progression in wild-type RAS metastatic colorectal cancer treated with cetuximab. *Oncotarget* **2015**, *6*, 38695–38704. [CrossRef] [PubMed]
34. Igarashi, H.; Kurihara, H.; Mitsuhashi, K.; Ito, M.; Okuda, H.; Kanno, S.; Naito, T.; Yoshii, S.; Takahashi, H.; Kusumi, T.; et al. Association of MicroRNA-31-5p with Clinical Efficacy of Anti-EGFR Therapy in Patients with Metastatic Colorectal Cancer. *Ann. Surg. Oncol.* **2015**, *22*, 2640–2648. [CrossRef] [PubMed]
35. Manceau, G.; Imbeaud, S.; Thiébaut, R.; Liébaert, F.; Fontaine, K.; Rousseau, F.; Génin, B.; Le Corre, D.; Didelot, A.; Vincent, M.; et al. Hsa-miR-31-3p Expression Is Linked to Progression-free Survival in Patients with KRAS Wild-type Metastatic Colorectal Cancer Treated with Anti-EGFR Therapy. *Clin. Cancer Res.* **2014**, *20*, 3338–3347. [CrossRef] [PubMed]
36. Lu, Y.; Zhao, X.; Liu, Q.; Li, C.; Graves-Deal, R.; Cao, Z.; Singh, B.; Franklin, J.L.; Wang, J.; Hu, H.; et al. LncRNA MIR100HG-derived miR-100 and miR-125b mediate cetuximab resistance via Wnt/β-catenin signaling. *Nat. Med.* **2017**, *23*, 1331–1341. [CrossRef]
37. Pichler, M.; Winter, E.; Ress, A.L.; Bauernhofer, T.; Gerger, A.; Kiesslich, T.; Lax, S.; Samonigg, H.; Hoefler, G. miR-181a is associated with poor clinical outcome in patients with colorectal cancer treated with EGFR inhibitor. *J. Clin. Pathol.* **2014**, *67*, 198–203. [CrossRef] [PubMed]

38. Schou, J.V.; Rossi, S.; Jensen, B.V.; Nielsen, D.L.; Pfeiffer, P.; Høgdall, E.; Yilmaz, M.; Tejpar, S.; Delorenzi, M.; Kruhøffer, M.; et al. miR-345 in metastatic colorectal cancer: A non-invasive biomarker for clinical outcome in Non-KRAS mutant patients treated with 3rd line cetuximab and irinotecan. *PLoS ONE* **2014**, *9*. [CrossRef]
39. Mussnich, P.; Rosa, R.; Bianco, R.; Fusco, A.; D'Angelo, D. MiR-199a-5p and miR-375 affect colon cancer cell sensitivity to cetuximab by targeting PHLPP1. *Expert Opin. Ther. Targets* **2015**, *19*, 1017–1026. [CrossRef]
40. Sun, L.; Fang, Y.; Wang, X.; Han, Y.; Du, F.; Li, C.; Hu, H.; Liu, H.; Liu, Q.; Wang, J.; et al. MiR-302a inhibits metastasis and cetuximab resistance in colorectal cancer by targeting NFIB and CD44. *Theranostics* **2019**, *9*, 8409–8425. [CrossRef]
41. Xing, Y.; Jing, H.; Zhang, Y.; Suo, J.; Qian, M. MicroRNA-141-3p affected proliferation, chemosensitivity, migration and invasion of colorectal cancer cells by targeting EGFR. *Int. J. Biochem. Cell Biol.* **2020**, *118*, 105643. [CrossRef]
42. Ellis, L.M.; Hicklin, D.J. VEGF-targeted therapy: Mechanisms of anti-tumour activity. *Nat. Rev. Cancer* **2008**, *8*, 579–591. [CrossRef]
43. Hurwitz, H.; Fehrenbacher, L.; Novotny, W.; Cartwright, T.; Hainsworth, J.; Heim, W.; Berlin, J.; Baron, A.; Griffing, S.; Holmgren, E.; et al. Bevacizumab plus irinotecan, fluorouracil, and leucovorin for metastatic colorectal cancer. *N. Engl. J. Med.* **2004**, *350*, 2335–2342. [CrossRef] [PubMed]
44. Giantonio, B.J.; Catalano, P.J.; Meropol, N.J.; O'Dwyer, P.J.; Mitchell, E.P.; Alberts, S.R.; Schwartz, M.A.; Benson, A.B., 3rd; Eastern Cooperative Oncology Group Study E3200. Bevacizumab in combination with oxaliplatin, fluorouracil, and leucovorin (FOLFOX4) for previously treated metastatic colorectal cancer: Results from the Eastern Cooperative Oncology Group Study E3200. *J. Clin. Oncol. Off. J. Am. Soc. Clin. Oncol.* **2007**, *25*, 1539–1544. [CrossRef] [PubMed]
45. Bennouna, J.; Sastre, J.; Arnold, D.; Österlund, P.; Greil, R.; Van Cutsem, E.; von Moos, R.; Viéitez, J.M.; Bouché, O.; Borg, C.; et al. Continuation of bevacizumab after first progression in metastatic colorectal cancer (ML18147): A randomised phase 3 trial. *Lancet Oncol.* **2013**, *14*, 29–37. [CrossRef]
46. Syed, Y.Y.; McKeage, K. Aflibercept: A Review in Metastatic Colorectal Cancer. *Drugs* **2015**, *75*, 1435–1445. [CrossRef]
47. Wilke, H.; Muro, K.; Van Cutsem, E.; Oh, S.C.; Bodoky, G.; Shimada, Y.; Hironaka, S.; Sugimoto, N.; Lipatov, O.; Kim, T.Y.; et al. Ramucirumab plus paclitaxel versus placebo plus paclitaxel in patients with previously treated advanced gastric or gastro-oesophageal junction adenocarcinoma (RAINBOW): A double-blind, randomised phase 3 trial. *Lancet Oncol.* **2014**, *15*, 1224–1235. [CrossRef]
48. Fuchs, C.S.; Shitara, K.; Di Bartolomeo, M.; Lonardi, S.; Al-Batran, S.E.; Van Cutsem, E.; Ilson, D.H.; Alsina, M.; Chau, I.; Lacy, J.; et al. Ramucirumab with cisplatin and fluoropyrimidine as first-line therapy in patients with metastatic gastric or junctional adenocarcinoma (RAINFALL): A double-blind, randomised, placebo-controlled, phase 3 trial. *Lancet Oncol.* **2019**, *20*, 420–435. [CrossRef]
49. Bergers, G.; Hanahan, D. Modes of resistance to anti-angiogenic therapy. *Nat. Rev. Cancer* **2008**, *8*, 592–603. [CrossRef]
50. Wang, S.; Aurora, A.B.; Johnson, B.A.; Qi, X.; McAnally, J.; Hill, J.A.; Richardson, J.A.; Bassel-Duby, R.; Olson, E.N. The endothelial-specific microRNA miR-126 governs vascular integrity and angiogenesis. *Dev. Cell* **2008**, *15*, 261–271. [CrossRef]
51. Boisen, M.K.; Dehlendorff, C.; Linnemann, D.; Nielsen, B.S.; Larsen, J.S.; Osterlind, K.; Nielsen, S.E.; Tarpgaard, L.S.; Qvortrup, C.; Pfeiffer, P.; et al. Tissue microRNAs as predictors of outcome in patients with metastatic colorectal cancer treated with first line capecitabine and oxaliplatin with or without bevacizumab. *PLoS ONE* **2014**, 9. [CrossRef]
52. Ulivi, P.; Canale, M.; Passardi, A.; Marisi, G.; Valgiusti, M.; Frassineti, G.L.; Calistri, D.; Amadori, D.; Scarpi, E. Circulating plasma levels of miR-20b, miR-29b and mir-155 as predictors of bevacizumab efficacy in patients with metastatic colorectal cancer. *Int. J. Mol. Sci.* **2018**, *19*, 307. [CrossRef]
53. Li, Y.; Cai, B.; Shen, L.; Dong, Y.; Lu, Q.; Sun, S.; Liu, S.; Ma, S.; Ma, P.X.; Chen, J. MiRNA-29b suppresses tumor growth through simultaneously inhibiting angiogenesis and tumorigenesis by targeting Akt3. *Cancer Lett.* **2017**, *397*, 111–119. [CrossRef] [PubMed]
54. Wong, P.F.; Jamal, J.; Tong, K.L.; Khor, E.S.; Yeap, C.E.; Jong, H.L.; Lee, S.T.; Mustafa, M.R.; Abubakar, S. Deregulation of hsa-miR-20b expression in TNF-α-induced premature senescence of human pulmonary microvascular endothelial cells. *Microvasc. Res.* **2017**, *114*, 26–33. [CrossRef] [PubMed]
55. Yang, D.; Wang, J.; Xiao, M.; Zhou, T.; Shi, X. Role of Mir-155 in Controlling HIF-1α Level and Promoting Endothelial Cell Maturation. *Sci. Rep.* **2016**, *6*, 35316. [CrossRef] [PubMed]
56. Rubin, I.; Yarden, Y. The basic biology of HER2. *Ann. Oncol.* **2001**, *12*, S3–S8. [CrossRef] [PubMed]
57. Vakiani, E. HER2 testing in gastric and gastroesophageal adenocarcinomas. *Adv. Anat. Pathol.* **2015**, *22*, 194–201. [CrossRef] [PubMed]
58. Cappellesso, R.; Fassan, M.; Hanspeter, E.; Bornschein, J.; d'Amore, E.S.; Cuorvo, L.V.; Mazzoleni, G.; Barbareschi, M.; Pizzi, M.; Guzzardo, V.; et al. HER2 status in gastroesophageal cancer: A tissue microarray study of 1040 cases. *Hum. Pathol.* **2015**, *46*, 665–672. [CrossRef] [PubMed]
59. Rüschoff, J.; Hanna, W.; Bilous, M.; Hofmann, M.; Osamura, R.Y.; Penault-Llorca, F.; van de Vijver, M.; Viale, G. HER2 testing in gastric cancer: A practical approach. *Mod. Pathol.* **2012**, *25*, 637–650. [CrossRef] [PubMed]
60. Van Cutsem, E.; Bang, Y.J.; Feng-Yi, F.; Xu, J.M.; Lee, K.W.; Jiao, S.C.; Chong, J.L.; López-Sanchez, R.I.; Price, T.; Gladkov, O.; et al. HER2 screening data from ToGA: Targeting HER2 in gastric and gastroesophageal junction cancer. *Gastric Cancer.* **2015**, *18*, 476–484. [CrossRef] [PubMed]
61. Nahta, R.; Esteva, F.J. Trastuzumab: Triumphs and tribulations. *Oncogene* **2007**, *26*, 3637–3643. [CrossRef]

62. Bang, Y.J.; Van Cutsem, E.; Feyereislova, A.; Chung, H.C.; Shen, L.; Sawaki, A.; Lordick, F.; Ohtsu, A.; Omuro, Y.; Satoh, T.; et al. Trastuzumab in combination with chemotherapy versus chemotherapy alone for treatment of HER2-positive advanced gastric or gastro-oesophageal junction cancer (ToGA): A phase 3, open-label, randomised controlled trial. *Lancet* **2010**, *376*, 687–697. [CrossRef]
63. Shimoyama, S. Unraveling trastuzumab and lapatinib inefficiency in gastric cancer: Future steps (Review). *Mol. Clin. Oncol.* **2014**, *2*, 175–181. [CrossRef]
64. Nagata, Y.; Lan, K.H.; Zhou, X.; Tan, M.; Esteva, F.J.; Sahin, A.A.; Klos, K.S.; Li, P.; Monia, B.P.; Nguyen, N.T.; et al. PTEN activation contributes to tumor inhibition by trastuzumab, and loss of PTEN predicts trastuzumab resistance in patients. *Cancer Cell* **2004**, *6*, 117–127. [CrossRef] [PubMed]
65. Eto, K.; Iwatsuki, M.; Watanabe, M.; Ida, S.; Ishimoto, T.; Iwagami, S.; Baba, Y.; Sakamoto, Y.; Miyamoto, Y.; Yoshida, N.; et al. The microRNA-21/PTEN pathway regulates the sensitivity of HER2-positive gastric cancer cells to trastuzumab. *Ann. Surg. Oncol.* **2014**, *21*, 343–350. [CrossRef] [PubMed]
66. Nakayama, K.I.; Nakayama, K. Ubiquitin ligases: Cell-cycle control and cancer. *Nat. Rev. Cancer* **2006**, *6*, 369–381. [CrossRef]
67. Eto, K.; Iwatsuki, M.; Watanabe, M.; Ishimoto, T.; Ida, S.; Imamura, Y.; Iwagami, S.; Baba, Y.; Sakamoto, Y.; Miyamoto, Y.; et al. The sensitivity of gastric cancer to trastuzumab is regulated by the miR-223/FBXW7 pathway. *Int. J. Cancer* **2015**, *136*, 1537–1545. [CrossRef]
68. Peggs, K.S.; Segal, N.H.; Allison, J.P. Targeting immunosupportive cancer therapies: Accentuate the positive, eliminate the negative. *Cancer Cell* **2007**, *12*, 192–199. [CrossRef] [PubMed]
69. Venturutti, L.; Cordo Russo, R.I.; Rivas, M.A.; Mercogliano, M.F.; Izzo, F.; Oakley, R.H.; Pereyra, M.G.; De Martino, M.; Proietti, C.J.; Yankilevich, P.; et al. MiR-16 mediates trastuzumab and lapatinib response in ErbB-2-positive breast and gastric cancer via its novel targets CCNJ and FUBP1. *Oncogene* **2016**, *35*, 6189–6202. [CrossRef] [PubMed]
70. Sui, M.; Jiao, A.; Zhai, H.; Wang, Y.; Wang, Y.; Sun, D.; Li, P. Upregulation of miR-125b is associated with poor prognosis and trastuzumab resistance in HER2-positive gastric cancer. *Exp. Ther. Med.* **2017**, *14*, 657–663. [CrossRef]
71. Wu, J.G.; Wang, J.J.; Jiang, X.; Lan, J.P.; He, X.J.; Wang, H.J.; Ma, Y.Y.; Xia, Y.J.; Ru, G.Q.; Ma, J.; et al. MiR-125b promotes cell migration and invasion by targeting PPP1CA-Rb signal pathways in gastric cancer, resulting in a poor prognosis. *Gastric Cancer* **2015**, *18*, 729–739. [CrossRef]
72. Voon, D.C.; Huang, R.Y.; Jackson, R.A.; Thiery, J.P. The EMT spectrum and therapeutic opportunities. *Mol. Oncol.* **2017**, *11*, 878–891. [CrossRef] [PubMed]
73. Zhou, X.; Men, X.; Zhao, R.; Han, J.; Fan, Z.; Wang, Y.; Lv, Y.; Zuo, J.; Zhao, L.; Sang, M.; et al. MiR-200c inhibits TGF-β-induced-EMT to restore trastuzumab sensitivity by targeting ZEB1 and ZEB2 in gastric cancer. *Cancer Gene Ther.* **2018**, *25*, 68–76. [CrossRef] [PubMed]
74. Hecht, J.R.; Bang, Y.J.; Qin, S.K.; Chung, H.C.; Xu, J.M.; Park, J.O.; Jeziorski, K.; Shparyk, Y.; Hoff, P.M.; Sobrero, A.; et al. Lapatinib in Combination with Capecitabine Plus Oxaliplatin in Human Epidermal Growth Factor Receptor 2-Positive Advanced or Metastatic Gastric, Esophageal, or Gastroesophageal Adenocarcinoma: TRIO-013/LOGiC–A Randomized Phase III Trial. *J. Clin. Oncol.* **2016**, *34*, 443–451. [CrossRef] [PubMed]
75. Yu, Y.; Yu, X.; Liu, H.; Song, Q.; Yang, Y. MiR-494 inhibits cancer-initiating cell phenotypes and reverses resistance to lapatinib by downregulating FGFR2 in HER2-positive gastric cancer. *Int. J. Mol. Med.* **2018**, *42*, 998–1007. [CrossRef]
76. Lemmon, M.A.; Schlessinger, J. Cell signaling by receptor tyrosine kinases. *Cell* **2010**, *141*, 1117–1134. [CrossRef]
77. Hojjat-Farsangi, M. Small-molecule inhibitors of the receptor tyrosine kinases: Promising tools for targeted cancer therapies. *Int. J. Mol. Sci.* **2014**, *15*, 13768–13801. [CrossRef]
78. Demetri, G.D.; von Mehren, M.; Blanke, C.D.; Van den Abbeele, A.D.; Eisenberg, B.; Roberts, P.J.; Heinrich, M.C.; Tuveson, D.A.; Singer, S.; Janicek, M.; et al. Efficacy and safety of imatinib mesylate in advanced gastrointestinal stromal tumors. *N. Engl. J. Med.* **2002**, *347*, 472–480. [CrossRef]
79. Antonescu, C.R.; Besmer, P.; Guo, T.; Arkun, K.; Hom, G.; Koryotowski, B.; Leversha, M.A.; Jeffrey, P.D.; Desantis, D.; Singer, S.; et al. Acquired resistance to imatinib in gastrointestinal stromal tumor occurs through secondary gene mutation. *Clin. Cancer Res.* **2005**, *11*, 4182–4190. [CrossRef]
80. Akçakaya, P.; Caramuta, S.; Åhlen, J.; Ghaderi, M.; Berglund, E.; Östman, A.; Bränström, R.; Larsson, C.; Lui, W.O. microRNA expression signatures of gastrointestinal stromal tumours: Associations with imatinib resistance and patient outcome. *Br. J. Cancer* **2014**, *111*, 2091–2102. [CrossRef]
81. Gao, X.; Shen, K.; Wang, C.; Ling, J.; Wang, H.; Fang, Y.; Shi, Y.; Hou, Y.; Qin, J.; Sun, Y.; et al. MiR-320a downregulation is associated with imatinib resistance in gastrointestinal stromal tumors. *Acta Biochim. Biophys. Sin.* **2014**, *46*, 72–75. [CrossRef]
82. Fan, R.; Zhong, J.; Zheng, S.; Wang, Z.; Xu, Y.; Li, S.; Zhou, J.; Yuan, F. microRNA-218 increase the sensitivity of gastrointestinal stromal tumor to imatinib through PI3K/AKT pathway. *Clin. Exp. Med.* **2015**, *15*, 137–144. [CrossRef]
83. Shi, Y.; Gao, X.; Hu, Q.; Li, X.; Xu, J.; Lu, S.; Liu, Y.; Xu, C.; Jiang, D.; Lin, J.; et al. PIK3C2A is a gene-specific target of microRNA-518a-5p in imatinib mesylate-resistant gastrointestinal stromal tumor. *Lab. Investig.* **2016**, *96*, 652–660. [CrossRef] [PubMed]
84. Oneyama, C.; Ikeda, J.; Okuzaki, D.; Suzuki, K.; Kanou, T.; Shintani, Y.; Morii, E.; Okumura, M.; Aozasa, K.; Okada, M. MicroRNA-mediated downregulation of mTOR/FGFR3 controls tumor growth induced by Src-related oncogenic pathways. *Oncogene* **2011**, *30*, 3489–3501. [CrossRef] [PubMed]

85. Zhang, X.; He, X.; Liu, Y.; Zhang, H.; Chen, H.; Guo, S.; Liang, Y. MiR-101-3p inhibits the growth and metastasis of non-small cell lung cancer through blocking PI3K/AKT signal pathway by targeting MALAT-1. *Biomed. Pharmacother.* **2017**, *93*, 1065–1073. [CrossRef] [PubMed]
86. Amirnasr, A.; Gits, C.M.M.; van Kuijk, P.F.; Smid, M.; Vriends, A.L.M.; Rutkowski, P.; Sciot, R.; Schöffski, P.; Debiec-Rychter, M.; Sleijfer, S.; et al. Molecular Comparison of Imatinib-Naïve and Resistant Gastrointestinal Stromal Tumors: Differentially Expressed microRNAs and mRNAs. *Cancers* **2019**, *11*, 882. [CrossRef]
87. Zhang, Z.; Jiang, N.Y.; Guan, R.Y.; Zhu, Y.K.; Jiang, F.Q.; Piao, D. Identification of critical microRNAs in gastrointestinal stromal tumor patients treated with Imatinib. *Neoplasma* **2018**, *65*, 683–692. [CrossRef]
88. Tang, W.; Chen, Z.; Zhang, W.; Cheng, Y.; Zhang, B.; Wu, F.; Wang, Q.; Wang, S.; Rong, D.; Reiter, F.P.; et al. The mechanisms of sorafenib resistance in hepatocellular carcinoma: Theoretical basis and therapeutic aspects. *Signal Transduct. Target. Ther.* **2020**, *5*, 87. [CrossRef]
89. Bai, S.; Nasser, M.W.; Wang, B.; Hsu, S.H.; Datta, J.; Kutay, H.; Yadav, A.; Nuovo, G.; Kumar, P.; Ghoshal, K. MicroRNA-122 inhibits tumorigenic properties of hepatocellular carcinoma cells and sensitizes these cells to sorafenib. *J. Biol. Chem.* **2009**, *284*, 32015–32027. [CrossRef]
90. Yang, F.; Li, Q.J.; Gong, Z.B.; Zhou, L.; You, N.; Wang, S.; Li, X.L.; Li, J.J.; An, J.Z.; Wang, D.S.; et al. MicroRNA-34a targets Bcl-2 and sensitizes human hepatocellular carcinoma cells to sorafenib treatment. *Technol. Cancer Res. Treat.* **2014**, *13*, 77–86. [CrossRef]
91. Shimizu, S.; Takehara, T.; Hikita, H.; Kodama, T.; Miyagi, T.; Hosui, A.; Tatsumi, T.; Ishida, H.; Noda, T.; Nagano, H.; et al. The let-7 family of microRNAs inhibits Bcl-xL expression and potentiates sorafenib-induced apoptosis in human hepatocellular carcinoma. *J. Hepatol.* **2010**, *52*, 698–704. [CrossRef]
92. Xu, H.; Zhao, L.; Fang, Q.; Sun, J.; Zhang, S.; Zhan, C.; Liu, S.; Zhang, Y. MiR-338-3p inhibits hepatocarcinoma cells and sensitizes these cells to sorafenib by targeting hypoxia-induced factor 1α. *PLoS ONE* **2014**, *9*, e115565. [CrossRef]
93. Xia, H.; Ooi, L.L.; Hui, K.M. MicroRNA-216a/217-induced epithelial-mesenchymal transition targets PTEN and SMAD7 to promote drug resistance and recurrence of liver cancer. *Hepatology* **2013**, *58*, 629–641. [CrossRef] [PubMed]
94. Ohta, K.; Hoshino, H.; Wang, J.; Ono, S.; Iida, Y.; Hata, K.; Huang, S.K.; Colquhoun, S.; Hoon, D.S. MicroRNA-93 activates c-Met/PI3K/Akt pathway activity in hepatocellular carcinoma by directly inhibiting PTEN and CDKN1A. *Oncotarget* **2015**, *6*, 3211–3224. [CrossRef] [PubMed]
95. He, C.; Dong, X.; Zhai, B.; Jiang, X.; Dong, D.; Li, B.; Jiang, H.; Xu, S.; Sun, X. MiR-21 mediates sorafenib resistance of hepatocellular carcinoma cells by inhibiting autophagy via the PTEN/Akt pathway. *Oncotarget* **2015**, *6*, 28867–28881. [CrossRef] [PubMed]
96. Salvi, A.; Conde, I.; Abeni, E.; Arici, B.; Grossi, I.; Specchia, C.; Portolani, N.; Barlati, S.; De Petro, G. Effects of miR-193a and sorafenib on hepatocellular carcinoma cells. *Mol. Cancer* **2013**, *12*, 162. [CrossRef]
97. Mao, K.; Zhang, J.; He, C.; Xu, K.; Liu, J.; Sun, J.; Wu, G.; Tan, C.; Zeng, Y.; Wang, J.; et al. Restoration of miR-193b sensitizes Hepatitis B virus-associated hepatocellular carcinoma to sorafenib. *Cancer Lett.* **2014**, *352*, 245–252. [CrossRef]
98. Braconi, C.; Valeri, N.; Gasparini, P.; Huang, N.; Taccioli, C.; Nuovo, G.; Suzuki, T.; Croce, C.M.; Patel, T. Hepatitis C virus proteins modulate microRNA expression and chemosensitivity in malignant hepatocytes. *Clin. Cancer Res.* **2010**, *16*, 957–966. [CrossRef]
99. Fornari, F.; Pollutri, D.; Patrizi, C.; La Bella, T.; Marinelli, S.; Casadei Gardini, A.; Marisi, G.; Baron Toaldo, M.; Baglioni, M.; Salvatore, V.; et al. In Hepatocellular Carcinoma miR-221 Modulates Sorafenib Resistance through Inhibition of Caspase-3-Mediated Apoptosis. *Clin. Cancer Res.* **2017**, *23*, 3953–3965. [CrossRef]
100. Sun, H.; Cui, C.; Xiao, F.; Wang, H.; Xu, J.; Shi, X.; Yang, Y.; Zhang, Q.; Zheng, X.; Yang, X.; et al. miR-486 regulates metastasis and chemosensitivity in hepatocellular carcinoma by targeting CLDN10 and CITRON. *Hepatol. Res.* **2015**, *45*, 1312–1322. [CrossRef]
101. Pollutri, D.; Patrizi, C.; Marinelli, S.; Giovannini, C.; Trombetta, E.; Giannone, F.A.; Baldassarre, M.; Quarta, S.; Vandewynckel, Y.P.; Vandierendonck, A.; et al. The epigenetically regulated miR-494 associates with stem-cell phenotype and induces sorafenib resistance in hepatocellular carcinoma. *Cell Death Dis.* **2018**, 9. [CrossRef]
102. Wei, X.; Tang, C.; Lu, X.; Liu, R.; Zhou, M.; He, D.; Zheng, D.; Sun, C.; Wu, Z. MiR-101 targets DUSP1 to regulate the TGF-β secretion in sorafenib inhibits macrophage-induced growth of hepatocarcinoma. *Oncotarget* **2015**, *6*, 18389–18405. [CrossRef] [PubMed]
103. Grothey, A.; Van Cutsem, E.; Sobrero, A.; Siena, S.; Falcone, A.; Ychou, M.; Humblet, Y.; Bouché, O.; Mineur, L.; Barone, C.; et al. Regorafenib monotherapy for previously treated metastatic colorectal cancer (CORRECT): An international, multicentre, randomised, placebo-controlled, phase 3 trial. *Lancet* **2013**, *381*, 303–312. [CrossRef]
104. Bruix, J.; Qin, S.; Merle, P.; Granito, A.; Huang, Y.H.; Bodoky, G.; Pracht, M.; Yokosuka, O.; Rosmorduc, O.; Breder, V.; et al. Regorafenib for patients with hepatocellular carcinoma who progressed on sorafenib treatment (RESORCE): A randomised, double-blind, placebo-controlled, phase 3 trial. *Lancet* **2017**, *389*, 56–66. [CrossRef]
105. Demetri, G.D.; Reichardt, P.; Kang, Y.K.; Blay, J.Y.; Rutkowski, P.; Gelderblom, H.; Hohenberger, P.; Leahy, M.; von Mehren, M.; Joensuu, H.; et al. Efficacy and safety of regorafenib for advanced gastrointestinal stromal tumours after failure of imatinib and sunitinib (GRID): An international, multicentre, randomised, placebo-controlled, phase 3 trial. *Lancet* **2013**, *381*, 295–302. [CrossRef]
106. Schirripa, M.; Borelli, B.; D'Aurizio, R.; Lubrano, S.; Cremolini, C.; Zucchelli, G.; Antoniotti, C.; Marmorino, F.; Prete, A.A.; Murgioni, S.; et al. Early modifications of circulating microRNAs levels in metastatic colorectal cancer patients treated with regorafenib. *Pharm. J.* **2019**, *19*, 455–464. [CrossRef] [PubMed]

107. Cai, M.H.; Xu, X.G.; Yan, S.L.; Sun, Z.; Ying, Y.; Wang, B.K.; Tu, Y.X. Regorafenib suppresses colon tumorigenesis and the generation of drug resistant cancer stem-like cells via modulation of miR-34a associated signaling. *J. Exp. Clin. Cancer Res.* **2018**, *37*, 151. [CrossRef]
108. Cheng, C.C.; Yang, B.L.; Chen, W.C.; Ho, A.S.; Sie, Z.L.; Lin, H.C.; Chang, C.C. STAT3 Mediated miR-30a-5p Inhibition Enhances Proliferation and Inhibits Apoptosis in Colorectal Cancer Cells. *Int. J. Mol. Sci.* **2020**, *21*, 7315. [CrossRef]
109. Teufel, M.; Seidel, H.; Köchert, K.; Meinhardt, G.; Finn, R.S.; Llovet, J.M.; Bruix, J. Biomarkers Associated with Response to Regorafenib in Patients with Hepatocellular Carcinoma. *Gastroenterology* **2019**, *156*, 1731–1741. [CrossRef]
110. Darvin, P.; Toor, S.M.; Sasidharan Nair, V.; Elkord, E. Immune checkpoint inhibitors: Recent progress and potential biomarkers. *Exp. Mol. Med.* **2018**, *50*, 1–11. [CrossRef] [PubMed]
111. Abdul-Latif, M.; Townsend, K.; Dearman, C.; Shiu, K.K.; Khan, K. Immunotherapy in gastrointestinal cancer: The current scenario and future perspectives. *Cancer Treat. Rev.* **2020**, *88*, 102030. [CrossRef] [PubMed]
112. André, T.; Shiu, K.K.; Kim, T.W.; Jensen, B.V.; Jensen, L.H.; Punt, C.; Smith, D.; Garcia-Carbonero, R.; Benavides, M.; Gibbs, P.; et al. Pembrolizumab in Microsatellite-Instability-High Advanced Colorectal Cancer. *N. Engl. J. Med.* **2020**, *383*, 2207–2218. [CrossRef] [PubMed]
113. Fuchs, C.S.; Doi, T.; Jang, R.W.; Muro, K.; Satoh, T.; Machado, M.; Sun, W.; Jalal, S.I.; Shah, M.A.; Metges, J.P.; et al. Safety and Efficacy of Pembrolizumab Monotherapy in Patients With Previously Treated Advanced Gastric and Gastroesophageal Junction Cancer: Phase 2 Clinical KEYNOTE-059 Trial. *JAMA Oncol.* **2018**, *4*, e180013. [CrossRef]
114. Leal, A.D.; Paludo, J.; Finnes, H.D.; Grothey, A. Response to pembrolizumab in patients with mismatch repair dficient (dMMR) colorectal cancer (CRC). *J. Clin. Oncol.* **2017**, *35*, 3558. [CrossRef]
115. Derks, S.; Liao, X.; Chiaravalli, A.M.; Xu, X.; Camargo, M.C.; Solcia, E.; Sessa, F.; Fleitas, T.; Freeman, G.J.; Rodig, S.J.; et al. Abundant PD-L1 expression in Epstein-Barr Virus-infected gastric cancers. *Oncotarget* **2016**, *7*, 32925–32932. [CrossRef]
116. Sangro, B.; Gomez-Martin, C.; de la Mata, M.; Iñarrairaegui, M.; Garralda, E.; Barrera, P.; Riezu-Boj, J.I.; Larrea, E.; Alfaro, C.; Sarobe, P.; et al. A clinical trial of CTLA-4 blockade with tremelimumab in patients with hepatocellular carcinoma and chronic hepatitis C. *J. Hepatol.* **2013**, *59*, 81–88. [CrossRef] [PubMed]
117. Hamid, O.; Robert, C.; Daud, A.; Hodi, F.S.; Hwu, W.J.; Kefford, R.; Wolchok, J.D.; Hersey, P.; Joseph, R.; Weber, J.S.; et al. Five-year survival outcomes for patients with advanced melanoma treated with pembrolizumab in KEYNOTE-001. *Ann. Oncol.* **2019**, *30*, 582–588. [CrossRef]
118. Wang, W.; Sun, J.; Li, F.; Li, R.; Gu, Y.; Liu, C.; Yang, P.; Zhu, M.; Chen, L.; Tian, W.; et al. A frequent somatic mutation in CD274 3′-UTR leads to protein over-expression in gastric cancer by disrupting miR-570 binding. *Hum. Mutat.* **2012**, *33*, 480–484. [CrossRef]
119. Trivedi, P.; Slack, F.J.; Anastasiadou, E. Epstein-Barr virus: From kisses to cancer, an ingenious immune evader. *Oncotarget* **2018**, *9*, 36411–36412. [CrossRef]
120. Anastasiadou, E.; Stroopinsky, D.; Alimperti, S.; Jiao, A.L.; Pyzer, A.R.; Cippitelli, C.; Pepe, G.; Severa, M.; Rosenblatt, J.; Etna, M.P.; et al. Epstein-Barr virus-encoded EBNA2 alters immune checkpoint PD-L1 expression by downregulating miR-34a in B-cell lymphomas. *Leukemia* **2019**, *33*, 132–147. [CrossRef]
121. Isosaka, M.; Niinuma, T.; Nojima, M.; Kai, M.; Yamamoto, E.; Maruyama, R.; Nobuoka, T.; Nishida, T.; Kanda, T.; Taguchi, T.; et al. A Screen for Epigenetically Silenced microRNA Genes in Gastrointestinal Stromal Tumors. *PLoS ONE* **2015**, *10*, e0133754. [CrossRef]
122. Zhao, L.; Yu, H.; Yi, S.; Peng, X.; Su, P.; Xiao, Z.; Liu, R.; Tang, A.; Li, X.; Liu, F.; et al. The tumor suppressor miR-138-5p targets PD-L1 in colorectal cancer. *Oncotarget* **2016**, *7*, 45370–45384. [CrossRef] [PubMed]
123. Martinez-Ciarpaglini, C.; Oltra, S.; Roselló, S.; Roda, D.; Mongort, C.; Carrasco, F.; Gonzalez, J.; Santonja, F.; Tarazona, N.; Huerta, M.; et al. Low miR200c expression in tumor budding of invasive front predicts worse survival in patients with localized colon cancer and is related to PD-L1 overexpression. *Mod. Pathol.* **2019**, *32*, 306–313. [CrossRef] [PubMed]
124. Ashizawa, M.; Okayama, H.; Ishigame, T.; Thar Min, A.K.; Saito, K.; Ujiie, D.; Murakami, Y.; Kikuchi, T.; Nakayama, Y.; Noda, M.; et al. miRNA-148a-3p Regulates Immunosuppression in DNA Mismatch Repair-Deficient Colorectal Cancer by Targeting PD-L1. *Mol. Cancer Res.* **2019**, *17*, 1403–1413. [CrossRef] [PubMed]

biomedicines

MDPI

Review

LAT1 and ASCT2 Related microRNAs as Potential New Therapeutic Agents against Colorectal Cancer Progression

Francisca Dias [1,†], Cristina Almeida [1,2,†], Ana Luísa Teixeira [1,*,†], Mariana Morais [1,3] and Rui Medeiros [1,2,3,4,5]

1 Molecular Oncology and Viral Pathology Group, IPO-Porto Research Center (CI-IPOP), Portuguese Oncology Institute of Porto (IPO-Porto), Research Center—LAB2, E Bdg 1st Floor, Rua Dr António Bernardino de Almeida, 4200-072 Porto, Portugal; Francisca.Carvalho.Dias@ipoporto.min-saude.pt (F.D.); cristinalmeida11@hotmail.com (C.A.); mariana.gomes.morais@ipoporto.min-saude.pt (M.M.); ruimedei@ipoporto.min-saude.pt (R.M.)
2 Research Department of the Portuguese League against Cancer Regional Nucleus of the North (LPCC-NRN), Estrada da Circunvalação 6657, 4200-177 Porto, Portugal
3 Institute of Biomedical Sciences Abel Salazar, University of Porto (ICBAS-UP), Rua Jorge Viterbo Ferreira 228, 4050-513 Porto, Portugal
4 Faculty of Medicine, University of Porto (FMUP), Alameda Prof. Hernâni Monteiro, 4200-319 Porto, Portugal
5 Biomedical Research Center (CEBIMED), Faculty of Health Sciences of Fernando Pessoa University (UFP), Praça 9 de Abril 349, 4249-004 Porto, Portugal
* Correspondence: ana.luisa.teixeira@ipoporto.min-saude.pt; Tel.: +351-225084000 (ext. 5410)
† Authors contributed equally to this work.

Abstract: The development and progression of colorectal cancer (CRC) have been associated with genetic and epigenetic alterations and more recently with changes in cell metabolism. Amino acid transporters are key players in tumor development, and it is described that tumor cells upregulate some AA transporters in order to support the increased amino acid (AA) intake to sustain the tumor additional needs for tumor growth and proliferation through the activation of several signaling pathways. LAT1 and ASCT2 are two AA transporters involved in the regulation of the mTOR pathway that has been reported as upregulated in CRC. Some attempts have been made in order to develop therapeutic approaches to target these AA transporters, however none have reached the clinical setting so far. MiRNA-based therapies have been gaining increasing attention from pharmaceutical companies and now several miRNA-based drugs are currently in clinical trials with promising results. In this review we combine a bioinformatic approach with a literature review in order to identify a miRNA profile with the potential to target both LAT1 and ASCT2 with potential to be used as a therapeutic approach against CRC.

Keywords: colorectal cancer; amino acid transporters; LAT1; ASCT2; miRNAs

Citation: Dias, F.; Almeida, C.; Teixeira, A.L.; Morais, M.; Medeiros, R. LAT1 and ASCT2 Related microRNAs as Potential New Therapeutic Agents against Colorectal Cancer Progression. *Biomedicines* **2021**, *9*, 195. https://doi.org/10.3390/biomedicines9020195

Received: 12 January 2021
Accepted: 12 February 2021
Published: 16 February 2021

1. Introduction

Colorectal cancer (CRC) is one of the most common cancers worldwide, with 1,849,518 new cases in 2018, being the third most common cancer [1]. Currently, CRC accounts for approximately 10% of all diagnosed cancers and it is the world's second most deadly cancer [2]. CRC is the second most common neoplasia diagnosed in women, and the third in men, being the incidence and mortality approximately 25% lower in woman [2]. CRC development can be modulated by several factors, being the high alcohol consumption, overweigh, physical inactivity, tobacco smoking, diabetes mellitus, age, personal or family history of CRC well established risk factors [3,4]. Although the mortality rates have declined due to the improvement in diagnosis and treatment, CRC still represents one of the most lethal cancer types [3]. Furthermore, metastasis is also found in, approximately, 15–25% of CRC cases at the diagnosis, and increase to 50% during the course of the disease [2,5]. The advances in the pathophysiological and molecular CRC knowledge allowed the increase of the treatment options, but these new therapeutic approaches were proven to be more effective in patients with non-metastatic disease [2]. Thus, it is imperative to clarify the mechanisms involved in disease progression, aggressiveness and metastasis formation in order to improve the patients' follow up and to identify new therapeutic approaches.

Recently, the literature showed that amino acid (AA) transporters, such as solute carrier transports (SLCs), are important players in tumor development, since tumor cells have an increased demand for AA to sustain their excessive proliferation rate [6]. In fact, it is described that tumor cells upregulate some AA transporters in order to support the increased AA demand and intake [7]. Moreover, SLCs are able to regulate the PI3K/Akt/mTORC1 signaling pathway, which is central in the regulation of CRC proliferation and aggressiveness and is also involved in metabolic reprograming [8,9]. In fact, there are some SLC dysfunctions associated with CRC, such as L-type amino acid transporter 1 (LAT1) and alanine-serine-cysteine transporter 2 (ASCT2) upregulation, that may have an impact on disease aggressiveness [9–11]. Given the growing evidence and interest in the impact of cancer metabolism in disease aggressiveness, it is imperative to further understand the regulatory mechanisms responsible for of LAT1 and ASCT2 modulation in CRC and study the potential of their inhibition as a therapeutic approach. Since these two AA transporters are frequently overexpressed in CRC cells, they have potential as drug targets because their inhibition or blockade could lead to cell cycle arrest and apoptosis [10,12–14].

1.1. Amino Acid Transporters Deregulation in CRC: The Impact of LAT1 and ASCT2

It has been nearly a century since the discovery that normal and tumor cells differ in energy metabolism, with tumor cells presenting a higher need of nutrients, being the AA bioavailability crucial to support cell proliferation and growth [15]. Amino acids can be classified into three groups: (1) essential AA (EAA), if the organism is not able to synthesize them and needs to acquire them from the diet; (2) non-essential AA, if they are synthesized in sufficient quantities by the organism or (3) conditional AA, if are usually nonessential, except in times of illness, trauma or stress were they become conditionally essential [16,17].

In addition to their need in protein synthesis, several amino acids have other roles in supporting cancer development. One example is glutamine, the most abundant AA that participates in energy production, redox homeostasis, macromolecular synthesis and cell signaling [18]. In fact, the commitment of glutamine in the these cell processes makes this AA conditionally essential in conditions characterized by a high proliferation rate, such as cancer, in which endogenous glutamine synthesis is not sufficient to satisfy the cell need [17].

Since AAs are hydrophilic, they need selective transport proteins in order to cross the plasma membrane of the cells. There are approximately two-dozen amino acid transporters in humans, and cancer cells must regulate one or more of these transporters to satisfy their nutrient demand [6]. LAT1 (SLC7A5) is a transmembrane transporter involved in the import of large and neutral AA such as leucine and phenylalanine, in exchange for intracellular AA, such as glutamine [10,13,19]. According to various studies, LAT1 is highly upregulated in multiple human cancers, including gastrointestinal cancers [10,19–21]. In fact, Hayase and coworkers found a higher expression of LAT1 in 72.4% of CRC cases when compared to colonic adenoma cases, concluding that LAT1 could be a marker for malignant lesions [10]. Furthermore, Zhang and colleagues also found an association of higher LAT1 expression levels to poorer outcomes and shorter survival in several types of cancer, including CRC [14]. The higher LAT1 expression in cancer cells shows the importance of this AA transporter in the maintenance of AA nutrition in cancer cells [6]. Studies conducted by Elorza and coworkers show that the upregulation of LAT1 is involved in the increase of mTORC1 activity through HIF2α activation, showing a relationship between the hypoxic microenvironment, HIF2α and LAT1 [22]. Furthermore, LAT1 mediates leucine uptake with high affinity, which is a key AA activator of the mTOR signaling pathway [23]. However, for mTOR activation, the functional LAT1 is coupled to ASCT2, another AA transporter involved in glutamine uptake [16].

The ASCT2 (SLC1A5) is expressed in most human tissues including the large intestine and CRC tumor cells, and is essentially responsible for the influx of glutamine inside the cells, inducing asparagine, serine and threonine efflux [24–26]. According to Liu and colleagues, ASCT2 expression levels can modulate the migration capacity of CRC cells, being the overexpression of this AA transportersassociated with a poorer patients' prognosis [1,27]. In fact, ASCT2 is upregulated in several cancers, including triple-negative breast cancer, CRC, lung cancer, melanoma, neuroblastoma, glioblastoma and prostate cancer [12]. Some studies in glioblastomas and neuroblastoma support the involvement of the activation of c-Myc, n-Myc oncogenes in the inducing of ASCT2 expression [28,29].

Metabolic reprogramming is a well-known hallmark of cancer that has been gaining increasing attention in the last few years due to its importance in cancer cells viability and growth [30]. Cancer associated metabolic reprogramming influences intracellular and extra-cellular availability of metabolites that will result in alterations in gene expression, cellular differentiation and also in the tumor microenvironment [31]. Glutamine is considered to be a crucial nutrient for cancer proliferation due to its ability to donate its nitrogen and carbon to several growth-promoting pathways [32]. In 2012, Mootha and colleagues reported that tumor cells have a high necessity of glutamine uptake compared to other AA and, consequently, a glutamine starvation can interfere with tumor metabolism inhibiting tumor proliferation and progression [32]. More recently, Varshavi and colleagues, described a molecular association between CRC that present oncogenic *KRAS* mutation and glutamine metabolism, since these cells exhibit special metabolic phenotypes, including differences in glycolysis, glutamine utilization and AA metabolism [33]. Furthermore, glutamine is described as a signaling factor in the uptake of AA for the activation of mTORC1 [34]. Thus, the upregulation of AA transporters have an important role in the support of the high-level protein synthesis for continuous cancer growth and proliferation [10,35]. The mTOR pathway is well described as deregulated in CRC, and the availability of AA functions as a regulator of this pathway, since a high AA microenvironmental bioavailability induces mTOR activity and consequent biological processes, such as protein translation [36]. Some studies report a relationship between LAT1 and ASCT2, with a two-step mechanism of these AAT being able to regulate mTOR pathway [37–39]. Firstly, ASCT2 regulates the intracellular concentration of glutamine, and in turn LAT1 uses this intracellular glutamine as an efflux substrate, in order to regulate the uptake of extracellular leucine, which will lead to an activation of mTOR signaling and consequent induction of cell growth and proliferation [40,41] (Figure 1). Furthermore, according to Rajasinghe and coworkers, the inhibition of glutamine uptake in proliferating cells, through the inhibition of glutamine transporters LAT1 and ASCT2, results in the inhibition of cell proliferation and induces apoptosis, through the downregulation of the mTOR pathway [38]. Thus, the inhibition of LAT1 and ASCT2 expression levels could represent a promising therapeutic approach for CRC since it would reduce the AA intake, consequently causing mTOR pathway inhibition and compromising cancer cell proliferation.

The use of pharmacologic approaches against LAT1 and ASCT2 in cancers with overexpression of these two AA transporters seems be a promising strategy. In fact, over the last few years there was investment in the development of drugs against LAT1 and ASCT2 [26,38,42,43]. The design of drugs against these two AA transporters usually follows an approach based on substrate analogues, which act as competitive inhibitors [26]. In the case of ASCT2 there are also been developed monoclonal antibodies against its cell surface domains [44]. The pharmacological inhibitors against LAT1 and ASCT2 reported in CRC are listed on Table 1.

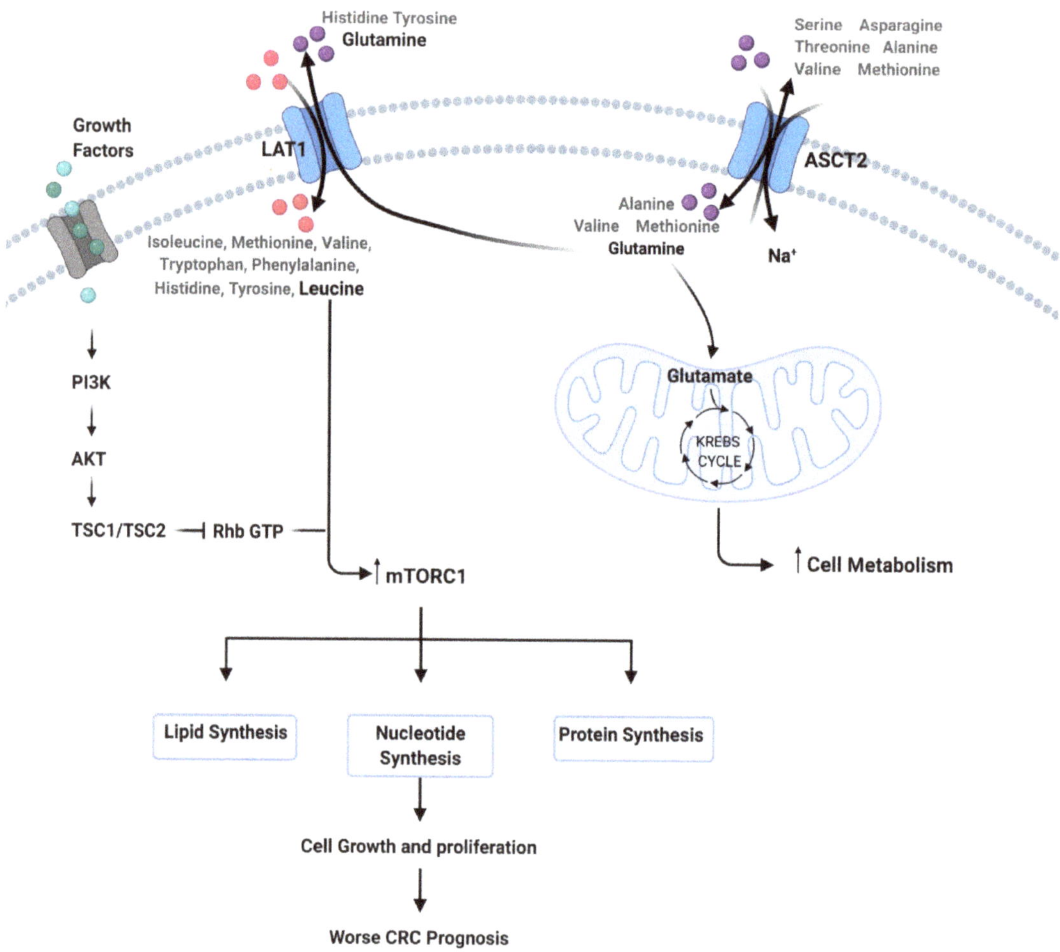

Figure 1. Representation of the interplay between ASCT2, LAT1 and mTOR pathway in colorectal cancer (CRC). This image was created using BioRender.

Table 1. Pharmacological inhibitors of LAT1 and ASCT2 reported in CRC.

AA Transporter	Inhibitor	Inhibitor Type	Reference
LAT1	JHP203	Tyrosine analog	[45,46]
ASCT2	MAb KM4008	Monoclinal antibodies against cell surface domains	[44]
	MAb KM4012		
	MAb KM4018		
	V-9302	Competitive antagonist	[47]

More recently, in a phase I study, Okano and coworkers observed that the JPH203 treatment was well tolerated by patients with CRC and biliary tract cancer (BTC). In fact, disease control was observed in two of the six CRC patients and in three of the five BTC patients [13]. Furthermore, a study from Toda and colleagues using two *KRAS*-mutated cells lines demonstrated a significant association between ASCT2 expression and *KRAS* mutation and, when the authors used siRNAs to silence KRAS, they observed a significant reduction of ASCT2 [48]. In addition to that, the authors also used specific inhibitors of Raf/MEK/ERK, and PI3K/Akt/mTOR pathways, and observed that both inhibitors presented the ability to reduce ASCT2 expression [48]. Moreover, studies using xenograft models demonstrated that the inhibition of ASCT2 expression is able to reduce the uptake of glutamine and inhibit tumor cell proliferation [49]. However, it is imperative to keep in mind that the block of AA transporters could be associated with the upregulation of compensatory and redundant pathways, being crucial an accurate overview of all network involved in the process [50]. In addition to that, there are some limitations in the use of pharmacological inhibitors due to the low affinity for the transporter and low selective capacity observed to cancer cells. Thus, these data highlight the need for a deeper understanding of other therapeutic approaches for the selective inhibition of LAT1 and ASCT2 in CRC.

1.2. Applicability of microRNAs as Therapeutic Agents

Over the years, advances in genomic technologies have led to an identification of a variety of epigenetic alterations believed to be strongly involved in cancer initiation and progression [51]. In fact, several studies revealed that the altered metabolic pathways in cancer are tightly regulated by microRNAs (miRNAs) [52–58]. MiRNAs are a family of short non-coding RNAs with a length of approximately 19–25 nucleotides that post-transcriptionally regulate gene expression, with an important role in several biological pathways, including cell proliferation and differentiation [59,60]. MiRNAs can regulate the expression of more than 50% of protein-coding genes by binding to their target mRNA transcript and causing its degradation or translation repression [61]. Furthermore, the downstream targets of several miRNAs are directly or indirectly connected to metabolic alterations [52].

Regarding their applicability in the clinical setting, a growing number of evidence suggests a significant utility of miRNAs as biomarkers for pathogenic conditions, modulators of drug resistance and as therapeutic agents for medical intervention in almost all human health-related conditions [62–65]. The pleiotropic nature of miRNAs makes them particularly attractive, both as drugs or drug targets, for diseases with a multifactorial origin and no current effective treatments [66,67]. In addition to that, circulating miRNAs present several advantages compared to other circulating nucleic acids, such as: protection from RNAse degradation, high stability in circulation through the body, and resistance to adverse conditions such as temperature or pH alterations [68,69]. Regarding miRNA therapeutics applicability, there are reports demonstrating clinical utility of miRNA mimics and miRNA repressors and miRNAs loaded in something. In fact, there are two major types of miRNA-based therapies: miRNA suppression therapy, when the goal is the target mRNA upregulation and miRNA replacement therapy, when the goal is the target mRNA downregulation.

Overall, the current evidence suggests a viable future for miRNA drugs in diseases with no current effective treatments, such as CRC. Hence, the scope of this review is to gather and systematize the information available regarding the impact of *LAT1* and *ASCT2* related miRNAs in CRC development and establish a profile with potential application to be used as a therapeutic agent through in silico analysis combined with a literature review (Figure 2).

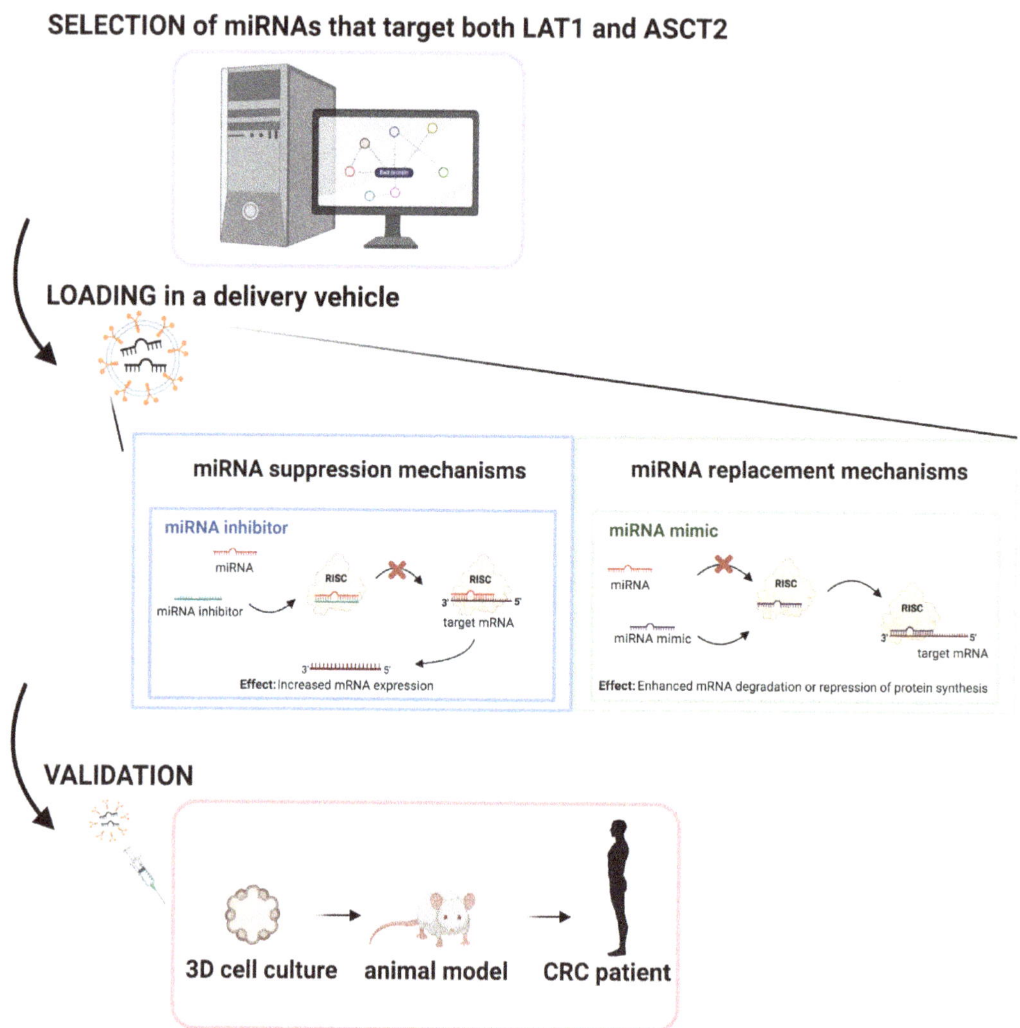

Figure 2. Schematic overview of the applicability of this review in the future development of miRNA-based therapies against LAT1 and ASCT2 in CRC. This image was created using BioRender.

2. Materials and Methods

2.1. MiRNA Selection and Literature Review

In order to select miRNAs that target both *LAT1* and *ASCT2* we used miRTarBase (version 8.0), the largest known online database of validated miRNA:mRNA interactions [70]. According to miRTarBase there are 267 miRNAs that target *LAT1* and 173 that target *ASCT2* mRNAs. Since one miRNA has multiple targets and the same mRNA can be regulated by several miRNAs, we went to see if there were miRNAs that targeted both *LAT1* and *ASCT2*. From the 440 miRNAs retrieved by miRTarBase, we observed that 33 targeted both *LAT1* and *ASCT2* (Figure 3).

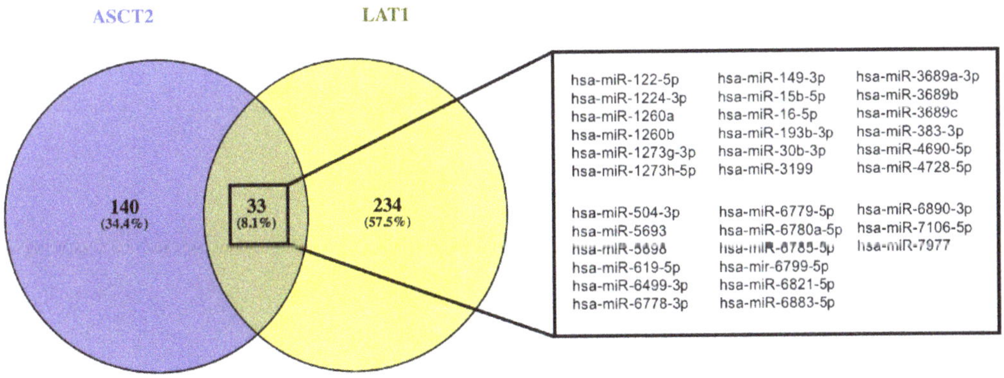

Figure 3. Venn diagram of the validated miRNAs that target ASCT2 and LAT1 obtained using Venny 2.1. (http://bioinfogp.cnb.csic.es/tools/venny, accessed on 21 December 2020) and the detailed list of the 33 miRNAs.

After retrieving the miRNAs that target both *LAT1* and *ASCT2* from miRTarBase, a literature search in PubMed and Google Academic was conducted using the search terms "colorectal cancer" plus one of the following 33 microRNAs: "miR-122-5p", "miR-1224-3p", "miR-1260a", "hsa-miR-1260b", "hsa-1273g-3p", hsa-miR-1273h-5p", "hsa-miR-149-3p", "hsa-miR-15b-5p", "miR-16-5p", "miR-193b-3p", "miR-30b-3p", "miR-3199", "miR-3689a-3p", "miR-3689b-3p", "miR-3689c", "miR-383-3p", "miR-4690-5p", "miR-4728-5p", "miR-504-3p", "miR-5693", "miR-5698", "miR-619-5p", "miR-6499", "miR-6778-3p", "miR-6799-5p", "miR-6780a-5p", "miR-6785", "miR-6799", "miR-6821", "miR-6883-5p", "miR-6890-3p", "miR-7106-5p" or "miR-7977". The articles were selected by relevance of their findings, namely, a significant association between these miRNAs and colorectal cancer. Literature analysis includes scientific papers published in the last 6 years (between 2014 and 2020). The obtained scientific papers were manually curated in order to determine associations between the miRNAs and CRC, giving a total of 28 selected papers. The exclusion criteria for the collected papers were as follows: (1) no significant association between the miRNAs and CRC; (2) association of the miRNAs with a benign tumor and (3) individual papers that were already included in meta-analysis. For each study, information was extracted concerning the following characteristics: the name of the miRNA, type of sample where the miRNA was studied, miRNA expression levels (upregulated and downregulated) and their effect on CRC (e.g., prognosis, therapy response or pathways regulation).

2.2. In Silico Analysis

The Search Tool for the Retrieval of Interacting Genes (STRING) database is an online tool that is used to develop protein–protein interaction (PPI) networks [71]. We used the STRINGapp of the Cytoscape software (v3.7.X) to construct and visualize the protein interaction network of the selected target genes. Those with a combined score of >0.4 were selected as significant. The functional enrichment analysis of Gene Ontology (GO), Kyoto Encyclopedia of Genes and Genomes (KEGG) and Reactome pathways was made with the STRING enrichment analysis tool, with a false discovery rate (FDR) of $p < 0.01$. The enrichment results were filtered, and redundant terms were removed according to the Jaccard index.

3. Results

3.1. miRNAs that Target Both LAT1 and ASCT2 and their Impact on CRC

From the 33 candidate miRNAs, only 16 have already been described in CRC (Table 2). However, in terms of the miRNA: mRNA target interaction with *LAT1* and *ASCT2*, none of the miRNAs have been yet validated for CRC.

Table 2. Selected miRNAs' impact on CRC.

miRNA	Expression	Sample Type	Effect	Reference
Hsa-miR-122-5p	Down	CRC Tissue and cells	Increase in cell proliferation, migration and invasion through the upregulation of CDC25A	Yin 2020 [72]
	Down	CRC Tissues	Upregulation of the PI3K/Akt pathway through upregulation of TRIM29	Asadi 2019 [73]
	Up	CRC liver metastatic tissues	Not described	Liu 2019 [74]
	Up	Serum and HT-29 and SW480 cell lines	Lymph node metastasis biomarker and cell migration inducer	Qu 2018 [75]
	Up	CRC Plasma	Worse prognosis in metastatic patients and shorter RFS and OS in non-metastatic patients	Maiertheler 2017 [76]
Hsa-miR-1224-3p	Up	CRC Tissues	Upregulated in E cadherin positive tissues	Lin 2017 [77]
Hsa-miR-1260a	Down	CRC Serum	Not described	Wang 2017 [78]
Hsa-miR-1260b	Up	HCT116 cells	Chemoresistance to 5-FU through upregulation of PDCD4	Zhao 2018 [79]
	Down	SW480 cells	Downregulated by STAT3-siRNA	Zhang 2014 [80]
	Up	Carcinoma vs adenoma (tissue)	Not described	Slattery 2016 [81]
	Down	CRC Serum	Not described	Zhang 2017 [82]
	Up	DKO-1 cells	Enriched in *KRAS* mutant cells	Cha 2015 [83]
Hsa-miR-1273g-3p	Up	LoVo cells	Proliferation, migration and invasion through activation of ERBB4/PIK3R3/mTOR/S6K2 pathway	Li 2018 [84]
Hsa-miR-1273h-5p	Up	CRC tissues	Not described	Du 2018 [85]
Hsa-miR-149-3p	Down	HCT-8 and HCT-116 cells	Chemoresistance to 5-FU through upregulation of PDK2	Liang 2020 [86]
Hsa-miR-15b-5p	Down	CRC tissues and cell lines	Chemoresistance to 5-FU through upregulation of XIAP	Zhao 2017 [87]
	Up	HT-29 cell line	Cell growth and inhibition of the proapoptotic pathway	Gasparello 2020 [88]
	Down	KRAS mutated CRC tissues vs wild type CRC tissues	Not described	Milanesi 2020 (82)
Hsa-miR-16-5p	Down	CRC tissues and cell lines	Upregulation of VEGFA	Wu 2020 [33]
Hsa-miR-193b-3p	Down	CRC tissues vs adjacent normal tissues	Shorter OS of CRC patients and upregulation of STMN1	Guo 2016 [89]
	Up	CRC tissues	Downregulation of RAD51	Kara 2015 [90]
Hsa-miR-3199	Down	SW620 cell line	Upregulation of SMAD4	Yan 2018 [91]
Hsa-miR-383-3p	Down	CRC tissues and HT-29 and LoVo cell lines	Upregulation of APRIL	Cui 2018 [92]
Hsa-miR-4690-5p	Down	CRC Stool	Not described	Ghanbari 2015 [93]
	Up	CRC tissues	Upregulated in CIMP high/MSI CRC tissues	Mullany 2016 [94]
Hsa-miR-619-5p	Down	CRC tissues vs adjacent normal tissues	Upregulation of MALAT1, lymphovascular invasion perineural invasion, shorter DFS and shorter OS	Qiu 2016 [95]
Hsa-miR-6821-5p	Down	SW480 CSCs vs SW480 wild-type	Not described	Zhou 2019 [96]
	Up	CRC tissues	Not described	Du 2018 [85]
Hsa-miR-6883-5p	Down	TCGA dataset and Cell lines	Upregulation of CDK4 and CDK6 and cell growth stimulus	Lulla 2017 [97]

Through the analysis of Table 1 we can observe that some of the miRNAs present opposite results regarding their expression levels, which may be related with the type of biological sample from which their expression levels are analyzed. Regarding their effects on CRC, the deregulation of miR-122-5p, miR-1273g-3p, miR-16-5p, miR-3199, miR-383-3p, miR-619-5p and miR-6883-5p was associated with the upregulation of important players of oncogenic pathways, such as TRIM29, CDC25A, PI3K/Akt, mTOR, VEGFA, MALAT1, SMAD4, STMN1, APRIL and CDK4, with an impact on cell proliferation, invasion and migration. In addition to that, miR-1260b, miR-149-3p and miR-15b-5p were reported as associated with resistance to 5′-FU treatment through the upregulation of PDCD4, PDK2 and XIAP, respectively. Moreover, only three miRNAs were associated with clinical endpoints. Higher plasmatic levels of hsa-miR-122-5p were associated with worse prognosis in metastatic patients and shorter RFS and OS in non-metastatic patients, while lower levels of CRC tissue hsa-miR-193b-3p and hsa-miR-619-5p were associated with shorter OS. Moreover, lower levels of CRC tissue hsa-miR-619-5p were also associated with shorter DFS, lymphovascular invasion and perineural invasion.

3.2. Functional Annotation and Pathway Enrichment Analysis

Since the downregulation of a miRNA usually leads to the upregulation of its mRNA targets, we focused on the 11 miRNAs that have been reported as downregulated in CRC cells and tissues and therefore could be implicated in the upregulation of *LAT1* and *ASCT2* (miR-122-5p, miR-1260b, miR-149-3p, miR-15b-5p, miR-16-5p, miR-193b-3p, miR-3199, miR-383-3p, miR-619-5p, miR-6821-5p and miR-6883-5p) and did an in silico analysis to obtain a deeper knowledge of their impact on CRC. We used miRTarBase v8.0 to retrieve the mRNA targets of the selected miRNAs that were validated with strong evidence methods in order to do the functional annotation and enrichment analysis. From the 11 miRNAs studied, only miR-15b-5p, miR-16-5p, miR-122-5p, miR-149-3p, miR-1260b, miR-193b-3p and miR-383-3p presented mRNA targets validated with strong evidence methods (Western blot, qRT-PCR or luciferase assay), which are listed on Table 3.

Table 3. Validated targets of miR-15b-5p, miR-16-5p, miR-122-5p, miR-1260b, miR-149-3p, miR-193b-3p and miR-383-3p.

miRNA	Target mRNA
miR-16-5p	*ZYX, YAP1, WNT4, WNT3A, WEE1, VEGFA, UNG, UCA1, TPPP3, TP53, SOX6, SOX5, SOCS3, SLC6A4, RPS6KB1, RICTOR, RECK, RAF1, PURA, PTGS2, PRDM4, PPM1D, PIM1, OPRM1, NCSTN, NCOR2, MYB, MTOR, METTL13, MAP7, KRAS, KDR, IL12B, IGF1R, IFNG, HMGA2, HMGA1, HGF, HDGF, GLS2, FGFR1, FGF2, CLDN2, CHUK, CHEK1, CDS2, CDK6, CCNE1, CCND3, CCND2, CCND1, CAPRIN1, CADM1, BRCA1, BMI1, BIRC5, BDNF, BCL2, BACE1, AXIN2, ARL2, ARHGDIA, APP, AKT3, ADORA2A, ACVR2A*
miR-15b-5p	*WEE1, VEGFA, TRIM29, TRIM14, TGFB1, TBR1, SOCS3, SMURF1, SMAD2, RECK, RAB1A, PURA, PPM1D, PEBP4, OIP5, MTSS1, MMP9, KDR, INSR, IFNG, HNF1A, FUT2, FOXO1, EIF4A1, CHEK1, CCNE1, CCND3, CCND1, BCL2, BAX, AXIN2, AKT3, AGO2*
miR-122-5p	*ZNF395, XPO6, WNT1, VEGFC, UBAP2, TRIB1, TPD52L2, TBX19, SRF, SPRY2, SOCS1, SLC7A11, SLC7A1, RHOA, RAC1, RAB6B, RAB11FIP1, PTPN1, PRKRA, PRKAB1, PKM, PEG10, PDK4, P4HA1, NUMBL, NT5C3A, NOD2, NFATC2IP, NCAM1, MEF2D, MECP2, MAPK11, LPIN1, IL1A, IGF1R, HMOX1, GYS1, GALNT10, G6PC3, FUT8, FUNDC2, FOXP1, FOXJ3, FAM117B, ENTPD4, EGLN3, EGFR, DUSP2, DSTYK, CYP7A1, CTDNEP1, CREB1, CLIC4, CDK4, CCNG1, BCL2L2, BAX, AXL, ATP1A2, AP3M2, ANXA11, ANK2, ALDOA, AKT3, ADAM17, ADAM10, AACS*
miR-1260b	SMAD4, SFRP1, DKK2
miR-193b-3p	*YWHAZ, SMAD3, SHMT2, RAD51, PRAP1, PLAU, NF1, MYB, MCL1, MAX, KRAS, KIT, ETS1, ESR1, CCND1, AKR1C2*
miR-383-3p	*PRPF31*
miR-149-3p	*WNT1, MYBL2, GPC1, FGFR1, E2F1, AKT1*

In order to explore the biological impact of these miRNA profiles in CRC, we analyzed their 186 validated targets with the STRINGapp Protein Query from Cytoscape software. A total of 168 of the 186 coding genes were filtered into a protein–protein interaction (PPI) network with 168 nodes and 1284 edges that presented a significant enrichment ($p = 1 \times 10^{-16}$).

We also applied a Markov clustering (MCL), which resulted in the clustering of the proteins into 11 clusters according to their STRING interaction score (Figure 4).

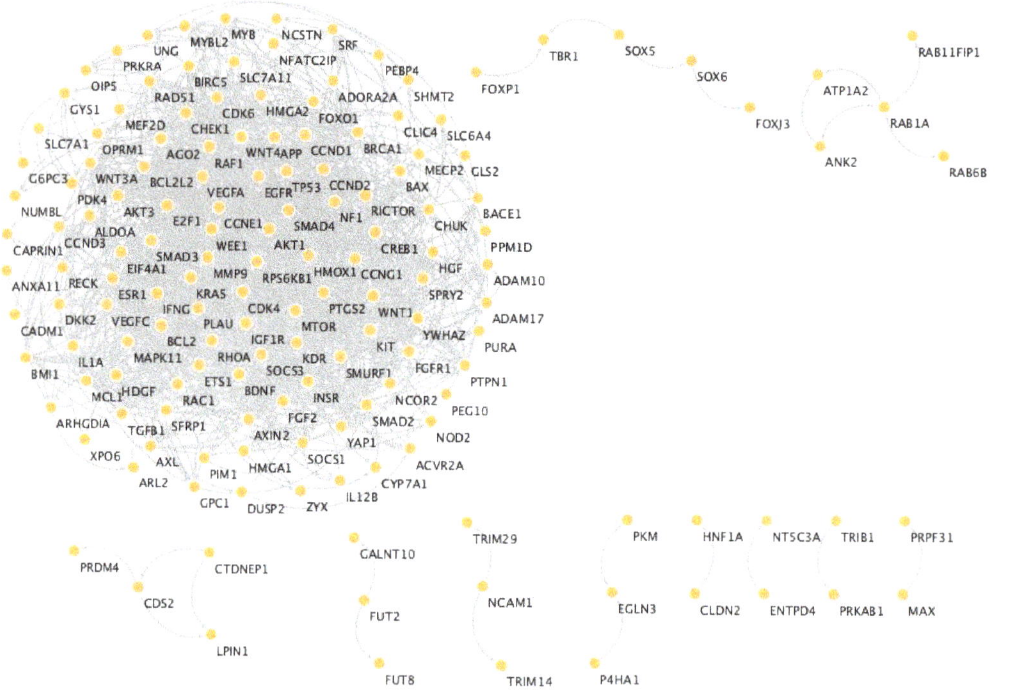

Figure 4. Protein–protein interaction (PPI) network. The proteins were clustered using the clusterMaker app from Cytoscape with an inflation value of 2.5 and a cutoff edge of 0.5. All the singletons were removed.

The functional enrichment analysis was made using an FDR threshold of $p < 0.01$, and the redundant terms were eliminated using a redundancy cutoff of 0.5, which resulted in a total of 892 enriched terms among the KEGG, Reactome and GO categories (Supplementary Tables S1–S5). The top 20 enriched terms for each category are represented on Figure 5. Among the functionally enriched terms in the KEGG and Reactome pathways we could find PI3K/Akt, MAPK, HIF-1, mTOR, VEGF and EGFR inhibitor resistant pathways, all of which are well established as involved in CRC development. Regarding the GO terms, if we focus on the molecular processes, we can observe that the two most enriched terms are the regulation of cell proliferation and the cellular response to organic substances, which may be related with the increase intake of nutrients as a consequence of the downregulation of this miRNA profile and consequent upregulation of AA transporters, such as LAT1 and ASCT2.

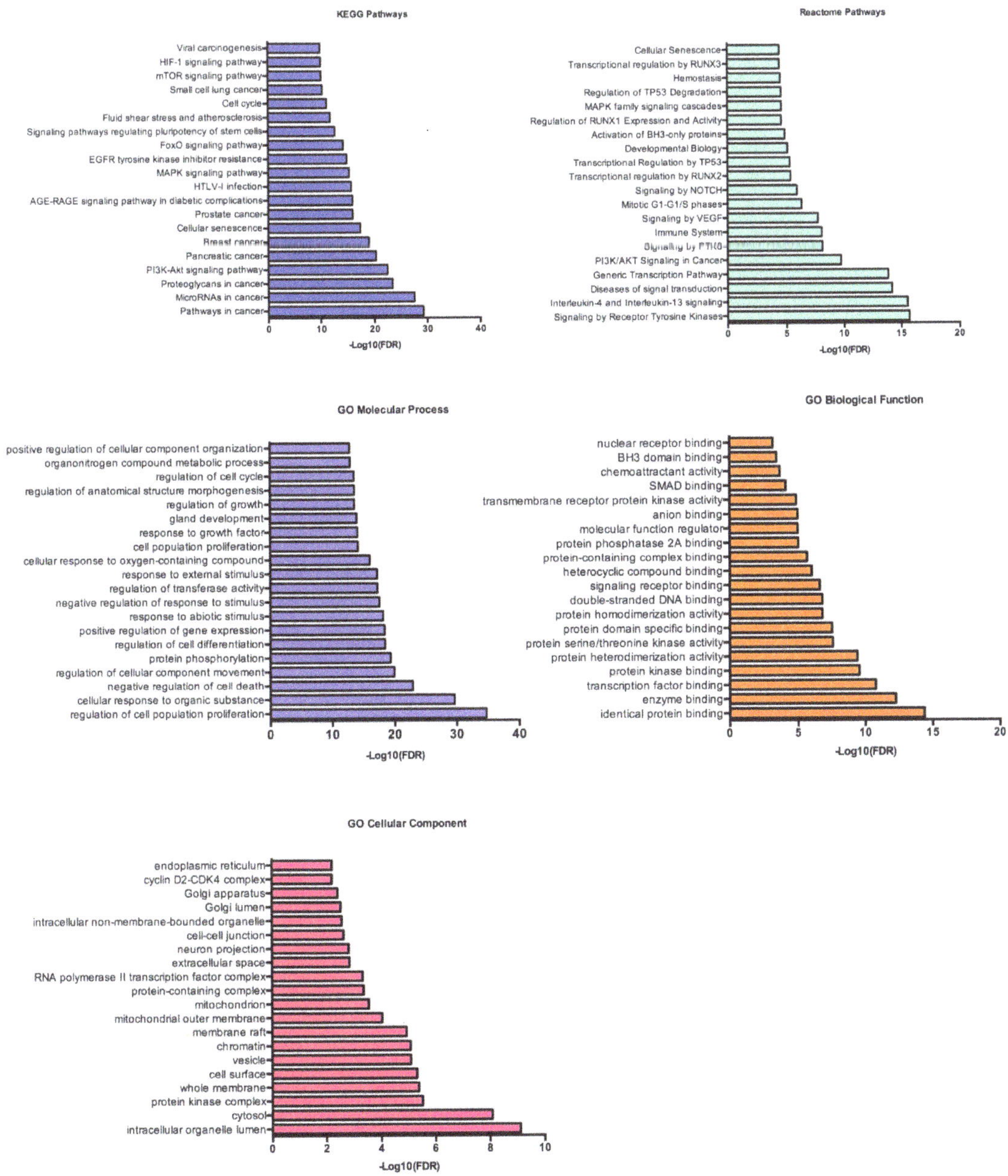

Figure 5. Kyoto Encyclopedia of Genes and Genomes (KEGG), Reactome and Gene Ontology (GO) analysis of the 186 selected target genes. The functional enrichment analysis was made using the STRINGapp from Cytoscape.

4. Discussion

CRC remains one of the most diagnosed cancers in the world, with a high metastatic potential and not enough therapeutic options. The previous underestimated metabolic alterations are now gaining more attention from the scientific community, and it is now known that metabolic cross-communication between tumor cells, immune cells, stromal cells and the gut microbiota are able to induce CRC proliferation, invasion and metastasis [98]. Among the metabolic alterations with potential to be targeted in order to develop new therapeutic approaches, the upregulation of AA transporters *LAT1* and *ASCT2* seems promising due to their impact in the regulation of the mTOR pathway. In addition to that, the liver is recognized as the most common metastatic site of CRC and cumulative evidences show that LAT1 and ASCT2 are overexpressed in hepatocellular carcinoma (HCC) and that these cells present a 10–20 fold increase in glutamine uptake, compared to normal hepatocytes [27,38,99–101]. Therefore, the definition of a new therapeutic approach involving the inhibition of these two AA transporters could be a promising strategy to control CRC proliferation and aggressiveness.

Recently, it has been suggested that modulation of miRNAs in cancer cells could be a potential tool for the improvement of cancer patients' therapies. In fact, by suppressing oncogenic miRNAs or substituting deficient tumor suppressive miRNAs, we are able to control cancer cell growth and progression. The world's first miRNA therapeutic, a short locked nucleic acid (LNA) antagonist for miR-122 named Miravirsen (produced by Roche/Santaris) was developed for the treatment of hepatitis C virus (HCV) infection [102]. Along with Miravirsen, all of the miRNA-based drugs are currently in clinical trials and none have yet reached the pharmaceutical breakthrough. However, acquisition of miRNA-based companies by famous pharmaceutical companies is sending a positive feedback on their potential [103]. Currently, there are several strategies used for miRNA-based therapies, which could include miRNA inhibition therapies that target oncomiRNAs, replacement therapies for tumor-suppressor miRNAs or miRNA-based delivery systems [104]. One example, of their applicability was the study performed by Callegari and coworkers that showed that the in vivo delivery of an anti-miR-221 caused a significant decrease in the size and number of tumor nodules, being established the promotor role of miR-221 in liver carcinogenesis [105]. On the other hand, Oshima and coworkers reported an effective delivery of miR-655-3p to CRC liver metastasis using nanoscale coordination polymers. The polymers used prolonged the miRNA distribution and miRNA-655-3p suppressed tumor growth when codelivered with oxaplatin, suggesting a synergistic effect of both therapeutic approaches [106].

Regarding the delivery mechanisms, miRNAs are delivered through the use of vectors that can be divided into two categories: viral vectors and nonviral vectors. The viral vectors used for miRNA delivery are mainly adenovirus vectors, adeno-associated virus vectors, retroviral vectors and lentivitus vectors. On the other hand, nonviral vectors include inorganic material-based delivery systems, lipid-based nanocarriers, polymeric vectors/dendrimer-based vectors, cell-derived membrane vesicles and 3D-Scaffold-based delivery systems [107]. The use of these delivery mechanisms improves targeting ability while protecting the miRNAs or miRNAs inhibitors from degradation. In fact, it was already demonstrated that, for cancer treatment, intratumoral injections of miRNA drugs directly into the tumor site are able to enhance target efficacy, specificity and minimize the side effects and there are also several ongoing clinical trials [62,108,109].

In the present review we combined a bioinformatic approach with a literature review to define a miRNA profile (miR-15b-5p, miR-16-5p, miR-122-5p, miR-1260b, miR-149-3p, miR-193b-3p and miR-383-3p) that has the potential to target both LAT1 and ASCT2 in CRC. The in silico approaches are very useful since they allow the simultaneous analysis of the interactions of hundreds of genes and, therefore, the creation of an integrative network that allows a deeper understanding of the biological processes regulated by them. Our in silico analysis result in a list of miRNAs that target both LAT1 and ASCT2 and the literature review allowed us to focus on the miRNAs that have already been studied in CRC and have been reported as downregulated. In addition to that, functional enrichment analysis showed that among the enriched terms derived from the miRNA profile targets, we can find PI3K/Akt, MAPK, HIF-1, mTOR, VEGF and EGFR inhibitor resistant pathways, all of which are well established as involved in CRC development.

The MAPK and PI3K/Akt signaling pathways are involved in cell proliferation and survival, and their deregulation confers proliferative advantages on cancer cells. In fact, KRAS, BRAF and PI3K mutations are frequent in CRC. Moreover, the increase of the PI3K/Akt pathway activation in CRC is also associated with the loss of the tumor suppressor PTEN, which is significantly associated with a worse prognosis [1]. In addition to that, according to Slatery and colleagues, approximately 41% of the genes of MAPK signaling are dysregulated in CRC [110]. These two signaling cascades can activate directly and indirectly the Ser/Thr protein kinase mTOR, respectively, being the mTOR involved in the regulation of cell proliferation and survival [111]. Similarly to other solid tumors, CRC is also characterized by a hypoxic microenvironment [112]. In fact, the cancer cells have the ability of adaptation to hypoxia through the regulation of the PI3K/AKT/mTOR pathway and by the transcription factors HIF-1α and HIF-2α, whose protein expression and transcriptional activity are also regulated by mTOR [113]. Furthermore, tumor hypoxia can also enhance cancer cells survival and proliferation through the upregulation of VEGF and its receptor VEGFR. VEGF promotes CRC growth through the stimulation of angiogenesis and its downstream signaling pathways are well characterized in cancer, with VEGF/VEGFR activation leading to the activation of MAPK/ERK, PI3K/Akt, PLC/PKC and other signaling pathways [114,115] .

Taking this information into consideration, we can conclude that the miRNA profile proposed in the present study plays an important role on CRC development and aggressiveness. However, despite promising, the results are still preliminary and require validation in CRC study models in order to assess the miRNA profile interaction with LAT1 and ASCT2 mRNAs, especially in terms of its inhibitory power. During the past few years, there has been a significant development of in vitro and in vivo preclinical research models, such as 3D cell culture of spheroids and organoids derived from several human tissues, which is helping in the translation of miRNAs into clinical practice [66,116,117]. In a recent study, Kawai and colleagues determined the culture conditions necessary to establish 3D cell culture models that mimic colon cancer heterogeneity [118]. In another study, Zoetemelk and colleagues established a robust, low-cost and reproducible short-term 3D colorectal cancer spheroids model to be used as a platform for screening the effect of combination therapies in CRC [119]. These enhanced research models are very useful for the study of miRNAs dynamics and for the development of the delivery systems for miRNA-based therapeutics [120,121]. Therefore, the next step should be focused on the delivery of the miRNA profile to CRC 3D culture models in order to see if it is sufficient to reverse the increased AA uptake caused by the increase of LAT1 and ASCT2 and inhibit cell proliferation.

Supplementary Materials: The following are available online at https://www.mdpi.com/2227-9059/9/2/195/s1, Table S1: GO Cellular Component enrichment results for terms with FDR $p < 0.01$, Table S2: GO Molecular Function enrichment results for terms with FDR $p < 0.01$, Table S3: GO Biological Process enrichment results for terms with FDR $p < 0.01$, Table S4: KEGG pathway enrichment results for terms with FDR $p < 0.01$, Table S5: Reactome pathways enrichment results for terms with FDR $p < 0.01$.

Author Contributions: Conceptualization, F.D.; methodology, F.D., C.A. and A.L.T.; software, F.D.; investigation, F.D., A.L.T.; data curation, F.D.; writing—original draft preparation, F.D., C.A. and A.L.T.; writing—review and editing, M.M., F.D., A.L.T.; supervision, R.M.; project administration, R.M.; funding acquisition, R.M., A.L.T. All authors have read and agreed to the published version of the manuscript.

Funding: This research was funded by the European Regional Development Fund (ERDF), grant number NORTE-01-0247-FEDER-033399.

Institutional Review Board Statement: Not applicable.

Informed Consent Statement: Not applicable.

Conflicts of Interest: The authors declare no conflict of interest. The funders had no role in the design of the study; in the collection, analyses, or interpretation of data; in the writing of the manuscript, or in the decision to publish the results.

References

1. Mármol, I.; Sánchez-De-Diego, C.; Dieste, A.P.; Cerrada, E.; Yoldi, M.J.R. Colorectal carcinoma: A general overview and future perspectives in colorectal cancer. *Int. J. Mol. Sci.* **2017**, *18*, 197. [CrossRef]
2. Dekker, E.; Tanis, P.J.; Vleugels, J.L.A.; Kasi, P.M.; Wallace, M.B. Colorectal cancer. *Lancet* **2019**, *394*, 1467–1480. [CrossRef]
3. Ahmed, M. Colon cancer: A clinician's perspective in 2019. *Gastroenterol. Res.* **2020**, *13*, 1–10. [CrossRef] [PubMed]
4. Karimi, N.; Feizi, M.A.H.; Safaralizadeh, R.; Hashemzadeh, S.; Baradaran, B.; Shokouhi, B.; Teimourian, S. Serum overexpression of miR-301a and miR-23a in patients with colorectal cancer. *J. Chin. Med Assoc.* **2019**, *82*, 215–220. [CrossRef]
5. Oh, B.Y.; Hong, H.K.; Lee, W.Y.; Cho, Y.B. Animal models of colorectal cancer with liver metastasis. *Cancer Lett.* **2017**, *387*, 114–120. [CrossRef]
6. Bhutia, Y.D.; Babu, E.; Ramachandran, S.; Ganapathy, V. Amino acid transporters in cancer and their relevance to "glutamine addiction": Novel targets for the design of a new class of anticancer drugs. *Cancer Res.* **2015**, *75*, 1782–1788. [CrossRef]
7. Bröer, S. Amino acid transporters as targets for cancer therapy: Why, where, when, and how. *Int. J. Mol. Sci.* **2020**, *21*, 6156. [CrossRef]
8. Bai, X.; Moraes, T.F.; Reithmeier, R.A.F. Structural biology of solute carrier (SLC) membrane transport proteins. *Mol. Membr. Biol.* **2017**, *34*, 1–32. [CrossRef] [PubMed]
9. Xie, J.; Zhu, X.Y.; Liu, L.M.; Meng, Z.Q. Solute carrier transporters: Potential targets for digestive system neoplasms. *Cancer Manag. Res.* **2018**, *10*, 153–166. [CrossRef] [PubMed]
10. Hayase, S.; Kumamoto, K.; Saito, K.; Kofunato, Y.; Sato, Y.; Okayama, H.; Miyamoto, K.; Ohki, S.; Takenoshita, S. L-type amino acid transporter 1 expression is upregulated and associated with cellular proliferation in colorectal cancer. *Oncol. Lett.* **2017**, *14*, 7410–7416. [CrossRef] [PubMed]
11. Nyquist, M.D.; Prasad, B.; Mostaghel, E.A. Harnessing solute carrier transporters for precision oncology. *Molecules* **2017**, *22*, 539. [CrossRef] [PubMed]
12. Bhutia, Y.D.; Ganapathy, V. Glutamine transporters in mammalian cells and their functions in physiology and cancer. *Biochim. Biophys. Acta* **1863**, 2531–2539. [CrossRef] [PubMed]
13. Scalise, M.; Pochini, L.; Console, L.; Losso, M.A.; Indiveri, C. The human SLC1A5 (ASCT2) amino acid transporter: From function to structure and role in cell biology. *Front. Cell Dev. Biol.* **2018**, *6*, 96. [CrossRef]
14. Zhang, C.; Xu, J.; Xue, S.; Ye, J. Prognostic Value of L-type amino acid transporter 1 (LAT1) in various cancers: A meta-analysis. *Mol. Diagn. Ther.* **2020**, *24*, 523–536. [CrossRef]
15. Ganapathy, V.; Thangaraju, M.; Prasad, P.D. Nutrient transporters in cancer: Relevance to Warburg hypothesis and beyond. *Pharmacol. Ther.* **2009**, *121*, 29–40. [CrossRef] [PubMed]
16. Choi, B.-H.; Coloff, J.L. The diverse functions of non-essential amino acids in cancer. *Cancers* **2019**, *11*, 675. [CrossRef]
17. Scalise, M.; Console, L.; Rovella, F.; Galluccio, M.; Pochini, L.; Indiveri, C. Membrane transporters for amino acids as players of cancer metabolic rewiring. *Cells* **2020**, *9*, 2028. [CrossRef] [PubMed]
18. Hensley, C.T.; Wasti, A.T.; DeBerardinis, R.J. Glutamine and cancer: Cell biology, physiology, and clinical opportunities. *J. Clin. Investig.* **2013**, *123*, 3678–3684. [CrossRef] [PubMed]
19. Singh, N.; Ecker, G.F. Insights into the structure, function, and ligand discovery of the large neutral amino acid transporter 1, LAT1. *Int. J. Mol. Sci.* **2018**, *19*, 1278. [CrossRef]
20. Häfliger, P.; Charles, R.-P. The L-type amino acid transporter LAT1—An emerging target in Cancer. *Int. J. Mol. Sci.* **2019**, *20*, 2428. [CrossRef]
21. Ogawa, H.; Kaira, K.; Motegi, Y.; Yokobori, T.; Takada, T.; Katoh, R.; Osone, K.; Takahashi, R.; Katayama, C.; Oyama, T.; et al. Role of amino acid transporter expression as a prognostic marker in patients with surgically resected colorectal cancer. *Anticancer Res.* **2019**, *39*, 2535–2543. [CrossRef] [PubMed]
22. Elorza, A.; Soro-Arnáiz, I.; Meléndez-Rodríguez, F.; Rodríguez-Vaello, V.; Marsboom, G.; De Cárcer, G.; Acosta-Iborra, B.; Albacete-Albacete, L.; Ordóñez, A.; Serrano-Oviedo, L.; et al. HIF2α acts as an mTORC1 activator through the amino acid carrier SLC7A5. *Mol. Cell* **2012**, *48*, 681–691. [CrossRef]
23. Taylor, P.M. Role of amino acid transporters in amino acid sensing. *Am. J. Clin. Nutr.* **2014**, *99*, 223S–230S. [CrossRef] [PubMed]
24. Liu, Y.; Zhao, T.; Li, Z.; Wang, L.; Yuan, S.; Sun, L. The role of ASCT2 in cancer: A review. *Eur. J. Pharmacol.* **2018**, *837*, 81–87. [CrossRef]
25. Cormerais, Y.; Massard, P.A.; Vucetic, M.; Giuliano, S.; Tambutté, E.; Durivault, J.; Vial, V.; Endou, H.; Wempe, M.F.; Parks, S.K.; et al. The glutamine transporter ASCT2 (SLC1A5) promotes tumor growth independently of the amino acid transporter LAT1 (SLC7A5). *J. Biol. Chem.* **2018**, *293*, 2877–2887. [CrossRef]
26. Jiang, H.; Zhang, N.; Tang, T.; Feng, F.; Sun, H.; Qu, W. Target the human alanine/serine/cysteine transporter 2(ASCT2): Achievement and future for novel cancer therapy. *Pharmacol. Res.* **2020**, *158*, 104844. [CrossRef]
27. Scalise, M.; Pochini, L.; Galluccio, M.; Console, L.; Indiveri, C. Glutamine transport and mitochondrial metabolism in cancer cell growth. *Front. Oncol.* **2017**, *7*, 306. [CrossRef] [PubMed]

28. Wise, D.R.; DeBerardinis, R.J.; Mancuso, A.; Sayed, N.; Zhang, X.-Y.; Pfeiffer, H.K.; Nissim, I.; Daikhin, E.; Yudkoff, M.; McMahon, S.B.; et al. Myc regulates a transcriptional program that stimulates mitochondrial glutaminolysis and leads to glutamine addiction. *Proc. Natl. Acad. Sci. USA* **2008**, *105*, 18782–18787. [CrossRef]
29. Ren, P.; Yue, M.; Xiao, D.; Xiu, R.; Gan, L.; Liu, H.; Qing, G. ATF4 and N-Myc coordinate glutamine metabolism in MYCN-amplified neuroblastoma cells through ASCT2 activation. *J. Pathol.* **2015**, *235*, 90–100. [CrossRef]
30. Hanahan, D.; Weinberg, R.A. Hallmarks of Cancer: The Next Generation. *Cell* **2011**, *144*, 646–674. [CrossRef]
31. Pavlova, N.N.; Thompson, C.B. The emerging hallmarks of cancer metabolism. *Cell Metab.* **2016**, *23*, 27–47. [CrossRef]
32. Liberti, M.V.; Locasale, J.W. The Warburg Effect: How does it benefit cancer cells? *Trends Biochem. Sci.* **2016**, *41*, 211–218. [CrossRef]
33. Wu, H.; Wei, M.; Jiang, X.; Tan, J.; Xu, W.; Fan, X.; Zhang, R.; Ding, C.; Zhao, F.; Shao, X.; et al. lncRNA PVT1 promotes tumorigenesis of colorectal cancer by stabilizing miR-16-5p and interacting with the VEGFA/VEGFR1/AKT axis. *Mol. Ther. Nucleic Acids* **2020**, *20*, 438–450. [CrossRef]
34. Ward, P.S.; Thompson, C.B. Metabolic reprogramming: A cancer hallmark even Warburg did not anticipate. *Cancer Cell* **2012**, *21*, 297–308. [CrossRef]
35. Yanagida, O.; Kanai, Y.; Chairoungdua, A.; Kim, D.K.; Segawa, H.; Nii, T.; Cha, S.H.; Matsuo, H.; Fukushima, J.-I.; Fukasawa, Y.; et al. Human L-type amino acid transporter 1 (LAT1): Characterization of function and expression in tumor cell lines. *Biochim. Biophys. Acta* **2001**, *1514*, 291–302. [CrossRef]
36. Bar-Peled, L.; Sabatini, D.M. Regulation of mTORC1 by amino acids. *Trends Cell Biol.* **2014**, *24*, 400–406. [CrossRef] [PubMed]
37. Toda, K.; Nishikawa, G.; Iwamoto, M.; Itatani, Y.; Takahashi, R.; Sakai, Y.; Kawada, K. Clinical role of ASCT2 (SLC1A5) in KRAS-mutated colorectal cancer. *Int. J. Mol. Sci.* **2017**, *18*, 1632. [CrossRef]
38. Rajasinghe, L.D.; Hutchings, M.; Gupta, S.V. Delta-tocotrienol modulates glutamine dependence by inhibiting ASCT2 and LAT1 transporters in non-small cell lung cancer (NSCLC) cells: A metabolomic approach. *Metabolites* **2019**, *9*, 50. [CrossRef]
39. Alves, M.J.F.; Uno, M.; da Silva, R.; Shinjo, S.M.; Marie, S.K.N. The expression of the aminoacid transporters ASCT2 (SLC1A5) and LAT1 (SLC7A5) in astrocytomas. *Med. Express* **2016**, *3*. [CrossRef]
40. Yazawa, T.; Shimizu, K.; Kaira, K.; Nagashima, T.; Ohtaki, Y.; Atsumi, J.; Obayashi, K.; Nagamori, S.; Kanai, Y.; Oyama, T.; et al. Clinical significance of coexpression of L-type amino acid transporter 1 (LAT1) and ASC amino acid transporter 2 (ASCT2) in lung adenocarcinoma. *Am. J. Transl. Res.* **2015**, *7*, 1126–1139.
41. Nicklin, P.; Bergman, P.; Zhang, B.; Triantafellow, E.; Wang, H.; Nyfeler, B.; Yang, H.; Hild, M.; Kung, C.; Wilson, C.; et al. Bidirectional transport of amino acids regulates mTOR and autophagy. *Cell* **2009**, *136*, 521–534. [CrossRef]
42. Foster, A.C.; Rangel-Diaz, N.; Yang, J.-Y.; Penjwini, M.; Viswanath, V.; Li, Y.-X.; Staubli, U. Phenylglycine analogs are inhibitors of the neutral amino acid transporters ASCT1 and ASCT2 and enhance NMDA receptor-mediated LTP in rat visual cortex slices. *Neuropharmacology* **2017**, *126*, 70–83. [CrossRef] [PubMed]
43. Hayashi, K.; Anzai, N. Novel therapeutic approaches targeting L-type amino acid transporters for cancer treatment. *World J. Gastrointest. Oncol.* **2017**, *9*, 21–29. [CrossRef] [PubMed]
44. Suzuki, M.; Toki, H.; Furuya, A.; Ando, H. Establishment of monoclonal antibodies against cell surface domains of ASCT2/SLC1A5 and their inhibition of glutamine-dependent tumor cell growth. *Biochem. Biophys. Res. Commun.* **2017**, *482*, 651–657. [CrossRef]
45. Muto, Y.; Furihata, T.; Kaneko, M.; Higuchi, K.; Okunushi, K.; Morio, H.; Reien, Y.; Uesato, M.; Matsubara, H.; Anzai, N. Different response profiles of gastrointestinal cancer cells to an L-type amino acid transporter inhibitor, JPH203. *Anticancer. Res.* **2019**, *39*, 159–165. [CrossRef]
46. Okunushi, K.; Furihata, T.; Morio, H.; Muto, Y.; Higuchi, K.; Kaneko, M.; Otsuka, Y.; Ohno, Y.; Watanabe, Y.; Reien, Y.; et al. JPH203, a newly developed anti-cancer drug, shows a preincubation inhibitory effect on L-type amino acid transporter 1 function. *J. Pharmacol. Sci.* **2020**, *144*, 16–22. [CrossRef]
47. Schulte, M.L.; Fu, A.; Zhao, P.; Allie, F.; Geng, L.; Smith, S.T.; Kondo, J.; Coffey, R.J.; Johnson, M.O.; Rathmell, J.C.; et al. Pharmacological blockade of ASCT2-dependent glutamine transport leads to antitumor efficacy in preclinical models. *Nat. Med.* **2018**, *24*, 194–202. [CrossRef] [PubMed]
48. Namikawa, M.; Kakizaki, S.; Kaira, K.; Tojima, H.; Yamazaki, Y.; Horiguchi, N.; Satoru, K.; Oriuchi, N.; Tominaga, H.; Sunose, Y.; et al. Expression of amino acid transporters (LAT1, ASCT2 and xCT) as clinical significance in hepatocellular carcinoma. *Hepatol. Res.* **2014**, *45*, 1014–1022. [CrossRef]
49. Song, W.; Li, D.; Tao, L.; Luo, Q.; Chen, L. Solute carrier transporters: The metabolic gatekeepers of immune cells. *Acta Pharm. Sin. B* **2020**, *10*, 61–78. [CrossRef]
50. Bröer, S. Amino acid transporters as disease modifiers and drug targets. *SLAS Discov.* **2018**, *23*, 303–320. [CrossRef] [PubMed]
51. Danese, E.; Montagnana, M. Epigenetics of colorectal cancer: Emerging circulating diagnostic and prognostic biomarkers. *Ann. Transl. Med.* **2017**, *5*, 279. [CrossRef] [PubMed]
52. Hatziapostolou, M.; Polytarchou, C.; Iliopoulos, D. miRNAs link metabolic reprogramming to oncogenesis. *Trends Endocrinol. Metab.* **2013**, *24*, 361–373. [CrossRef]
53. Shu, S.; Yang, Y.; Allen, C.L.; Maguire, O.; Minderman, H.; Sen, A.; Ciesielski, M.J.; Collins, K.A.; Bush, P.J.; Singh, P.; et al. Metabolic reprogramming of stromal fibroblasts by melanoma exosome microRNA favours a premetastatic microenvironment. *Sci. Rep.* **2018**, *8*, 12905. [CrossRef] [PubMed]

54. Sun, Y.; Oravecz-Wilson, K.; Bridges, S.; McEachin, R.; Wu, J.; Kim, S.H.; Taylor, A.; Zajac, C.; Fujiwara, H.; Peltier, D.C.; et al. miR-142 controls metabolic reprogramming that regulates dendritic cell activation. *J. Clin. Investig.* **2019**, *129*, 2029–2042. [CrossRef]
55. Yan, W.; Wu, X.; Zhou, W.; Fong, M.Y.; Cao, M.; Liu, J.; Liu, X.; Chen, C.H.; Fadare, O.; Pizzo, D.P.; et al. Cancer-cell-secreted exosomal miR-105 promotes tumour growth through the MYC-dependent metabolic reprogramming of stromal cells. *Nat. Cell Biol.* **2018**, *20*, 597–609. [CrossRef]
56. Nogueira, I.; Dias, F.; Morais, M.; Teixeira, A.L.; Medeiros, R. Everolimus resistance in clear cell renal cell carcinoma: miRNA-101 and HIF-2α as molecular triggers? *Futur. Oncol.* **2019**, *15*, 2361–2370. [CrossRef] [PubMed]
57. Nogueira, I.; Dias, F.; Teixeira, A.L.; Medeiros, R. miRNAs as potential regulators of mTOR pathway in renal cell carcinoma. *Pharmacogenomics* **2018**, *19*, 249–261. [CrossRef]
58. Morais, M.; Dias, F.; Teixeira, A.L.; Medeiros, R. MicroRNAs and altered metabolism of clear cell renal cell carcinoma: Potential role as aerobic glycolysis biomarkers. *Biochim. Biophys. Acta. Gen. Subj.* **2017**, *1861*, 2175–2185. [CrossRef] [PubMed]
59. Acunzo, M.; Romano, G.; Wernicke, D.; Croce, C.M. MicroRNA and cancer—A brief overview. *Adv. Biol. Regul.* **2015**, *57*, 1–9. [CrossRef] [PubMed]
60. Khan, S.; Ayub, H.; Khan, T.; Wahid, F. MicroRNA biogenesis, gene silencing mechanisms and role in breast, ovarian and prostate cancer. *Biochimie* **2019**, *167*, 12–24. [CrossRef]
61. Tovar-Camargo, O.A.; Toden, S.; Goel, A. Exosomal microRNA biomarkers: Emerging frontiers in colorectal and other human cancers. *Expert Rev. Mol. Diagn.* **2016**, *16*, 553–567. [CrossRef] [PubMed]
62. Hanna, J.; Hossain, G.S.; Kocerha, J. The potential for MicroRNA therapeutics and clinical research. *Front. Genet.* **2019**, *10*, 478. [CrossRef] [PubMed]
63. Dias, F.; Teixeira, A.L.; Nogueira, I.; Morais, M.; Maia, J.; Bodo, C.; Ferreira, M.; Silva, A.; Vilhena, M.; Lobo, J.; et al. Extracellular vesicles enriched in hsa-miR-301a-3p and hsa-miR-1293 dynamics in clear cell renal cell carcinoma patients: Potential biomarkers of metastatic disease. *Cancers* **2020**, *12*, 1450. [CrossRef] [PubMed]
64. Adem, B.F.; Bastos, N.R.A.; Dias, F.; Teixeira, A.L.; Medeiros, R. miRNAs: Mediators of ErbB family targeted therapy resistance. *Pharmacogenomics* **2016**, *17*, 1175–1187. [CrossRef]
65. Mollaei, H.; Safaralizadeh, R.; Rostami, Z. MicroRNA replacement therapy in cancer. *J. Cell Physiol.* **2019**, *234*, 12369–12384. [CrossRef]
66. Chakraborty, C.; Sharma, A.R.; Sharma, G.; Doss, C.G.P.; Lee, S.-S. Therapeutic miRNA and siRNA: Moving from Bench to Clinic as Next Generation Medicine. *Mol. Ther. Nucleic Acids* **2017**, *8*, 132–143. [CrossRef]
67. Van Rooij, E.; Purcell, A.L.; Levin, A.A. Developing MicroRNA therapeutics. *Circ. Res.* **2012**, *110*, 496–507. [CrossRef] [PubMed]
68. Glinge, C.; Clauss, S.; Boddum, K.; Jabbari, R.; Jabbari, J.; Risgaard, B.; Tomsits, P.; Hildebrand, B.; Kääb, S.; Wakili, R.; et al. Stability of circulating blood-based MicroRNAs—Pre-analytic methodological considerations. *PLoS ONE* **2017**, *12*, e0167969. [CrossRef]
69. Sohel, M.H. Extracellular/circulating MicroRNAs: Release mechanisms, functions and challenges. *Achiev. Life Sci.* **2016**, *10*, 175–186. [CrossRef]
70. Huang, H.Y.; Lin, Y.C.; Li, J.; Huang, K.Y.; Shrestha, S.; Hong, H.C.; Tang, Y.; Chen, Y.G.; Jin, C.N.; Yu, Y.; et al. miRTarBase 2020: Updates to the experimentally validated microRNA-target interaction database. *Nucleic Acids Res.* **2020**, *48*, D148–D154. [CrossRef]
71. Szklarczyk, D.; Gable, A.L.; Lyon, D.; Junge, A.; Wyder, S.; Huerta-Cepas, J.; Simonovic, M.; Doncheva, N.T.; Morris, J.H.; Bork, P.; et al. STRING v11: Protein–protein association networks with increased coverage, supporting functional discovery in genome-wide experimental datasets. *Nucleic Acids Res.* **2019**, *47*, D607–D613. [CrossRef] [PubMed]
72. Yin, W.; Xu, J.; Li, C.; Dai, X.; Wu, T.; Wen, J. Circular RNA circ_0007142 facilitates colorectal cancer progression by modulating CDC25A expression via miR-122-5p. *OncoTargets Ther.* **2020**, *13*, 3689–3701. [CrossRef]
73. Asadi, M.; Talesh, S.T.; Gjerstorff, M.F.; Shanehbandi, D.; Baradaran, B.; Hashemzadeh, S.; Zafari, V. Identification of miRNAs correlating with stage and progression of colorectal cancer. *Color. Cancer* **2019**, *8*, CRC06. [CrossRef]
74. Liu, J.; Li, H.; Sun, L.; Shen, S.; Zhou, Q.; Yuan, Y.; Xing, C. Epigenetic alternations of MicroRNAs and DNA methylation contribute to liver metastasis of colorectal cancer. *Dig. Dis. Sci.* **2019**, *64*, 1523–1534. [CrossRef] [PubMed]
75. Qu, A.; Yang, Y.; Zhang, X.; Wang, W.; Liu, Y.; Zheng, G.; Du, L.; Wang, C. Development of a preoperative prediction nomogram for lymph node metastasis in colorectal cancer based on a novel serum miRNA signature and CT scans. *EBioMedicine* **2018**, *37*, 125–133. [CrossRef]
76. Maierthaler, M.; Benner, A.; Hoffmeister, M.; Surowy, H.; Jansen, L.; Knebel, P.; Chang-Claude, J.; Brenner, H.; Burwinkel, B. Plasma miR-122 and miR-200 family are prognostic markers in colorectal cancer. *Int. J. Cancer* **2017**, *140*, 176–187. [CrossRef] [PubMed]
77. Lin, M.; Duan, B.; Hu, J.; Yu, H.; Sheng, H.; Gao, H.; Huang, J. Decreased expression of miR-193a-3p is associated with poor prognosis in colorectal cancer. *Oncol. Lett.* **2017**, *14*, 1061–1067. [CrossRef]
78. Wang, Y.-N.; Chen, Z.-H.; Chen, W.-C. Novel circulating microRNAs expression profile in colon cancer: A pilot study. *Eur. J. Med. Res.* **2017**, *22*, 1–11. [CrossRef]
79. Zhao, J.; Cao, J.; Zhou, L.; Du, Y.; Zhang, X.; Yang, B.; Gao, Y.; Wang, Y.; Ma, N.; Yang, W. MiR-1260b inhibitor enhances the chemosensitivity of colorectal cancer cells to fluorouracil by targeting PDCD4/IGF. *Oncol. Lett.* **2018**, *16*, 5131–5139. [CrossRef]

80. Zhang, J.; Luo, X.; Li, H.; Deng, L.; Wang, Y. Genome-wide uncovering of STAT3-mediated miRNA expression profiles in colorectal cancer cell lines. *BioMed Res. Int.* **2014**, *2014*, 1–11. [CrossRef]
81. Slattery, M.L.; Herrick, J.S.; Pellatt, D.F.; Stevens, J.R.; Mullany, L.E.; Wolff, E.; Hoffman, M.D.; Samowitz, W.S.; Wolff, R.K. MicroRNA profiles in colorectal carcinomas, adenomas and normal colonic mucosa: Variations in miRNA expression and disease progression. *Carcinogenesis* **2016**, *37*, 245–261. [CrossRef] [PubMed]
82. Zhang, Y.; Li, M.; Ding, Y.; Fan, Z.; Zhang, J.; Zhang, H.; Jiang, B.; Zhu, Y. Serum MicroRNA profile in patients with colon adenomas or cancer. *BMC Med Genom.* **2017**, *10*, 23. [CrossRef] [PubMed]
83. Cha, D.J.; Franklin, J.L.; Dou, Y.; Liu, Q.; Higginbotham, J.N.; Beckler, M.D.; Weaver, A.M.; Vickers, K.C.; Prasad, N.; Levy, S.; et al. KRAS-dependent sorting of miRNA to exosomes. *eLife* **2015**, *4*, e07197. [CrossRef]
84. Li, M.; Qian, X.; Zhu, M.; Li, A.; Fang, M.; Zhu, Y.; Zhang, J. miR-1273g-3p promotes proliferation, migration and invasion of LoVo cells via cannabinoid receptor 1 through activation of ERBB4/PIK3R3/mTOR/S6K2 signaling pathway. *Mol. Med. Rep.* **2018**, *17*, 4619–4626. [CrossRef]
85. Du, B.; Wu, D.; Yang, X.; Wang, T.; Shi, X.; Lv, Y.; Zhou, Z.; Liu, Q.; Zhang, W. The expression and significance of microRNA in different stages of colorectal cancer. *Medicine* **2018**, *97*, e9635. [CrossRef] [PubMed]
86. Liang, Y.; Hou, L.; Li, L.; Li, L.; Zhu, L.; Wang, Y.; Huang, X.; Hou, Y.; Zhu, D.; Zou, H.; et al. Dichloroacetate restores colorectal cancer chemosensitivity through the p53/miR-149-3p/PDK2-mediated glucose metabolic pathway. *Oncogene* **2020**, *39*, 469–485. [CrossRef] [PubMed]
87. Zhao, C.; Zhao, Q.; Zhang, C.; Wang, G.; Yao, Y.; Huang, X.; Zhan, F.; Zhu, Y.; Shi, J.; Chen, J.; et al. miR-15b-5p resensitizes colon cancer cells to 5-fluorouracil by promoting apoptosis via the NF-κB/XIAP axis. *Sci. Rep.* **2017**, *7*, 4194. [CrossRef] [PubMed]
88. Gasparello, J.; Gambari, L.; Papi, C.; Rozzi, A.; Manicardi, A.; Corradini, R.; Gambari, R.; Finotti, A. High levels of apoptosis are induced in the human colon cancer HT-29 cell line by co-administration of sulforaphane and a peptide nucleic acid targeting miR-15b-5p. *Nucleic Acid Ther.* **2020**, *30*, 164–174. [CrossRef] [PubMed]
89. Guo, F.; Luo, Y.; Mu, Y.-F.; Qin, S.-L.; Qi, Y.; Qiu, Y.-E.; Zhong, M. miR-193b directly targets STMN1 and inhibits the malignant phenotype in colorectal cancer. *Am. J. Cancer Res.* **2016**, *6*, 2463–2475.
90. Kara, M.; Yumrutas, O.; Ozcan, O.; Celik, O.I.; Bozgeyik, E.; Bozgeyik, I.; Tasdemir, S. Differential expressions of cancer-associated genes and their regulatory miRNAs in colorectal carcinoma. *Gene* **2015**, *567*, 81–86. [CrossRef]
91. Yan, W.; Liu, Z.; Yang, W.; Wu, G. miRNA expression profiles in Smad4-positive and Smad4-negative SW620 human colon cancer cells detected by next-generation small RNA sequencing. *Cancer Manag. Res.* **2018**, *10*, 5479–5490. [CrossRef]
92. Cui, Y.; Chen, L.; Yao, H.; Zhang, J.; Ding, K. Upregulation of microRNA-383 inhibits the proliferation, migration and invasion of colon cancer cells. *Oncol. Lett.* **2017**, *15*, 1184–1190. [CrossRef]
93. Ghanbari, R.; Mosakhani, N.; Sarhadi, V.K.; Armengol, G.; Nouraee, N.; Mohammadkhani, A.; Khorrami, S.; Arefian, E.; Paryan, M.; Malekzadeh, R.; et al. Simultaneous underexpression of let-7a-5p and let-7f-5p microRNAs in plasma and stool samples from early stage colorectal carcinoma. *Biomark. Cancer* **2015**, *7*, BIC.S25252. [CrossRef] [PubMed]
94. Mullany, L.E.; Herrick, J.S.; Wolff, R.K.; Stevens, J.R.; Slattery, M.L. Association of cigarette smoking and microRNA expression in rectal cancer: Insight into tumor phenotype. *Cancer Epidemiol.* **2016**, *45*, 98–107. [CrossRef] [PubMed]
95. Qiu, G.; Zhang, X.-B.; Zhang, S.-Q.; Liu, P.-L.; Wu, W.; Zhang, J.-Y.; Dai, S.-R. Dysregulation of MALAT1 and miR-619-5p as a prognostic indicator in advanced colorectal carcinoma. *Oncol. Lett.* **2016**, *12*, 5036–5042. [CrossRef]
96. Zhou, J.-M.; Hu, S.-Q.; Jiang, H.; Chen, Y.-L.; Feng, J.-H.; Chen, Z.-Q.; Wen, K.-M. OCT4B1 Promoted EMT and regulated the self-renewal of CSCs in CRC: Effects associated with the balance of miR-8064/PLK1. *Mol. Ther. Oncolytics* **2019**, *15*, 7–20. [CrossRef] [PubMed]
97. Lulla, A.R.; Slifker, M.J.; Zhou, Y.; Lev, A.; Einarson, M.B.; Dicker, D.T.; El-Deiry, W.S. miR-6883 family miRNAs target CDK4/6 to induce G(1) phase cell-cycle arrest in colon cancer cells. *Cancer Res.* **2017**, *77*, 6902–6913. [CrossRef]
98. Brown, R.E.; Short, S.P.; Williams, C.S. Colorectal cancer and metabolism. *Curr. Color. Cancer Rep.* **2018**, *14*, 226–241. [CrossRef]
99. Verrey, F. System L: Heteromeric exchangers of large, neutral amino acids involved in directional transport. *Pflügers Arch.* **2003**, *445*, 529–533. [CrossRef]
100. Fuchs, B.C.; Bode, B.P. Amino acid transporters ASCT2 and LAT1 in cancer: Partners in crime? *Semin. Cancer Biol.* **2005**, *15*, 254–266. [CrossRef]
101. Hur, K.; Toiyama, Y.; Okugawa, Y.; Ide, S.; Imaoka, H.; Boland, C.R.; Goel, A. Circulating microRNA-203 predicts prognosis and metastasis in human colorectal cancer. *Gut* **2017**, *66*, 654–665. [CrossRef] [PubMed]
102. Gebert, L.F.R.; Rebhan, M.A.E.; Crivelli, S.E.M.; Denzler, R.; Stoffel, M.; Hall, J. Miravirsen (SPC3649) can inhibit the biogenesis of miR-122. *Nucleic Acids Res.* **2014**, *42*, 609–621. [CrossRef] [PubMed]
103. Bonneau, E.; Neveu, B.; Kostantin, E.; Tsongalis, G.J.; De Guire, V. How close are miRNAs from clinical practice? A perspective on the diagnostic and therapeutic market. *EJIFCC* **2019**, *30*, 114–127. [PubMed]
104. Sharad, S.; Kapur, S. MiRNA-based therapeutics in oncology, realities and challenges (2018). In *Antisense Therapy*; Balacescu, O., Visan, S., Baldasici, O., Balacescu, L., Vlad, C., Achimas-Cadariu, P., Eds.; IntechOpen: London, UK.
105. Callegari, E.; Elamin, B.K.; Giannone, F.; Milazzo, M.; Altavilla, G.; Fornari, F.; Giacomelli, L.; D'Abundo, L.; Ferracin, M.; Bassi, C.; et al. Liver tumorigenicity promoted by microRNA-221 in a mouse transgenic model. *Hepatology* **2012**, *56*, 1025–1033. [CrossRef] [PubMed]

106. Oshima, G.; Guo, N.; He, C.; Stack, M.E.; Poon, C.; Uppal, A.; Wightman, S.C.; Parekh, A.; Skowron, K.B.; Posner, M.C.; et al. In vivo delivery and therapeutic effects of a MicroRNA on colorectal liver metastases. *Mol. Ther.* **2017**, *25*, 1588–1595. [CrossRef]
107. Fu, Y.; Chen, J.; Huang, Z. Recent progress in microRNA-based delivery systems for the treatment of human disease. *ExRNA* **2019**, *1*, 24. [CrossRef]
108. Chen, Y.; Gao, D.-Y.; Huang, L. In vivo delivery of miRNAs for cancer therapy: Challenges and strategies. *Adv. Drug Deliv. Rev.* **2015**, *81*, 128–141. [CrossRef]
109. Mercatelli, N.; Coppola, V.; Bonci, D.; Miele, F.; Costantini, A.; Guadagnoli, M.; Bonanno, E.; Muto, G.; Frajese, G.V.; De Maria, R.; et al. The inhibition of the highly expressed Mir-221 and Mir-222 impairs the growth of prostate carcinoma xenografts in mice. *PLoS ONE* **2008**, *3*, e4029. [CrossRef] [PubMed]
110. Slattery, M.L.; Mullany, L.E.; Sakoda, L.C.; Wolff, R.K.; Samowitz, W.S.; Herrick, J.S. The MAPK-signaling pathway in colorectal cancer: Dysregulated genes and their association with MicroRNAs. *Cancer Inform.* **2018**, *17*, 1176935118766522. [CrossRef]
111. Reita, D.; Bour, C.; BenBrika, R.; Groh, A.; Pencreach, E.; Guérin, E.; Guenot, D. Synergistic anti-tumor effect of mTOR inhibitors with irinotecan on colon cancer cells. *Cancers* **2019**, *11*, 1581. [CrossRef]
112. Mi, Y.; Mu, L.; Huang, K.; Hu, Y.; Yan, C.; Zhao, H.; Ma, C.; Li, X.; Tao, D.; Qin, J. Hypoxic colorectal cancer cells promote metastasis of normoxic cancer cells depending on IL-8/p65 signaling pathway. *Cell Death Dis.* **2020**, *11*, 1–11. [CrossRef] [PubMed]
113. Agani, F.; Jiang, B.H. Oxygen-independent regulation of HIF-1: Novel involvement of PI3K/AKT/mTOR pathway in cancer. *Curr. Cancer Drug Targets* **2013**, *13*, 245–251. [CrossRef]
114. Ntellas, P.; Mavroeidis, L.; Gkoura, S.; Gazouli, I.; Amylidi, A.-L.; Papadaki, A.; Zarkavelis, G.; Mauri, D.; Karpathiou, G.; Kolettas, E.; et al. Old player-new tricks: Non angiogenic effects of the VEGF/VEGFR pathway in cancer. *Cancers* **2020**, *12*, 3145. [CrossRef]
115. Ahluwalia, A.; Jones, M.K.; Szabo, S.; Tarnawski, A.S. Aberrant, ectopic expression of VEGF and VEGF receptors 1 and 2 in malignant colonic epithelial cells. Implications for these cells growth via an autocrine mechanism. *Biochem. Biophys. Res. Commun.* **2013**, *437*, 515–520. [CrossRef]
116. Boo, L.; Ho, W.Y.; Ali, N.M.; Yeap, S.K.; Ky, H.; Chan, K.G.; Yin, W.F.; Satharasinghe, D.A.; Liew, W.C.; Tan, S.W.; et al. MiRNA transcriptome profiling of spheroid-enriched cells with cancer stem cell properties in human breast MCF-7 cell line. *Int. J. Biol. Sci.* **2016**, *12*, 427–445. [CrossRef]
117. Park, H.; Hwang, S.; Jeong, J.-Y.; Jung, S.G.; Choi, M.C.; Joo, W.D.; Song, S.H.; Lee, C.; An, H.J. Integrative analysis of transcription factors and microRNAs in ovarian cancer cell spheroids. *J. Ovarian Res.* **2020**, *13*, 1–10. [CrossRef] [PubMed]
118. Kawai, S.; Yamazaki, M.; Shibuya, K.; Yamazaki, M.; Fujii, E.; Nakano, K.; Suzuki, M. Three-dimensional culture models mimic colon cancer heterogeneity induced by different microenvironments. *Sci. Rep.* **2020**, *10*, 1–11. [CrossRef] [PubMed]
119. Zoetemelk, M.; Rausch, M.; Colin, D.J.; Dormond, O.; Nowak-Sliwinska, P. Short-term 3D culture systems of various complexity for treatment optimization of colorectal carcinoma. *Sci. Rep.* **2019**, *9*, 7103. [CrossRef] [PubMed]
120. Zhang, L.; Yang, X.; Lv, Y.; Xin, X.; Qin, C.; Han, X.; Yang, L.; He, W.; Yin, L. Cytosolic co-delivery of miRNA-34a and docetaxel with core-shell nanocarriers via caveolae-mediated pathway for the treatment of metastatic breast cancer. *Sci. Rep.* **2017**, *7*, 46186. [CrossRef]
121. Morgan, R.G.; Chambers, A.C.; Legge, D.N.; Coles, S.J.; Greenhough, A.; Williams, A.C. Optimized delivery of siRNA into 3D tumor spheroid cultures in situ. *Sci. Rep.* **2018**, *8*, 1–10. [CrossRef]

biomedicines

MDPI

Systematic Review

Role of Circulating miRNAs in Therapeutic Response in Epithelial Ovarian Cancer: A Systematic Revision

Gloria Ravegnini [1,*], Pierandrea De Iaco [2,3,4], Francesca Gorini [1], Giulia Dondi [2], Isabella Klooster [5], Eugenia De Crescenzo [2], Alessandro Bovicelli [2], Patrizia Hrelia [1], Anna Myriam Perrone [2,3,4,†] and Sabrina Angelini [1,†]

1 Department of Pharmacy and Biotechnology, University of Bologna, 40126 Bologna, Italy; francesca.gorini3@unibo.it (F.G.); patrizia.hrelia@unibo.it (P.H.); s.angelini@unibo.it (S.A.)
2 Division of Oncologic Gynecology, IRCCS Azienda Ospedaliero-Universitaria di Bologna, 40138 Bologna, Italy; pierandrea.deiaco@unibo.it (P.D.I.); giulia.dondi@aosp.bo.it (G.D.); eugenia.decrescenzo2@unibo.it (E.D.C.); alessandro.bovicelli@unibo.it (A.B.); myriam.perrone@aosp.bo.it (A.M.P.)
3 Department of Medical and Surgical Sciences, DIMEC, University of Bologna, 40138 Bologna, Italy
4 Centro di Studio e Ricerca delle Neoplasie Ginecologiche, University of Bologna, 40138 Bologna, Italy
5 Department of Pathology, Brigham and Women's Hospital, 75 Francis Street, Boston, MA 02115, USA; iklooster@bwh.harvard.edu
* Correspondence: gloria.ravegnini2@unibo.it
† These authors contributed equally to this work.

Citation: Ravegnini, G.; De Iaco, P.; Gorini, F.; Dondi, G.; Klooster, I.; De Crescenzo, E.; Bovicelli, A.; Hrelia, P.; Perrone, A.M.; Angelini, S. Role of Circulating miRNAs in Therapeutic Response in Epithelial Ovarian Cancer: A Systematic Revision. *Biomedicines* **2021**, *9*, 1316. https://doi.org/10.3390/biomedicines9101316

Academic Editor: Veronique Baud

Received: 23 August 2021
Accepted: 20 September 2021
Published: 26 September 2021

Abstract: Epithelial ovarian cancer (EOC) is one of the most lethal cancers worldwide, mostly due to nonspecific symptoms and a lack of screening tests, which, taken together, contribute to delayed diagnosis and treatment. The current clinical biomarker is serum CA-125, which allows the identification of most advanced primary and relapsed disease and correlates with disease burden; however, as well highlighted in the literature, CA-125 often lacks sensitivity and specificity, and is not helpful in monitoring chemotherapeutic response or in predicting the risk of relapse. Given that, the identification of novel biomarkers able to foster more precise medical approaches and the personalization of patient management represents an unmet clinical requirement. In this context, circulating miRNAs may represent an interesting opportunity as they can be easily detected in all biological fluids. This is particularly relevant when looking for non-invasive approaches that can be repeated over time, with no pain and stress for the oncological patient. Given that, the present review aims to describe the circulating miRNAs currently identified as associated with therapeutic treatments in OC and presents a complete overview of the available evidence.

Keywords: ovarian cancer; epithelial ovarian cancer; liquid biopsy; circulating miRNAs; drug response; personalized medicine; chemotherapy

1. Introduction

1.1. Epithelial Ovarian Cancer

Ovarian cancer represents the gynecological malignancy responsible for the highest number of deaths each year in western countries [1]. Ovarian cancers are a heterogeneous group of tumors including separate entities and are divided into epithelial (about 90% of cases), germ cell (3%), and sex cord–stromal (2%) [2]. Epithelial ovarian cancers (EOCs) are the most common and are, in turn, divided into serous ovarian carcinoma (SOC), endometrioid carcinoma (EMOC), clear cell carcinoma (CCOC), mucinous carcinoma (MCOC), Brenner tumors, undifferentiated, and carcinosarcomas. These cancers are grouped into a dualistic model, type I, and type II, which reflects different clinical–molecular features. Usually, type I tumors have indolent behavior and they are often limited to the ovary at the time of diagnosis; they have a stable genome with no *TP53* mutations, even if somatic alterations can be frequently detected in different genes, such as

BRAF, *KRAS*, and *CTNNB1*. Type I tumors include low-grade SOC, EMOC, MCOC, CCOC, and Brenner tumors. On the other hand, type II tumors are more aggressive, usually identified at an advanced stage, and genetically highly unstable; the majority of them harbor *TP53* mutations, and a good portion of the cases have mutations or dysfunction of *BRCA-1/2* genes. High-grade serous EOC is the predominant histotype in type II cancers [3,4]. EOC is a relatively rare condition with the highest incidence rates in western countries such as in Europe and North America (8 cases per 100,000 population) [5,6]. This cancer is the most lethal and silent gynecological tumor with diagnosis in an advanced stage in about 80% of cases and a 5-year relative survival of only 20–30%. Primary tumors originate from the epithelium of the ovary, fallopian tube, or peritoneum and then spread to the peritoneal surface and to the viscera of the pelvis and abdomen (carcinosis). The standard approach is surgical cytoreduction followed by standard first-line chemotherapy with platinum and taxane compounds. When surgery is not feasible due to the extent of the disease, neoadjuvant chemotherapy is an option to reduce the burden of the disease and give the patient the opportunity for cytoreduction [7]. Despite optimal surgery and proper chemotherapy, approximately 70% to 80% of patients will develop a recurrent disease and gradually patients susceptible to platinum experience shorter intervals without illness, with the development of platinum resistance and poor prognosis [8–11]. In the last decade, targeted therapies including vascular endothelial growth factor (*VEGF*) inhibitors and poly (adenosine diphosphate-ribose) polymerase (*PARP*) inhibitors have been introduced with positive outcomes in clinical trials, but their role in OC therapeutic algorithms is still under debate [12].

Currently, serum markers (CA-125 and HE4) represent the only weapon to assess response to therapy, tumor progression, and disease recurrence. HE4, although of great promise, is not yet widely used in clinical practice because it has a clinical value overlapping with CA-125 [13]. Thus, serum CA-125 is the only biomarker available, but it lacks sensitivity and specificity that do not allow routine use for early diagnosis even in combination with other tools such as ultrasound [14]. It should be noted that a part of ovarian carcinomas is also CA-125 silent. It remains useful in the gross evaluation of response to therapy and its increase in treated patients is often a sign of disease recurrence. However, CA-125 assay does not correlate with the prediction of platinum sensitivity or resistance. In the absence of specific markers, presently, response therapy is evaluated surgically by diagnostic laparoscopy on the accuracy of laparoscopy to assess peritoneal spread in ovarian cancer [15].

1.2. Liquid Biopsy

Currently, the gold standard approach for the histological diagnosis and genetic–molecular characterization of EOC is tissue biopsy; however, standard biopsy is an invasive procedure that provides a static picture of the disease, strictly related to the portion of tissue analyzed. Given its invasiveness, it cannot be repeated easily over time, thus, not providing it does not provide a reliable, dynamic image of tumor evolution. An appealing alternative approach attempting to overcome these limits is liquid biopsy, which allows the detection of circulating molecules directly released by the tumor mass in body fluids. Not surprisingly, in recent years there has been a growing research interest in this field. Liquid biopsy allows access, through a non-invasive approach, to molecular information or identifies specific biomarkers in biological fluids (including, but not limited to, blood, ascitic fluid, urine, saliva), which could be very helpful in better characterizing a cancer patient [16,17].

In the last decade, research advances have boosted several steps forwards, promoting in 2016 the FDA approval of the first diagnostic test based on liquid biomarkers for non-small cell lung cancer (NSCLC) [18,19]. This test can detect specific *EGFR* mutations in the blood of NSCLC patients, who would not be able to provide a tumor biopsy for conventional EGFR testing due to advanced tumor stage, comorbidities, or tissue inadequacy. This fosters the choice of the most suitable treatment, maximizing the benefits for those patients. As demonstrated by this, much progress has been directed to diagnosis; however, another important potential application of liquid biopsy is therapeutic monitoring in order to achieve more personalized treatment [20–22].

1.3. microRNAs

Body fluids contain several types of molecules including circulating tumor cells (CTCs), circulating nucleic acids (both DNA and RNA), and extracellular vesicles (EVs) [23–25]. Among those, in particular, circulating microRNAs (miRNAs) have attracted research interest due to their extraordinary stability in body fluids [26]. MiRNAs are small non-coding RNAs (snRNA) that play an important role in gene regulation [27]. miRNAs modulate gene expression by binding a complementary sequence of a target mRNA [28,29]. Of note, a single miRNA can regulate the expression of hundreds of mRNAs and, conversely, an mRNA may present numerous sequences that can guarantee interaction with multiple miRNAs [30]. When taking into consideration the involvement of miRNAs in several biological processes, it is clear that they may play a role in many diseases, including cancer [31,32]. An aberrant miRNA profile is indeed associated with tumor development, progression, metastasis process, and chemotherapy response, suggesting their possible use not only as diagnostic biomarkers but also as predictive biomarkers of therapeutic response [33]. In particular, in recent years, growing interest has been paid to circulating miRNAs, which are detected in body fluid as complexed with other RNA binding proteins or enclosed in EVs. In both cases, the miRNA is protected from enzymatic degradation, ensuring that it can carry out its function [34,35]. Over the years, many studies have identified different miRNAs as potential diagnostic and prognostic biomarkers in EOC. However, in most of these, deregulation was observed when comparing the tumor with a normal counterpart or healthy tissue. On the contrary, the studies that analyzed miRNA expression in relation to pharmacological response are limited and with a small consensus. Given these premises, the aim of this review is to provide a picture of the current knowledge on circulating miRNAs identified to be significantly associated with EOC clinical response.

2. Methods

Systematic Review of Studies Investigating Circulating miRNAs in Therapeutic Response in EOC Patients

To this purpose, we systematically searched for papers analyzing expression of circulating miRNA in EOC in relation to prognostic and molecular classifications.

The systematic review was conducted in accordance with the PRISMA Statement principles [36]. The research question was "can miRNAs be used as biomarkers to monitor clinical response in EOC?", and it was determined using the PICOS process (population, intervention, comparison, outcomes, study design) [37]. PubMed, Web of Knowledge, and Scopus databases were systematically searched for original articles analyzing the circulating miRNAs associated with drug response in EOC (last updated search 1 August 2021). The papers included in this revision are summarized in Table 1. Relevant studies were selected using the Boolean combination of the following key terms: "miR OR miRNA or miRNAs or microRNA" AND "circulating OR plasma OR whole blood OR serum OR ascites OR effusions OR exosome OR exosomes OR exosomal" AND "ovarian cancer OR tumor OR tumour OR neoplasia OR carcinoma OR tumors OR tumours OR cancers OR carcinomas" AND "adjuvant OR neoadjuvant OR clinical response OR chemotherapy OR treatment response". Additionally, the reference lists of reviews, meta-analyses, and all original studies were hand-searched to acquire further relevant studies missed from the initial electronic search (Figure 1).

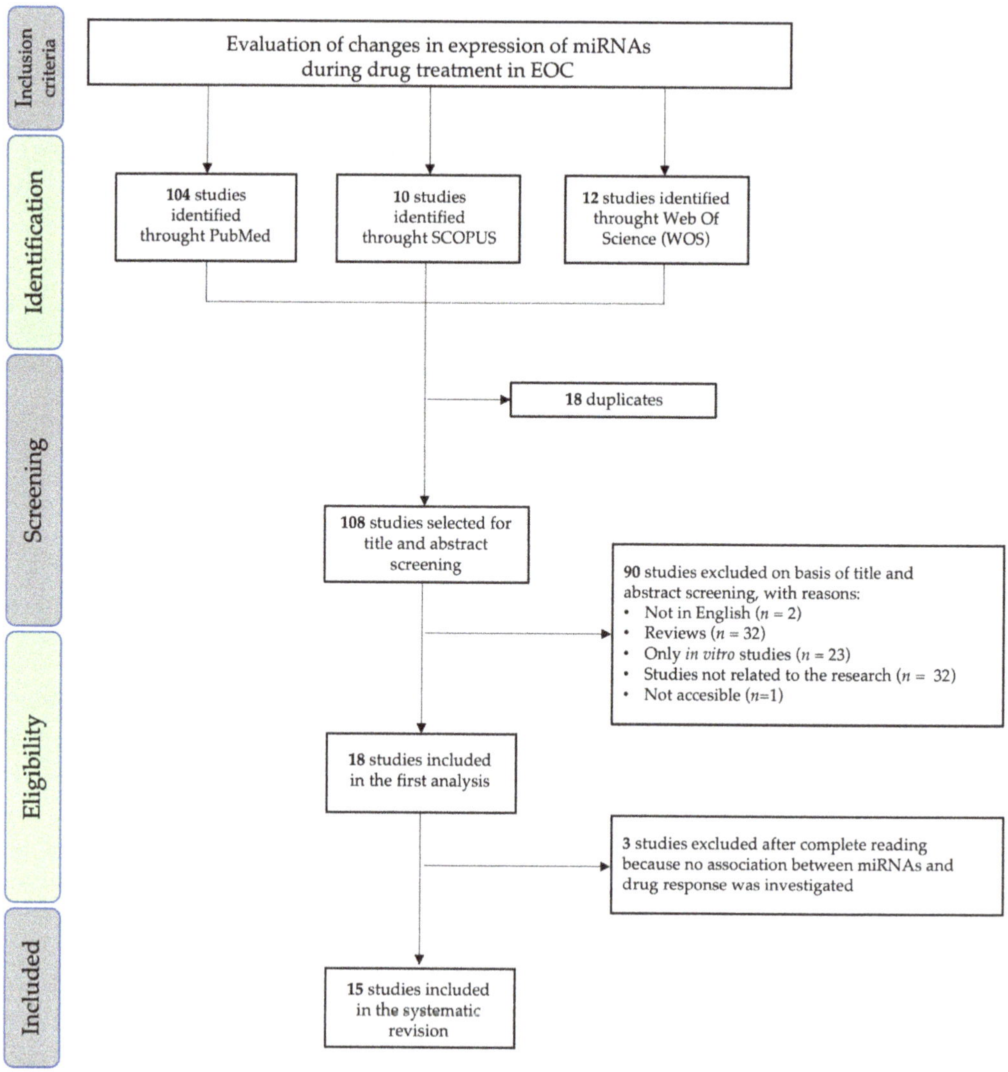

Figure 1. Workflow of the systematic review.

Eligible studies were required to meet the following inclusion criteria: studies evaluating circulating miRNAs in relation to therapy in EOC. Exclusion criteria were: (i) meta-analyses, reviews, and editorials; (ii) non-human studies; (iii) in vitro studies; (iv) non-English articles.

After removing duplicate studies, two investigators (GR and FG) independently checked titles and abstracts of the retrieved articles and judged their eligibility. One study was not accessible [38]. Then, the entire text of potentially eligible studies was evaluated to assess appropriateness of inclusion in this systematic review. The same two authors independently extracted the following data from the selected papers: (1) first author, publication year, and aim; (2) sample size; (3) type of drug; (4) type of body fluid (plasma/serum/exosomes), techniques used, and validations; (5) type of association between circulating miRNA and clinical outcome. The results are reported in Table 1.

Table 1. Studies included in the systematic review.

Author, Year, [ref.]	Aim of the Study	Number of Patients	Additional Details and Histology (If Reported)	Therapy	Biological Matrix	Technique/s Used	Validation of the Results	Most Important Findings
miRNAs expression in chemotherapy resistant and sensitive OC patients								
Li et al., 2021 [39]	To characterize the expression of hsa-miR-105 in PTX-resistant EOC	105 EOC pts: $n = 59$ resistant, $n = 56$ sensitive to chemo	Primary diagnosis EOC	TX-based chemo	Plasma	qRT-PCR	Cell lines and xenograft models	↓ miR-105 in PTX-resistant EOC, to PTX-responsive EOC ($p < 0.0001$).
Chen et al., 2020 [40]	To investigate serum miR-125b as a biomarker for diagnosis and prediction of treatment response in EOC	83 EOC pts: $n = 35$ resistant, $n = 48$ sensitive to chemo	Primary diagnosis EOC	PT and TX-based chemo	Serum	qRT-PCR	\	↓ miR-125b in PT-resistant EOC pts
Biamonte et al., 2019 [41]	To explore the functional roles of let-7g in EOC	17 EOC pts: $n = 9$ resistant, $n = 8$ sensitive to chemo	Primary diagnosis HGSOC (Stage IIIc–IV)	PT + TX + BVZ	Serum	qRT-PCR	Cell lines	↓ let-7g in resistant EOC pts
Kuhlmann et al., 2019 [42]	To explore the signature of EV-associated miRNAs in PT-resistant EOCs	30 EOC pts: $n = 15$ resistant, $n = 15$ sensitive to chemo	SOC	PT and TX-based chemo	Exosomes from plasma	Illumina NGS	\	12 miRNAs (miR-181a-2-3p, miR-1908-5p, miR-1304-3p, miR-486-3p, miR-21-3p, miR-548o-3p, miR-1185-1-3p, miR-223-5p, miR-664-5p, miR-345-5p, miR-625-3p, miR-443b-3p); however, after adjustment, no significance maintained
Fukagawa et al., 2017 [43]	To identify candidate circulating miRNAs as biomarkers, and potential therapeutic targets	Profiling in 12 EOC pts: $n = 6$ resistant, $n = 6$ sensitive to chemo. Validation in 98 sera	Primary diagnosis EOC (Profiling: $n = 7$ SOC, $n = 7$ EMOC, $n = 1$ CCOC; Validation: $n = 34$ SOC, $n = 16$ EMOC, $n = 6$ MCOC, $n = 28$ CCOC, $n = 14$ other)	PT and TX-based chemo	Serum	Agilent Microarray; qRT-PCR	Independent cohort of pts; cell lines and xenograft models	↑ miR-135a-3p associated with ↑ OS

Table 1. *Cont.*

Author, Year, [ref.]	Aim of the Study	Number of Patients	Additional Details and Histology (If Reported)	Therapy	Biological Matrix	Technique/s Used	Validation of the Results	Most Important Findings
Longitudinal analysis of miRNA levels to monitor chemotherapy response								
Robelin et al., 2020 [44]	To identify specific circulating miRNAs to monitor disease burden and guide clinicians in decision making for EOC pts	Profiling in 8 EOC pts; validation in 111 OC pts	Primary diagnosis EOC (Profiling: $n = 8$ SOC; Validation: $n = 98$ SOC, $n = 1$ EMOC, $n = 1$ CCOC, $n = 1$ MCOC, $n = 3$ Undifferentiated, $n = 7$ NA)	PT and TX-based chemotherapy +/− nintedanib and debulking surgery	Plasma	miScript miRNA PCR Array (Qiagen); qRT-PCR	Independent cohort of pts	The longitudinal kinetics of miRNA expressions were highly inconsistent and there was no relation with the CA-125 dynamics
Zhu et al., 2019 [45]	To analyze the correlation between exosomal miR-223 and recurrence	12 relapsed EOC pts. 2 time points: at the time of surgery and after recurrence	SOC (stage IIIC–V)	PT and TX-based chemo	Exosomes from serum	qRT-PCR	Cell lines and xenograft models	↑ miR-223 at recurrence vs. time of surgery
Kobayashi et al., 2018 [46]	To identify circulating miRNAs as potential diagnostic and prognostic biomarkers in HGSOCs	16 EOC pts. 2 time points: before surgery and after the first post-surgical chemo cycle (about 28 days after surgery)	Primary diagnosis HGSOC	PT and TX-based chemo	Serum	qRT-PCR	Cell lines	↓ miR-1290 after debulking surgery and chemo
Grabosch et al., 2017 [47]	To confirm the feasibility of collecting serial peritoneal samples from implanted catheters in EOC pts receiving IP chemo	13 EOC pts. 3 time points: after surgery, before chemo (T0) and after the first (T1) and second (T2) cycles of chemo	Primary diagnosis EOC ($n = 9$ SOC, $n = 3$ EMOC, $n = 1$ CCOC)	PT + TX + BVZ-based IP chemo	Plasma ($n = 9$) and PW/PF ($n = 4$)	NanoString nCounter miRNA Expression Assay	\	In plasma, T0 vs. T1: 55 miRNAs deregulated; T1 vs. T2: 33 miRNAs deregulated In PW/PF, T0 vs. T1: 12 miRNAs deregulated; T1 vs. T2: 33 miRNAs deregulated

Table 1. *Cont.*

Author, Year, [ref.]	Aim of the Study	Number of Patients	Additional Details and Histology (If Reported)	Therapy	Biological Matrix	Technique/s Used	Validation of the Results	Most Important Findings
Benson et al., 2015 [48]	To identify alterations in circulating miRNAs associated with decitabine followed by carboPT chemo treatment	14 EOC pts. 2 time points: at baseline and on day 29 after first cycle of chemo	EOC, progressed to previous PT-based chemo	Decitabine followed by carboplatin chemo	Plasma	qRT-PCR miRNA OpenArrays (Thermo)	\	In the overall cohort, T0 vs. T1: ↓ miR-193a-5p and miR-375 after chemotherapy; In the non-responder pts, T0 vs. T1: ↑ miR-339-3p, miR-340-5p, miR-133a, and miR-10a, ↓ miR-375, miR-25-3p, and miR-148b-5p. In decitabine, sensitive vs. resistant pts: ↑ miR-616, miR-532-3p, and miR-148b-5p after first cycle of chemo
Kapetanakis et al., 2015 [49]	To assess the plasma levels of miR-200b in EOCs in a longitudinal study	33 EOC pts: $n = 9$ unresectable tumors treated with chemo, $n = 14$ debulking after chemo, $n = 10$ direct debulking. 2 time points: pre- and post-chemo	HGSOC	PT and TX-based chemo	Plasma	qRT-PCR	\	Pre vs. post-chemotherapy: ↓ miR-200b in 33% of unresectable tumors versus in 54% for tumors resectable immediately or after neoadjuvant chemo
Kuhlmann et al., 2014 [50]	To identify deregulated miRNAs/snRNAs in sera of EOC pts and investigate their potential in therapy monitoring	69 EOC pts. 2 time points: before surgery ($n = 63$) and after post-surgical chemo ($n = 56$)	Primary diagnosis EOC ($n = 45$ SOC, $n = 5$ MCOC, $n = 5$ EOC, $n = 3$ CCOC, $n = 4$ mixed, $n = 7$ other)	PT-based chemotherapy	Serum	Agilent Microarray; qRT-PCR	\	↑ RNU2-1f in pts with residual abdominal tumor mass after chemotherapy and PT resistance. In 50 pts with available paired serum samples before surgery and after adjuvant chemotherapy: pts with persistently RNU2-1f-positive levels had ↓ PFS and OS

Table 1. *Cont.*

Author, Year, [ref.]	Aim of the Study	Number of Patients	Additional Details and Histology (If Reported)	Therapy	Biological Matrix	Technique/s Used	Validation of the Results	Most Important Findings
Shapira et al., 2014 [51]	To analyze circulating miRNAs as potential biomarkers for EOC detection and outcome	5 EOC pts. 2 time points: before surgery and after post-surgical chemotherapy	Primary diagnosis EOC	PT-based chemotherapy (not clearly indicated)	Plasma	qRT-PCR miRNA OpenArrays (Thermo)	\	↓ miR-1274a, miR-1274b, and miR-1290 after treatment; ↑ miR-19b, miR-25, miR-195, and miR-16 in post-chemotherapy samples
Association between miRNAs and clinical response								
Vigneron et al., 2020 [52]	To assess the predictive value of circulating miR-622 prior to first-line chemotherapy and at relapse	130 EOC pts (n = 65: prospective cohort, n = 65 retrospective cohort; additional n = 35 at relapse, from the retrospective cohort)	Newly diagnosed HGSOC (stages III–IV)	PT and TX-based chemotherapy	Serum	qRT-PCR	Independent cohort of pts (prospective and retrospective)	↑ miR-622 in pts with ↓ PFS
Halvorsen et al., 2017 [53]	To identify circulating miRNAs able to identify EOC pts at high risk for relapse	207 EOC pts: Profiling in 91 EOC pts; validation in 116 EOC pts	Primary diagnosis EOC (Profiling: n = 58 SOC, n = 6 EMOC, n = 2 MCOC, n = 13 CCOC, n = 8 mixed, n = 4 other; validation: n = 79 SOC, n = 6 EMOC, n = 0 MCOC, n = 14 CCOC, n = 13 mixed, n = 4 other)	PT and TX or PT and TX-based chemotherapy + BVZ	Plasma	Taqman miRNA low density array (Thermo); qRT-PCR	Independent cohort of pts	↓ miR-200c in pts with ↑ OS treated with BVZ

BVZ: bevacizumab; Chemo: chemotherapy; CCOC: clear cell ovarian carcinoma; EOC: epithelial ovarian cancer; HGSOC: high-grade serous ovarian cancinoma; EMOC: endometrioid ovarian carcinoma; IP: intraperitoneal; MCOC: mucinous ovarian carcinoma; PF: peritoneal fluid; PW: peritoneal washing; pts: patients; PTX: paclitaxel; PT: platinum; SOC: serous ovarian carcinoma; TX: taxane; ↑: higher; ↓: lower; \: information not available.

3. Results

We included in the final review a total of 15 works. The majority of the studies analyzed miRNAs in plasma or serum, a small portion (n = 3) investigated exosomal miRNAs, and one analyzed peritoneal washing (PW) and fluid (PF). Overall, the studies retrieved can be divided in three different groups based on the main goal (Figure 2): (i) comparing miRNA expression in chemotherapy-resistant and -sensitive OC patients; (ii) longitudinal analysis of miRNA levels to monitor chemotherapy response; (iii) identifying potential associations between miRNAs and chemotherapy response (i.e., in terms of progression-free survival (PFS) or overall survival (OS)).

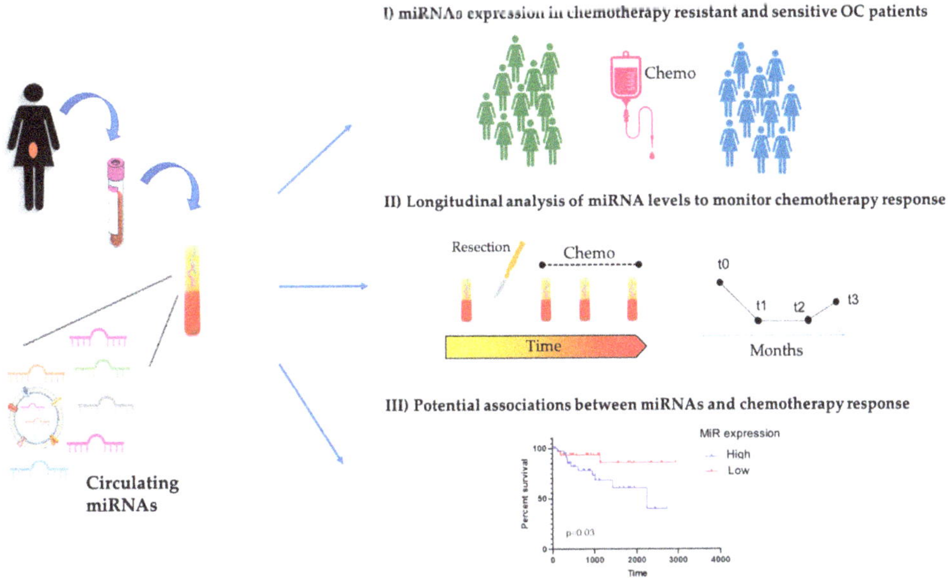

Figure 2. Main types of studies investigating circulating miRNAs and therapeutic response in EOC.

3.1. miRNA Expression in Chemotherapy-Resistant and -Sensitive EOC Patients

The first study analyzing the expression of circulating miRNAs in chemotherapy-sensitive and -resistant EOC patients was published in 2017 [43]. Resistance was defined as relapse occurring ≤6 months following the completion of chemotherapy. The authors first analyzed the miRNAs' global expression profile in 12 EOC patients, of which six were platinum-resistant (had recurrence within 6 months after completion of platinum and taxane-based treatment) and six showed platinum sensitivity. Based on this comparison, the authors identified three deregulated miRNAs (miR-135a-3p, miR-630, and miR-1207), which were further validated in 98 EOC sera. In particular, after having stratified the patients based on the median value for each miRNA, they showed that EOCs with higher miR-135a-3p had significantly improved OS compared to the patients with lower miRNA levels. To provide clinical insights in EOC, miR-135a-3p expression in sera was compared with the one in peritoneal fluid and tissue samples of patients with EOC, ovarian cysts, normal ovaries, or endometrial cancer. In all these comparisons, the biological matrix related to EOC patients showed lower miR-135a expression. Finally, functional studies demonstrated that in OC cell lines (SKOV-3 and ES-2), enhanced miR-135a-3p expression was able to promote cisplatin and paclitaxel sensitivity and suppress cell proliferation and xenograft tumor growth. Subsequently, Kuhlmann et al. evaluated the exosomal miRNAs in 30 EOC patients by Illumina NGS [42]; among those, 15 patients recurred within 6 months after the adjuvant platinum-based chemotherapy, whereas 15 remained platinum-sensitive. In addition, the authors compared different EV-enrichment strategies for optimizing the miRNA isolation and library preparation. The results showed the deregulation of 12 miRNAs (hsa-miR-181a-2-3p, hsa-miR-1908-5p, hsa-miR-1304-3p, hsa-miR-486-3p, hsa-miR-21-3p, hsa-miR-548o-3p, hsa-miR-1185-1-3p, has-miR-223-5p, hsa-miR-664-5p, hsa-miR-345-5p, hsa-miR-625-3p, and hsa-miR-443b-3p); however, after adjustment, the findings did not maintain statistical significance. However, the results are of potential interest considering that among these miRNAs, a few (miR-181a, miR-1908, miR-21, miR-486, and miR-223) were previously reported in EOC [54–57]. Besides these two papers investigating large profiles of miRNAs, the other works published in the literature explored single miRNAs from previous evidence on different cancer types. Biamonte and colleagues explored the role of let-7g in EOC and chemoresistance. The analysis started from an in vitro evaluation in two OC cell lines showing that let-7g acts as a tumor suppressor in EOC and that its enhanced expression promotes higher sensitivity to cisplatin treatment. To further corroborate the results, let-7g levels were evaluated in the tissue and serum of 17 EOC patients, highlighting that in both cases let-7g was expressed at a significantly lower level in chemotherapy-resistant cases (n = 9) compared to chemo-sensitive cases (n = 8). Another example of a single miRNA investigated in EOC and chemoresistance is miR-125b. This miRNA was previously characterized in EOC specimens as markedly poorly expressed [58,59], but its correlation with therapeutic response had not been investigated. In a recent work, Chen and colleagues [40] first compared circulating miR-125b in sera from EOC (n = 152), healthy controls (n = 42), and benign and borderline tumors (n = 30 and n = 35, respectively) and confirmed that lower levels were detected in EOC patients. In this cohort of EOC cases, miR-125b was also correlated with FIGO stage and lymph node metastasis. With regard to chemotherapy resistance, the authors showed that sensitive patients had miR-125b upregulation compared to the non-sensitive patients. More recently, Li et al. deepened the understanding of the role of miR-105 in EOC starting by data mining publicly available datasets comprising the miRNA profiling of EOC cells and their PTX-resistant sublines [39]. Based on that, miR-105 was significantly downregulated in PTX-resistant cell lines compared to parental ones, and this deregulation was further confirmed by the same authors by generating a set of two additional PTX-resistant models (exposing PTX-sensitive cells to increasing doses of PTX) and their matched xenograft models. In both cells and xenografts, lower miR-105 expression was significantly associated with PTX resistance. To further test these findings, tissue and sera clinical specimens from 105 EOC patients were analyzed. The results revealed that miR-105 was significantly decreased in both tissue and sera derived from PTX-resistant patients compared with the PTX-responsive cases. With regard to the circulating miR-105 in particular, high plasmatic levels were associated with improved responsiveness to PTX. All the studies applied the same 6-month cut off to define resistance.

3.2. Longitudinal Analysis of miRNA Levels to Monitor Chemotherapy Response

Longitudinal analysis of miRNA levels, through the collection of multiple blood samples over time, is particularly interesting because it may display the peculiar deregulation of certain miRNAs potentially correlated with poor or good response to specific drugs, including chemotherapy.

The first studies in EOC with this purpose analyzed a small sample size of patients.

Shapira et al. investigated plasma samples of 42 EOC patients; however, the association between miRNAs and therapeutic response was evaluated in only five cases, with OS > 4 years, for whom blood samples were collected both before surgical resection and after chemotherapy [51]. The comparison showed seven differentially expressed miRNAs between presurgical and post-chemotherapy time points; in particular, miR-1274a, miR-1274b, and miR-1290 were decreased after treatment, whereas miR-19b, miR-25, miR-195, and miR-16 displayed over-expression in post-chemotherapy samples. Comparison between plasma collected before and within 2 weeks from the surgical resection did not show any difference. Similarly, Kuhlmann and colleagues started their analysis by profiling miRNA expression in five EOC patients and five healthy controls [50]; based on the results, one snRNA, *RNU2-1f*, was selected for further validation in 69 sera, of which $n = 63$ were collected before surgery and $n = 56$ after adjuvant platinum-based chemotherapy. The detection of *RNU2-1f* within the profiling was made by two probes (miR-1246 and miR-1290) that were previously shown to be specific for *RNU2-1* since they detect fragmented forms of *RNU2-1* [60,61]. While the expression of *RNU2-1* was confirmed to be higher in EOC patients versus the healthy controls (independently by the specific time points), no differences were observed between preoperative circulating *RNU2-1f* and after adjuvant regimen. Interestingly, for a subset of 15 patients with suboptimal primary debulking, radiographic reports on restaging after chemotherapy were available; of these, 10 were defined platinum-sensitive and five resistant. The levels of *RNU2-1f* were significantly higher in patients with residual abdominal tumor mass after chemotherapy and platinum resistance. Finally, for 50 patients, for whom paired serum samples before surgery and after adjuvant chemotherapy were available, *RNU2-1f* abundance dynamics were evaluated. Kaplan–Meier analysis highlighted that the patients who had persistently *RNU2-1f*-positive levels at primary diagnosis and after chemotherapy showed significantly shorter PFS and OS than the other patients.

Kapetanakis et al. evaluated the expression of miR-200b in 33 patients, with blood samples collected before a diagnostic laparoscopy and at the end of the primary treatment (treatment including chemotherapy and debulking surgery when feasible), 4-8 months after the initial laparoscopy [49]. The authors also evaluated the association between miR-200b and the serum marker CA-125. CA-125 levels returned to normal plasma concentrations within the first months of the treatment, even among patients with unresectable tumors. On the contrary, expression levels of miR-200b were quite heterogeneous among the different types of EOCs. In general, the proportion of patients with decreasing concentrations of miR-200b was 33% for unresectable tumors versus 54% for patients with resectable tumors treated with adjuvant or neoadjuvant chemotherapy. For 24 out of 33 EOCs, follow-up longer than 10 months was available and the miR-200b level was analyzed in association with PFS. Patients with a miR-200b-negative variation pre- and post-chemotherapy showed significantly longer PFS, compared with the remaining patients, even after adjustment for multiple variables. All these data, taken together, suggest that specific miRNAs could be more sensitive liquid biomarkers than CA-125, which is currently widely used in clinical management. Similarly, Kobayashi et al. evaluated miR-1290 with a longitudinal approach [46]. The work originated from an miRNA profiling from in vitro models of OC and normal ovary cells, which was then validated in clinical specimens, confirming a higher expression of miR-1290 in EOC patients compared with healthy control sera. The same miRNA was also evaluated before and after the first cycle of adjuvant chemotherapy in 16 patients. In line with the other results, miR-1290 expression was significantly decreased after debulking surgery and chemotherapy, suggesting that circulating miR-1290 may be directly related to tumor burden. Similarly, Zhu and collaborators characterized the role of exosomal miR-223 in chemoresistance, starting with a careful in vitro study in cell lines and xenograft models [45]. The authors demonstrated that exosomal miR-223 derived from macrophages was able to foster drug resistance in EOC cells and that its upregulation is directly associated with a chemoresistant phenotype. To further test this hypothesis, the authors compared sera of 12 patients collected before and after resistance occurrence, confirming an increased miR-223 expression at the time of recurrence.

More recently, Robelin et al. published a longitudinal report including a large number of patients ($n = 119$), which is, so far, the widest series investigated [44]. The enrolled patients received standard neoadjuvant and adjuvant chemotherapy (three to four cycles) before and after cytoreductive surgery, followed by a maintenance treatment with nintedanib/placebo for up to 2 years. In total, the authors were able to assess 756 serial blood samples. From a profiling of 84 miRNAs in eight patients, and from literature data, 11 miRNAs (iR-15b-5p, miR-16-5p, miR-20a-5p, miR-21-5p, miR-93-5p, miR-122-5p, miR-150-5p, miR-195-5p, miR-200b-3p, miR-148b-5p, and miR-34a-5p) were selected to be further tested in serial blood samples derived from 111 EOC cases. However, even with the good clinical design of the study, the results were mainly negative; indeed, as clearly highlighted by the same authors, the longitudinal kinetics of the 11 miRNA expressions were highly inconsistent, and no relation with CA-125 dynamics was identified. The miRNA changes during neoadjuvant treatment were not found to be associated with RECIST tumor response or outcomes. The conclusion of this study indicates, therefore, a lack of assessable longitudinal prognostic or predictive kinetic profiles for the selected miRNAs, which cannot be automatically applied to other miRNAs.

Finally, among the papers investigating circulating miRNAs and therapeutic response, two are different but deserve to be included in this list. In particular, the first, by Benson et al., evaluated plasmatic miRNA levels in EOC patients treated with a regimen of low dose decitabine—a DNA methyltransferase inhibitor—and carboplatin [48]. The second, published by Grabosch and colleagues, investigated circulating miRNAs in serial peritoneal samples in women receiving intraperitoneal (IP) chemotherapy [47]. The analysis by Benson and collaborators included 14 EOC patients enrolled in the previously described open label phase II clinical trial [62]. This study is particularly relevant because, among the works herein described, it is the only one focused on platinum-resistant, recurrent patients treated with an alternative drug. The aim of this report was to characterize the alterations in circulating miRNAs associated with decitabine followed by a carboplatin chemotherapy regimen and clinical response. To this purpose, plasma samples were collected before treatment and after the completion of the first cycle of treatment (day 29). Among the 14 patients, $n = 8$ showed tumor progression prior to six cycles of chemotherapy and were considered non-responders, whereas the remaining six were considered responders. By simultaneously analyzing 93 miRNAs, the authors identified 10 miRNAs related to response to decitabine followed by carboplatin chemotherapy. In detail, miR-193a-5p and miR-375 decreased after chemotherapy; moreover, in the non-responder patients, four miRNAs (miR-339-3p, miR-340-5p, miR-133a, and miR-10a) displayed increased levels, while three miRNAs (miR-375, miR-25-3p, and miR-148b-5p) showed a significant decrease. MiRNA expression was compared also in resistant and sensitive patients at the post-treatment timepoint; in this regard, the authors observed three miRNAs (miR-616, miR-532-3p, and miR-148b-5p) that were significantly increased in responders. Finally, Kaplan–Meier analysis was applied to evaluate if any of the miRNA alterations were able to predict treatment response. In this case, four patients were excluded due to their progression within the first cycle of chemotherapy. The remaining 10 were divided into two groups (high and low expression) based on the median value of each miRNA, showing that a lower concentration of miR-148b-5p on day 29 was associated with disease progression.

The second previously mentioned study aimed to assess miRNA expression in serial peritoneal samples from implanted catheters in women receiving IP chemotherapy. The analysis involved 13 women, and, besides miRNAs, other potential biomarkers were evaluated, including, but not limited to, immune genes and cytokines. miRNAs were profiled in plasma ($n = 9$), peritoneal fluid (PF, $n = 1$), and peritoneal wash (PW, $n = 3$) at three time points (T0: after surgery, before chemo; T1: after the first cycle of chemo; T2: after the second cycle of chemo) using the NanoString nCounter miRNA Expression Assay. In plasma, after the first round of chemo (T0 vs. T1) and after the second cycle of chemotherapy (T1 vs. T2), 51 and 33 miRNAs were deregulated, respectively, eight of which were in common. When altered, the miRNA tended to remain expressed in the same direction (up or downregulated from baseline). On the contrary, in PW, a larger number of miRNAs were deregulated after the second cycle of chemo (T0 vs. T1: 12 miRNAs; T1 vs. T2: 33 miRNAs). As suggested by the authors, plasma miRNAs may be modulated by early changes due to systemic effects of chemotherapy. In contrast, PW miRNAs can be related to later local tumor changes. Interestingly, observing the deregulated miRNAs in plasma and PW, no overlap was detected, implying that the alterations of miRNAs happening in PF/PW (at local level) could be not detected by analyzing plasma miRNAs. In this context, PF/PW evaluation could be particularly appealing to accurately monitor molecular changes, assess response to therapy, or to develop more personalized therapeutic approaches.

3.3. Association between miRNAs and Clinical Response

The first work to assess the association between miRNAs and clinical response—in terms of PFS or OS—in EOC dates to 2017 [53]. Specifically, Halvorsen et al. enrolled 207 EOC patients, under standard chemotherapy or in association with bevacizumab, aiming at identifying circulating miRNAs able to discriminate patients at high risk for relapse. The discovery step assessed the levels of 754 miRNAs in 91 sera. The remaining 116 patients were included in the validation cohort; patients were stratified based on treatment type and survival length (in long or short PFS). Four miRNAs (miR-1274a, miR-141, miR-200b, and miR-200c) were shown to be significantly associated with survival. In the validation set, miR-141 and miR-200b confirmed the prognostic association. Considering the treatment, no difference in PFS related to miRNAs was observed in the discovery set; however, in the validation set, low levels of miR-200c were associated with significantly better survival in patients treated with bevacizumab (with 5-month prolongation of PFS) compared to standard chemotherapy. No additional associations were reported. Vigneron et al. analyzed the ability of miR-622 to predict platinum response [52].

This miRNA has been reported to be involved in the homologous recombination repair system, which plays a role in the platinum mechanism of action [63,64]. The authors analyzed miR-622 in two distinct cohorts of 65 HGSOC patients (one prospective and one retrospective) treated with adjuvant platinum and taxane-based chemotherapy. The sera were collected before the first cycle of chemotherapy; moreover, for 35 patients included in the retrospective cohort, an additional serum sample was available at the time of relapse. Each cohort was sorted into miR-622 low and high expression based on a cut-off value. In the prospective group, the high expression of miR-622 group was associated with significantly lower PFS compared with the patients showing lower miR-622 levels; similarly, high miR-622 expression was correlated with lower OS. In the validation, in the retrospective cohort, applying the same cut-off value, high miR-622 expression was correlated with lower OS; however, in the multivariate analysis, this did not maintain statistical significance. With regard to the predictive value at relapse, the 35 patients were divided into short-term (<12 months) and long-term (>12 months) survivors according to the OS and a new cut-off value was calculated by an ROC curve. Once again, high miR-622 levels were correlated with lower OS compared with patients with lower miR-622. The new cut-off value was re-applied to the retrospective cohort, and this time the correlation of high miR-622 expression/lower OS maintained statistical significance even in the multivariate analysis. All these results together showed that miR-622 was an independent predictive factor of PFS and OS in the prospective cohort, prior to first-line chemotherapy; in the retrospective cohort, miR-622 was a predictive factor of OS before first-line chemotherapy and at the time of relapse.

4. Discussion

Ovarian carcinoma is one of the most lethal cancers worldwide; this is mostly due to its unspecific symptoms and the lack of screening tests, which, taken together, contribute to delaying diagnosis and treatment. The current serum biomarker, CA-125, lacks sensitivity and specificity. It is useful in identifying primary and relapsed disease and correlates with disease burden, but is inadequate in the response to chemotherapy and risk of relapse [10,11].

Given that, the identification of novel biomarkers able to foster more precise medical approaches and the personalization of patients' management represents an unmet clinical requirement. In this context, circulating miRNAs may represent an interesting opportunity as they are highly stable and can be easily detected in all biological fluids, including blood samples. This is particularly relevant when looking for non-invasive approaches that can be repeated over time, with no pain and stress for the oncological patient. Based on this, it is reasonable to think that miRNAs could potentially be integrated into the existing prognostic outline and promote a better patient management. In this regard, the present review aimed to describe the circulating miRNAs currently reported as associated with therapeutic treatments in EOC. Considering that most of reports have investigated tissue miRNAs, the available literature results are limited and we were able to identify only 15 studies focused on our topic. Among those, the majority analyzed serum/plasma miRNAs, three exosomal miRNAs, and one evaluated PF/PB. Eight of 15 used large profiling to simultaneously screen multiple miRNAs, whereas the remaining adopted RT-PCR as the main technique to evaluate a limited number of miRNAs; four studies had an independent cohort of patients to validate their preliminary findings and five described functional validations in cell lines and/or animal models.

Overall, as previously mentioned, the available reports can be divided according to their main goals, thus identifying three main groups (Figure 2); however, even considering the specific aims, the consensus among the studies remains very limited. With all aspects taken together, it is understandable that no clinical translation has happened, and it seems that further extensive research will be needed to define reliable miRNAs as candidate biomarkers. In addition, the lack of standardized protocols, including sample collection, the type of biological fluid, RNA extraction, and techniques, makes it challenging to compare the results between independent studies. We should also be aware that it would be particularly difficult to identify one or a few miRNAs that are able, by themselves, to accurately monitor therapeutic response in EOC patients based on molecular or clinical features. The best approach would be combining multiple variables (including, but not limited to, miRNAs, any DNA mutations, and clinical parameters).

Recently, advances in therapeutic monitoring in EOC have been made with circulating tumor DNA (ctDNA) providing important evidence about its utility in determining outcome and individualizing cancer therapy in patients with EOC [65–67]; on the contrary, the role of circulating miRNAs in EOC clinical monitoring needs to be further investigated in order to obtain a larger concordance between the results from independent investigators. Of note, we should bear in mind that ctDNA represents a sort of barcode originating directly from the tumor, but "liquid" miRNAs are not derived uniquely from the cancerous mass. Indeed, miRNAs are also physiologically released by other, normal cells and this makes the general landscape more complex to decipher. As a consequence, the research on cancer liquid biomarkers is still in its embryonal phase and no reliable miRNA candidates to accurately follow the treatment response "in real-time" have been identified yet.

Based on the data reported in our work, the most appealing miRNAs in EOC belong to the miR-200 family. Indeed, two independent works have identified miR-200b and miR-200c as potential biomarkers. Given that, the miR-200 family could have a role as a non-invasive biomarker in EOC. This family has already been reported as of potential interest in gynecological cancers, particularly in endometrial cancer [68]. The above-mentioned correlation could be due to the involvement of miR-200s in the epithelial–mesenchymal transition (EMT) process, which is known to play a key role in EOC progression, metastases, and recurrence and to be one of the cancer escape routes to medical treatments [69].

5. Conclusions

So far, the role of circulating miRNAs in therapeutic monitoring in EOC remains to be clarified given the inconsistent findings reported by different studies. This could be in part due to the limited number of analyses, the small sample size, and the lack of a standardized procedure to properly assess the miRNAs' contribution. Nevertheless, circulating miRNAs have potential as novel non-invasive and highly useful biomarkers in EOC.

Further studies with standardized protocols and larger cohorts of patients are warranted to foster the identification of circulating miRNAs of potential clinical significance in EOC.

Author Contributions: Conceptualization, G.R. and F.G.; data curation, G.R. and F.G.; writing—original draft preparation, G.R., F.G., G.D., E.D.C. and A.M.P.; writing—review and editing, G.R., A.M.P., I.K., S.A., P.H., P.D.I. and A.B. All authors have read and agreed to the published version of the manuscript.

Funding: This research received no external funding.

Institutional Review Board Statement: Not applicable.

Informed Consent Statement: Not applicable.

Acknowledgments: G.R. was supported by a Fulbright Research Scholarship (year 2021); F.G. is supported by a fellowship from the Fondazione Cassa di Risparmio in Bologna (CARISBO).

Conflicts of Interest: The authors declare no conflict of interest.

References

1. Kurnit, K.C.; Fleming, G.F.; Lengyel, E. Updates and New Options in Advanced Epithelial Ovarian Cancer Treatment. *Obstet. Gynecol.* **2020**, *137*, 108–121. [CrossRef]
2. De Leo, A.; Santini, D.; Ceccarelli, C.; Santandrea, G.; Palicelli, A.; Acquaviva, G.; Chiarucci, F.; Rosini, F.; Ravegnini, G.; Pession, A.; et al. What Is New on Ovarian Carcinoma: Integrated Morphologic and Molecular Analysis Following the New 2020 World Health Organization Classification of Female Genital Tumors. *Diagnostics* **2021**, *11*, 697. [CrossRef]
3. Koshiyama, M.; Matsumura, N.; Konishi, I. Recent Concepts of Ovarian Carcinogenesis: Type I and Type II. *BioMed Res. Int.* **2014**, *2014*, 934261. [CrossRef] [PubMed]
4. Salazar, C.; Campbell, I.G.; Gorringe, K.L. When Is "Type I" Ovarian Cancer Not "Type I"? Indications of an Out-Dated Dichotomy. *Front. Oncol.* **2018**, *8*, 654. [CrossRef] [PubMed]
5. Global Cancer Observatory. 2020. Available online: https://gco.iarc.fr/ (accessed on 23 August 2021).
6. Gaona-Luviano, P.; Medina-Gaona, L.A.; Magaña-Pérez, K. Epidemiology of ovarian cancer. *Chin. Clin. Oncol.* **2020**, *9*, 47. [CrossRef] [PubMed]
7. Fagotti, A.; De Iaco, P.; Fanfani, F.; Vizzielli, G.; Perelli, F.; Pozzati, F.; Perrone, A.M.; Turco, L.C.; Scambia, G. Systematic Pelvic and Aortic Lymphadenectomy in Advanced Ovarian Cancer Patients at the Time of Interval Debulking Surgery: A Double-Institution Case–Control Study. *Ann. Surg. Oncol.* **2012**, *19*, 3522–3527. [CrossRef] [PubMed]
8. Kim, S.; Han, Y.; Kim, S.I.; Kim, H.-S.; Kim, S.J.; Song, Y.S. Tumor evolution and chemoresistance in ovarian cancer. *NPJ Precis. Oncol.* **2018**, *2*, 20. [CrossRef]
9. Guerra, F.; Paiano, A.; Migoni, D.; Girolimetti, G.; Perrone, A.M.; De Iaco, P.; Fanizzi, F.P.; Gasparre, G.; Bucci, C. Modulation of RAB7A Protein Expression Determines Resistance to Cisplatin through Late Endocytic Pathway Impairment and Extracellular Vesicular Secretion. *Cancers* **2019**, *11*, 52. [CrossRef]
10. Karst, A.M.; Drapkin, R. Ovarian Cancer Pathogenesis: A Model in Evolution. *J. Oncol.* **2010**, *2010*, 932371. [CrossRef]
11. Schorge, J.O.; Modesitt, S.C.; Coleman, R.L.; Cohn, D.E.; Kauff, N.D.; Duska, L.R.; Herzog, T.J. SGO White Paper on Ovarian Cancer: Etiology, Screening and Surveillance. *Gynecol. Oncol.* **2010**, *119*, 7–17. [CrossRef] [PubMed]
12. Nero, C.; Ciccarone, F.; Pietragalla, A.; Duranti, S.; Daniele, G.; Salutari, V.; Carbone, M.; Scambia, G.; Lorusso, D. Ovarian Cancer Treatments Strategy: Focus on PARP Inhibitors and Immune Check Point Inhibitors. *Cancers* **2021**, *13*, 1298. [CrossRef] [PubMed]
13. Han, Y.; Jiang, L.; Liu, K.; Ouyang, L.; Li, Y. Predictive Value of HE4 in Platinum-Based Chemotherapy for Ovarian Cancer: A Systematic Review. *Front. Oncol.* **2021**, *11*, 703949. [CrossRef] [PubMed]
14. Grandi, G.; Perrone, A.M.; Toss, A.; Vitagliano, A.; Friso, S.; Facchinetti, F.; Cortesi, L.; Cascinu, S.; De Iaco, P. The generally low sensitivity of CA 125 for FIGO stage I ovarian cancer diagnosis increases for endometrioid histotype. *Minerva Med.* **2020**, *111*, 133–140. [CrossRef] [PubMed]
15. Fagotti, A.; Vizzielli, G.; De Iaco, P.; Surico, D.; Buda, A.; Mandato, V.D.; Petruzzelli, F.; Ghezzi, F.; Garzarelli, S.; Mereu, L.; et al. A multicentric trial (Olympia–MITO 13) on the accuracy of laparoscopy to assess peritoneal spread in ovarian cancer. *Am. J. Obstet. Gynecol.* **2013**, *209*, 462.e1–462.e11. [CrossRef] [PubMed]
16. Jiang, P.; Lo, Y.D. The Long and Short of Circulating Cell-Free DNA and the Ins and Outs of Molecular Diagnostics. *Trends Genet.* **2016**, *32*, 360–371. [CrossRef]
17. Ilié, M.; Hofman, P. Pros: Can tissue biopsy be replaced by liquid biopsy? *Transl. Lung Cancer Res.* **2016**, *5*, 420–423. [CrossRef]
18. Bernabé, R.; Hickson, N.; Wallace, A.; Blackhall, F.H. What do we need to make circulating tumour DNA (ctDNA) a routine diagnostic test in lung cancer? *Eur. J. Cancer* **2017**, *81*, 66–73. [CrossRef]
19. Malapelle, U.; Sirera, R.; Jantus-Lewintre, E.; Reclusa, P.; Calabuig, S.; Blasco, A.; Pisapia, P.; Rolfo, C.; Camps, C. Profile of the Roche cobas® EGFR mutation test v2 for non-small cell lung cancer. *Expert Rev. Mol. Diagn.* **2017**, *17*, 209–215. [CrossRef]
20. Heidrich, I.; Ačkar, L.; Mohammadi, P.M.; Pantel, K. Liquid biopsies: Potential and challenges. *Int. J. Cancer* **2020**, *148*, 528–545. [CrossRef]
21. Kilgour, E.; Rothwell, D.; Brady, G.; Dive, C. Liquid Biopsy-Based Biomarkers of Treatment Response and Resistance. *Cancer Cell* **2020**, *37*, 485–495. [CrossRef]

22. Oellerich, M.; Schütz, E.; Beck, J.; Walson, P.D. Circulating Cell-Free DNA—Diagnostic and Prognostic Applications in Personalized Cancer Therapy. *Ther. Drug Monit.* **2019**, *41*, 115–120. [CrossRef]
23. Marrugo-Ramírez, J.; Mir, M.; Samitier, J. Blood-Based Cancer Biomarkers in Liquid Biopsy: A Promising Non-Invasive Alternative to Tissue Biopsy. *Int. J. Mol. Sci.* **2018**, *19*, 2877. [CrossRef]
24. Yang, M.; Forbes, M.; Bitting, R.; O'Neill, S.; Chou, P.C.; Topaloglu, U.; Miller, L.; Hawkins, G.; Grant, S.; DeYoung, B.; et al. Incorporating blood-based liquid biopsy information into cancer staging: Time for a TNMB system? *Ann. Oncol.* **2017**, *29*, 311–323. [CrossRef] [PubMed]
25. Wu, J.; Hu, S.; Zhang, L.; Xin, J.; Sun, C.; Wang, L.; Ding, K.; Wang, B. Tumor circulome in the liquid biopsies for cancer diagnosis and prognosis. *Theranostics* **2020**, *10*, 4544–4556. [CrossRef] [PubMed]
26. Mitchell, P.; Parkin, R.K.; Kroh, E.M.; Fritz, B.R.; Wyman, S.K.; Pogosova-Agadjanyan, E.L.; Peterson, A.; Noteboom, J.; O'Briant, K.C.; Allen, A.; et al. Circulating microRNAs as stable blood-based markers for cancer detection. *Proc. Natl. Acad. Sci. USA* **2008**, *105*, 10513–10518. [CrossRef] [PubMed]
27. Ravegnini, G.; Sammarini, G.; Hrelia, P.; Angelini, S. Key Genetic and Epigenetic Mechanisms in Chemical Carcinogenesis. *Toxicol. Sci.* **2015**, *148*, 2–13. [CrossRef]
28. Baer, C.; Claus, R.; Plass, C. Genome-Wide Epigenetic Regulation of miRNAs in Cancer. *Cancer Res.* **2013**, *73*, 473–477. [CrossRef]
29. Simonson, B. MicroRNA Therapeutics: The Next Magic Bullet? *Mini-Rev. Med. Chem.* **2015**, *15*, 467–474. [CrossRef]
30. Oliveto, S.; Mancino, M.; Manfrini, N.; Biffo, S. Role of microRNAs in translation regulation and cancer. *World J. Biol. Chem.* **2017**, *8*, 45–56. [CrossRef]
31. Babar, A.I.; Slack, F.J.; Weidhaas, J.B. miRNA modulation of the cellular stress response. *Futur. Oncol.* **2008**, *4*, 289–298. [CrossRef]
32. Khan, S.; Ayub, H.; Khan, T.; Wahid, F. MicroRNA biogenesis, gene silencing mechanisms and role in breast, ovarian and prostate cancer. *Biochimie* **2019**, *167*, 12–24. [CrossRef]
33. Iorio, M.; Croce, C.M. MicroRNA dysregulation in cancer: Diagnostics, monitoring and therapeutics. A comprehensive review. *EMBO Mol. Med.* **2012**, *4*, 143–159. [CrossRef]
34. Swarup, V.; Rajeswari, M. Circulating (cell-free) nucleic acids—A promising, non-invasive tool for early detection of several human diseases. *FEBS Lett.* **2007**, *581*, 795–799. [CrossRef]
35. Sun, Z.; Shi, K.; Yang, S.; Liu, J.; Zhou, Q.; Wang, G.; Song, J.; Li, Z.; Zhang, Z.; Yuan, W. Effect of exosomal miRNA on cancer biology and clinical applications. *Mol. Cancer* **2018**, *17*, 1–19. [CrossRef]
36. Moher, D.; PRISMA-P Group; Shamseer, L.; Clarke, M.; Ghersi, D.; Liberati, A.; Petticrew, M.; Shekelle, P.; Stewart, L.A. Preferred reporting items for systematic review and meta-analysis protocols (PRISMA-P) 2015 statement. *Syst. Rev.* **2015**, *4*, 1. [CrossRef]
37. Stone, P.W. Popping the (PICO) question in research and evidence-based practice. *Appl. Nurs. Res.* **2002**, *15*, 197–198. [CrossRef]
38. Zuberi, M.; Mir, R.; Khan, I.; Javid, J.; Guru, S.A.; Bhat, M.; Sumi, M.P.; Ahmad, I.; Masroor, M.; Yadav, P.; et al. The Promising Signatures of Circulating microRNA-145 in Epithelial Ovarian Cancer Patients. *MicroRNA* **2019**, *9*, 49–57. [CrossRef]
39. Li, M.; Zhang, S.; Ma, Y.; Yang, Y.; An, R. Role of hsa-miR-105 during the pathogenesis of paclitaxel resistance and its clinical implication in ovarian cancer. *Oncol. Rep.* **2021**, *45*, 84. [CrossRef]
40. Chen, Z.; Guo, X.; Sun, S.; Lu, C.; Wang, L. Serum miR-125b levels associated with epithelial ovarian cancer (EOC) development and treatment responses. *Bioengineered* **2020**, *11*, 311–317. [CrossRef]
41. Biamonte, F.; Santamaria, G.; Sacco, A.; Perrone, F.M.; Di Cello, A.; Battaglia, A.M.; Salatino, A.; Di Vito, A.; Aversa, I.; Venturella, R.; et al. MicroRNA let-7g acts as tumor suppressor and predictive biomarker for chemoresistance in human epithelial ovarian cancer. *Sci. Rep.* **2019**, *9*, 5668. [CrossRef]
42. Kuhlmann, J.D.; Chebouti, I.; Kimmig, R.; Buderath, P.; Reuter, M.; Puppel, S.-H.; Wimberger, P.; Kasimir-Bauer, S. Extracellular vesicle-associated miRNAs in ovarian cancer—Design of an integrated NGS-based workflow for the identification of blood-based biomarkers for platinum-resistance. *Clin. Chem. Lab. Med.* **2018**, *57*, 1053–1062. [CrossRef] [PubMed]
43. Fukagawa, S.; Miyata, K.; Yotsumoto, F.; Kiyoshima, C.; Nam, S.O.; Anan, H.; Katsuda, T.; Miyahara, D.; Murata, M.; Yagi, H.; et al. MicroRNA-135a-3p as a promising biomarker and nucleic acid therapeutic agent for ovarian cancer. *Cancer Sci.* **2017**, *108*, 886–896. [CrossRef] [PubMed]
44. Robelin, P.; Tod, M.; Colomban, O.; Lachuer, J.; Ray-Coquard, I.; De Rauglaudre, G.; Joly, F.; Chevalier-Place, A.; Combe, P.; Lortholary, A.; et al. Comparative analysis of predictive values of the kinetics of 11 circulating miRNAs and of CA125 in ovarian cancer during first line treatment (a GINECO study). *Gynecol. Oncol.* **2020**, *159*, 256–263. [CrossRef] [PubMed]
45. Zhu, X.; Shen, H.; Yin, X.; Yang, M.; Wei, H.; Chen, Q.; Feng, F.; Liu, Y.; Xu, W.; Li, Y. Macrophages derived exosomes deliver miR-223 to epithelial ovarian cancer cells to elicit a chemoresistant phenotype. *J. Exp. Clin. Cancer Res.* **2019**, *38*, 1–14. [CrossRef] [PubMed]
46. Kobayashi, M.; Sawada, K.; Nakamura, K.; Yoshimura, A.; Miyamoto, M.; Shimizu, A.; Ishida, K.; Nakatsuka, E.; Kodama, M.; Hashimoto, K.; et al. Exosomal miR-1290 is a potential biomarker of high-grade serous ovarian carcinoma and can discriminate patients from those with malignancies of other histological types. *J. Ovarian Res.* **2018**, *11*, 81. [CrossRef] [PubMed]
47. Grabosch, S.; Tseng, G.; Edwards, R.P.; Lankes, H.A.; Moore, K.; Odunsi, K.; Vlad, A.; Ma, T.; Strange, M.; Brozick, J.; et al. Multiplex profiling identifies distinct local and systemic alterations during intraperitoneal chemotherapy for ovarian cancer: An NRG Oncology/Gynecologic Oncology Group Study. *Gynecol. Oncol.* **2017**, *146*, 137–145. [CrossRef]
48. Benson, E.A.; Skaar, T.C.; Liu, Y.; Nephew, K.P.; Matei, D. Carboplatin with Decitabine Therapy, in Recurrent Platinum Resistant Ovarian Cancer, Alters Circulating miRNAs Concentrations: A Pilot Study. *PLoS ONE* **2015**, *10*, e0141279. [CrossRef]

49. Kapetanakis, N.-I.; Uzan, C.; Jimenez-Pailhes, A.-S.; Gouy, S.; Bentivegna, E.; Morice, P.; Caron, O.; Gourzones-Dmitriev, C.; Le Teuff, G.; Busson, P. Plasma miR-200b in ovarian carcinoma patients: Distinct pattern of pre/post-treatment variation compared to CA-125 and potential for prediction of progression-free survival. *Oncotarget* **2015**, *6*, 36815–36824. [CrossRef]
50. Kuhlmann, J.D.; Baraniskin, A.; Hahn, S.; Mosel, F.; Bredemeier, M.; Wimberger, P.; Kimmig, R.; Kasimir-Bauer, S. Circulating U2 Small Nuclear RNA Fragments as a Novel Diagnostic Tool for Patients with Epithelial Ovarian Cancer. *Clin. Chem.* **2014**, *60*, 206–213. [CrossRef]
51. Shapira, I.; Oswald, M.; Lovecchio, J.; Khalili, H.; Menzin, A.; Whyte, J.; Dos Santos, L.; Liang, S.; Bhuiya, T.; Keogh, M.; et al. Circulating biomarkers for detection of ovarian cancer and predicting cancer outcomes. *Br. J. Cancer* **2013**, *110*, 976–983. [CrossRef]
52. Vigneron, N.; Vernon, M.; Meryet-Figuière, M.; Lambert, B.; Briand, M.; Louis, M.-H.; Krieger, S.; Joly, F.; Lheureux, S.; Blanc-Fournier, C.; et al. Predictive Relevance of Circulating miR-622 in Patients with Newly Diagnosed and Recurrent High-Grade Serous Ovarian Carcinoma. *Clin. Chem.* **2020**, *66*, 352–362. [CrossRef]
53. Halvorsen, A.R.; Kristensen, G.; Embleton, A.; Adusei, C.; Barretina-Ginesta, M.P.; Beale, P.; Helland, A. Evaluation of Prognostic and Predictive Significance of Circulating MicroRNAs in Ovarian Cancer Patients. *Dis. Markers* **2017**, *2017*, 3098542. [CrossRef]
54. Feng, Y.-H.; Tsao, C.-J. Emerging role of microRNA-21 in cancer. *Biomed. Rep.* **2016**, *5*, 395–402. [CrossRef]
55. Echevarría-Vargas, I.M.; Valiyeva, F.; Vivas-Mejía, P.E. Upregulation of miR-21 in Cisplatin Resistant Ovarian Cancer via JNK-1/c-Jun Pathway. *PLoS ONE* **2014**, *9*, e97094. [CrossRef]
56. Parikh, A.; Lee, C.; Joseph, P.; Marchini, S.; Baccarini, A.; Kolev, V.; Romualdi, C.; Fruscio, R.; Shah, H.; Wang, F.; et al. microRNA-181a has a critical role in ovarian cancer progression through the regulation of the epithelial–mesenchymal transition. *Nat. Commun.* **2014**, *5*, 2977. [CrossRef]
57. Laios, A.; O'Toole, S.; Flavin, R.; Martin, C.; Kelly, L.; Ring, M.; Finn, S.P.; Barrett, C.; Loda, M.; Gleeson, N.; et al. Potential role of miR-9 and miR-223 in recurrent ovarian cancer. *Mol. Cancer* **2008**, *7*, 1–14. [CrossRef]
58. Ying, X.; Wei, K.; Lin, Z.; Cui, Y.; Ding, J.; Chen, Y.; Xu, B. MicroRNA-125b Suppresses Ovarian Cancer Progression via Suppression of the Epithelial-Mesenchymal Transition Pathway by Targeting the SET Protein. *Cell. Physiol. Biochem.* **2016**, *39*, 501–510. [CrossRef] [PubMed]
59. Guan, Y.; Yao, H.; Zheng, Z.; Qiu, G.; Sun, K. MiR-125b targets BCL3 and suppresses ovarian cancer proliferation. *Int. J. Cancer* **2010**, *128*, 2274–2283. [CrossRef] [PubMed]
60. Mazieres, J.; Catherinne, C.; Delfour, O.; Gouin, S.; Rouquette, I.; Delisle, M.-B.; Prévot, G.; Escamilla, R.; Didier, A.; Persing, D.H.; et al. Alternative Processing of the U2 Small Nuclear RNA Produces a 19–22nt Fragment with Relevance for the Detection of Non-Small Cell Lung Cancer in Human Serum. *PLoS ONE* **2013**, *8*, e60134. [CrossRef] [PubMed]
61. Baraniskin, A.; Nöpel-Dünnebacke, S.; Ahrens, M.; Jensen, S.G.; Zöllner, H.; Maghnouj, A.; Wos, A.; Mayerle, J.; Munding, J.; Kost, D.; et al. Circulating U2 small nuclear RNA fragments as a novel diagnostic biomarker for pancreatic and colorectal adenocarcinoma. *Int. J. Cancer* **2012**, *132*, E48–E57. [CrossRef] [PubMed]
62. Matei, D.; Fang, F.; Shen, C.; Schilder, J.; Arnold, A.; Zeng, Y.; Berry, W.A.; Huang, T.; Nephew, K.P. Epigenetic Resensitization to Platinum in Ovarian Cancer. *Cancer Res.* **2012**, *72*, 2197–2205. [CrossRef] [PubMed]
63. Damia, G.; Broggini, M. Platinum Resistance in Ovarian Cancer: Role of DNA Repair. *Cancers* **2019**, *11*, 119. [CrossRef] [PubMed]
64. Deans, A.; West, S.C. DNA interstrand crosslink repair and cancer. *Nat. Rev. Cancer* **2011**, *11*, 467–480. [CrossRef]
65. Thusgaard, C.F.; Korsholm, M.; Koldby, K.M.; Kruse, T.A.; Thomassen, M.; Jochumsen, K.M. Epithelial ovarian cancer and the use of circulating tumor DNA: A systematic review. *Gynecol. Oncol.* **2021**, *161*, 884–895. [CrossRef]
66. Charo, L.M.; Eskander, R.N.; Okamura, R.; Patel, S.P.; Nikanjam, M.; Lanman, R.B.; Piccioni, D.E.; Kato, S.; McHale, M.T.; Kurzrock, R. Clinical implications of plasma circulating tumor DNA in gynecologic cancer patients. *Mol. Oncol.* **2020**, *15*, 67–79. [CrossRef]
67. Sharbatoghli, M.; Vafaei, S.; Es, H.A.; Asadi-Lari, M.; Totonchi, M.; Madjd, Z. Prediction of the treatment response in ovarian cancer: A ctDNA approach. *J. Ovarian Res.* **2020**, *13*, 1–12. [CrossRef]
68. Lee, J.-W.; Park, Y.-A.; Choi, J.-J.; Lee, Y.Y.; Kim, C.-J.; Choi, C.; Kim, T.-J.; Lee, N.W.; Kim, B.-G.; Bae, D.-S. The expression of the miRNA-200 family in endometrial endometrioid carcinoma. *Gynecol. Oncol.* **2011**, *120*, 56–62. [CrossRef]
69. Loret, N.; Denys, H.; Tummers, P.; Berx, G. The Role of Epithelial-to-Mesenchymal Plasticity in Ovarian Cancer Progression and Therapy Resistance. *Cancers* **2019**, *11*, 838. [CrossRef] [PubMed]

MDPI
St. Alban-Anlage 66
4052 Basel
Switzerland
Tel. +41 61 683 77 34
Fax +41 61 302 89 18
www.mdpi.com

Biomedicines Editorial Office
E-mail: biomedicines@mdpi.com
www.mdpi.com/journal/biomedicines

www.ingramcontent.com/pod-product-compliance
Lightning Source LLC
Chambersburg PA
CBHW040752250926
51850CB00007BB/1478
9783036530888